WAVES ON FLUID INTERFACES

Publication No. 50
of the Mathematics Research Center
The University of Wisconsin–Madison

Academic Press Rapid Manuscript Reproduction

WAVES ON FLUID INTERFACES

Edited by

RICHARD E. MEYER

Mathematics Research Center
The University of Wisconsin–Madison
Madison, Wisconsin

Proceedings of a Symposium Conducted by the
Mathematics Research Center
The University of Wisconsin–Madison
October 18–20, 1982

1983

ACADEMIC PRESS

A Subsidiary of Harcourt Brace Jovanovich, Publishers

New York London

Paris San Diego San Francisco São Paulo Sydney Tokyo Toronto

ACADEMIC PRESS, INC.
111 Fifth Avenue, New York, New York 10003

United Kingdom Edition published by
ACADEMIC PRESS, INC. (LONDON) LTD.
24/28 Oval Road, London NW1 7DX

Library of Congress Cataloging in Publication Data
Main entry under title:

Waves on fluid interfaces.

(Publication of the Mathematics Research Center, the University of Wisconsin--Madison ; no. 50)
Includes index.
1. Wave-motion, Theory of--Congresses. I. Meyer, Richard E., Date II. University of Wisconsin--Madison. Mathematics Research Center. III. Title: Fluid interfaces. IV. Series.
QA3.U45 no. 50 [QA927] 510s [532'.059] 83-11797
ISBN 0-12-493220-7 (alk. paper)

PRINTED IN THE UNITED STATES OF AMERICA

83 84 85 86 9 8 7 6 5 4 3 2 1

CONTENTS

Senior Contributors vii
Preface ix

Finite-amplitude Interfacial Waves 1
P. G. Saffman

Instability of Finite-amplitude Interfacial Waves 17
H. C. Yuen

Finite-amplitude Water Waves with Surface Tension 41
J.-M. Vanden-Broeck

Generalized Vortex Methods for Free-Surface Flows 53
G. R. Baker

A Review of Solution Methods for Viscous Flow in the Presence of Deformable Boundaries 83
L. G. Leal

On Existence Criteria for Fluid Interfaces in the Absence of Gravity 113
P. Concus

Capillary Waves and Interfacial Structure 123
H. T. Davis

Interfacial and Critical Phenomena in Microemulsions 151
M. W. Kim, J. Bock, and J. S. Huang

Electrohydrodynamic Surface Waves 167
J. Melcher

Solitary Waves on Density Interfaces 201
T. Maxworthy

Interfacial Instabilities Caused by Air Flow over a Thin Liquid Layer 221
T. J. Hanratty

Film Waves 261
S. P. Lin

Rupture of Thin Liquid Films 291
S. Davis

The Moving Contact Line 303
E. B. Dussan V.

The Endothelial Interface between Tissue and Blood 325
S. Weinbaum

Index *355*

Senior Contributors

Numbers in parentheses indicate the pages on which the authors' contributions begin.

GREGORY R. BAKER (53), *Department of Mathematics, University of Arizona, Tucson, Arizona 85721*

PAUL CONCUS (113), *Lawrence Berkeley Laboratory, University of California, Berkeley, California 94720*

H. TED DAVIS (123), *Department of Chemical Engineering and Materials Science, University of Minnesota, Minneapolis, Minnesota 55455*

STEPHEN DAVIS (291), *Division of Applied Mathematics and Engineering Science, Northwestern University, Evanston, Illinois 60201*

ELIZABETH B. DUSSAN V. (303), *Department of Chemical Engineering, University of Pennsylvania, Philadelphia, Pennsylvania 19104*

THOMAS J. HANRATTY (221), *Department of Chemical Engineering, University of Illinois, Urbana, Illinois 61801*

MAHN WON KIM (151), *Science Laboratories, Exxon Research and Engineering Company, Linden, New Jersey 07036*

L. GARY LEAL (83), *Chemical Engineering Department, California Institute of Technology, Pasadena, California 91125*

SUNG P. LIN (261), *Department of Mechanical and Industrial Engineering, Clarkson College, Potsdam, New York 13676*

TONY MAXWORTHY (201), *Department of Mechanical Engineering, University of Southern California, Los Angeles, California 90007*

JAMES R. MELCHER (167), *Department of Electrical Engineering, Massachusetts Institute of Technology, Cambridge, Massachusetts 02139*

PHILIP G. SAFFMAN (1), *Applied Mathematics Department, California Institute of Technology, Pasadena, California 91125*

JEAN-MARC VANDEN-BROECK (41), *Mathematics Research Center, University of Wisconsin, Madison, Wisconsin 53706*

SHELDON WEINBAUM (325), *Department of Mechanical Engineering, City College of the City University of New York, New York, New York 10031*

HENRY C. YUEN (17), *Defense and Space Systems Group, TRW Inc., Redondo Beach, California 90278*

PREFACE

This volume collects the invited addresses given at a symposium on fluid interfaces of the Mathematics Research Center of the University of Wisconsin in Madison in October 1982. The articles survey many different aspects of recent research developments, from nonlinear instabilities of classical interfaces to the physical structure of real interfaces and the new challenges they pose to our intuition about fluids. They concern theory and experiment, and touch on many applications of acute interest in technology and medicine. Collectively, they illuminate a multifaceted subject in rapid progress.

I am greatly indebted to the authors for the excellence of their articles outlining so many recent advances in which they have played a decisive part. The Mathematics Research Center also wishes to thank the United States Army, which sponsored the conference under its Contract No. DAAG29-80-C-0041, the National Science Foundation, which supported it by Grants MEA-8212157 and MCS-7927062(2) of its Mechanical Engineering and Applied Mechanics and its Mathematical and Computer Sciences Divisions, and to the United States Department of Energy, which supported it by Grant DE-FG02-82ER13020 of its Applied Mathematical Sciences Division in the Office of Basic Energy Sciences. My personal thanks go to Gladys Moran for the expert handling of yet another symposium and to Elaine DuCharme for assembling the volume and index.

Richard E. Meyer

FINITE-AMPLITUDE INTERFACIAL WAVES

P. G. Saffman

1. INTRODUCTION.

We consider gravity waves at the interface between two uniform, unbounded fluids of different densities in the presence of a current or relative horizontal velocity U. The fluids are supposed to be immiscible, incompressible and inviscid, and the motion is assumed to be irrotational. We are concerned with the properties and existence of finite amplitude two-dimensional, periodic waves of permanent form which propagate steadily without change of shape. By two-dimensional, we mean that the flow field depends only on the horizontal direction of propagation, which will be the x-axis, and the vertical y-direction. In the field of surface gravity waves, which is the limit of the present study when the density of the upper fluid is zero, it has been found recently that three-dimensional waves of permanent form exist and are observed experimentally (see e.g. [5]). It is expected that such waves will also exist and be important for interfacial waves, but they will not be considered in the present work.

For the purpose of calculating steady waves, there is no loss of generality in taking the speed of propagation c parallel to the current U, as an arbitrary constant transverse velocity may be linearly superposed on any

ISBN 0-12-493220-7

two-dimensional steady wave without affecting its properties (the stability characteristics would, however, be affected). The wave can be reduced to rest by choosing a frame of reference moving with the wave. The problem is then to calculate steady irrotational solutions of the Euler equations which satisfy continuity of pressure across a common streamline. It follows from dimensional analysis that apart from scaling factors all flow variables will depend upon three dimensionless parameters:

$$\frac{h}{L}, \quad \frac{\rho_2}{\rho_1}, \quad \frac{\rho_2 U^2}{\rho_1 g L}, \tag{1.1}$$

where h is the height of the wave defined as the vertical distance between crest and trough, L is the wavelength (the horizontal distance over which the flow field repeats itself which in the present work will be the distance between crests), ρ_2 and ρ_1 are the densities of the upper and lower fluid respectively, and g is the acceleration due to gravity. For example, the speed of the waves is given by

$$c = (gL/2\pi)^{1/2} C\left(\frac{h}{L}, \frac{\rho_2}{\rho_1}, \frac{\rho_2 U^2}{\rho_1 g L}\right) \tag{1.2}$$

where C is a dimensionless function of its arguments. For surface waves, where $\rho_2 = 0$ and there is dependence on only one parameter, namely h/L, it is known that many interesting and unexpected phenomena exist, especially when the wave steepness becomes large. When there is dependence on three parameters, it is to be expected that many more phenomena are likely. However, in the absence of exact solutions for large h/L, it is a highly non-trivial task to search a three-dimensional parameter space. The results to be presented below are limited to those phenomena which seem currently to be of the most interest.

In contrast to the voluminous work on surface waves, relatively little seems to have been done on interfacial waves of permanent form, and that work seems to have been confined to the case of zero current, i.e. $U = 0$. Tsuji and Nagata [7] calculated Stokes type expansions to order $(h/L)^5$, and Holyer [3] used the computer to compute the coefficients in such an expansion to order $(h/L)^{37}$, and then used Padé approximants to estimate the behavior for large h/L. We are not aware of any work for waves with current.

For the mathematical formulation, there is no loss of generality in taking $g = 1$, $L = 2\pi$, and $\rho_1 = 1$. The mathematical problem is to determine the x-periodic velocity potentials and stream functions, for the lower and upper fluid respectively, which satisfy Laplace's equation and are harmonic conjugate pairs, so that at the unknown interface $y = Y(x)$,

$$\psi_1(x,Y(x)) = 0, \quad \psi_2(x,Y(x)) = 0 , \tag{1.3}$$

$$\frac{1}{2}(\nabla\phi_1)^2 + Y(x) + b = \frac{1}{2}\rho_2(\nabla\phi_2)^2 + \rho_B Y(x) . \tag{1.4}$$

In general, $\rho_B = \rho_2$, but we allow for the possibility of Boussinesq waves (in which the inertia of the two fluids is the same and density differences only matter when multiplied by g) by setting $\rho_2 = 1$ and $\rho_B = 0$. Surface tension is neglected throughout. The quantity b is the Bernoulli constant, which by suitable choice of the origin of pressure may be set equal to zero in the lower fluid. Infinitely far from the interface, we have

$$\phi_1 \sim -Cx, \quad \phi_2 \sim (U - c)x . \tag{1.5}$$

The vertical origin is set by requiring that the mean elevation of the interface is zero and the horizontal origin can be fixed by placing the crest at $x = 0$. This problem now appears to be free of arbitrary constants and the wave is determined by the crest to trough height h. It is expected that isolated families of solutions exist in connected regions in (h,ρ_2,U) space, although this does not yet appear to have been proved.

One question of considerable interest is the domain of parameter space in which solutions exist. Suppose that we consider a fixed value of ρ_2 and vary h and U. It is found that as U increases with h kept constant the system of equations describing steady solutions fails to have a solution, even though the 'limiting' wave profile is smooth and exhibits no singular properties. For $U > 0$ there are, when solutions exist, at least two physically distinct waves corresponding to the two wave speeds for propagation with and against the current. As U increases, the wave propagating against the current is 'entrained' by the current and at a certain value of U, which depends on h and ρ_2, the two waves become identical and for larger U there are no real solutions of the equations. Mathematically, this is like the disappearance of roots of a quadratic). We shall term this factor which limits existence a 'dynamical limit'.

The second factor is what we term a 'geometrical limit'. The mathematical formulation remains well-behaved but the solutions cease to make physical sense as the wave profiles cross themselves. This occurs for fixed U and increasing h. Examples of this phenomenon are found in pure capillary and capillary-gravity waves [2,1] for which the wave profile crosses itself at a critical value of h. If $U \neq 0$, this limit is going to be different for the two solutions of waves moving with and against the current. In the case of surface waves, this limit corresponds to a 120° cusp. It is easy to see that except for two special cases (see §4), this cannot happen for interfacial waves. Holyer [3] identified the geometrical limit for $U = 0$ with the existence of a vertical tangent. We shall present evidence that waves can exist with a vertical tangent and significant overhang, and the evidence indicates that the geometrical limit is associated with the wave crossing itself when it is sufficiently high for $U > 0$.

2. WEAKLY NONLINEAR WAVES.

The properties of weakly nonlinear steady waves may be obtained by using the Stokes expansion in which all variables are expanded as power series in h/L. However, the algebra can be simplified somewhat by using Whitham's variational approach. Proceeding in the usual manner, one finds after some algebra that the average Lagrangian is

$$L = \frac{1}{4}(\rho_B - 1)(\frac{1}{4}h^2 + a_2^2)$$

$$+ \frac{h^2}{16k}[\omega^2 + \rho_2(Uk-\omega)^2] - (\frac{a_2^2}{2k} + \frac{kh^4}{256})[\omega^2 + \rho_2(Uk-\omega)^2]$$

$$- \frac{h^2 a_2}{16}[\omega^2 - \rho_2(Uk-\omega)^2] + O(h^6) \qquad (2.1)$$

for the wave with interface shape

$$Y(x) = \frac{h}{2}\cos(kx - \omega t) + a_2 \cos 2(kx - \omega t) \qquad (2.2)$$

The value of a_2 is found from $\partial L/\partial a_2 = 0$ to be

$$a_2 = \frac{h^2}{8(1-\rho_B)}[C^2 - \rho_2(U - C)^2] + O(h^4) ,$$

where $C = \omega/k$ is the phase speed. The dispersion relation for the weakly nonlinear wave then follows from $\partial L/\partial h = 0$:

$$C^2 + \rho_2(U - C)^2$$

$$= (1 - \rho_B)[1 + \frac{h^2}{8}(\frac{2C^2}{1-\rho_B} - 1)^2 + \frac{h^2}{8}] + O(h^4) . \qquad (2.4)$$

For $U = 0$, the values of C agree with those in [7]

The values of the energy, momentum and action densities and fluxes follow from the expression (2.1) for L in the usual way. In particular, the total energy density E is given by

$$E = kCL_\omega - L . \qquad (2.5)$$

It is to be noted that for $U > 0$, the energy is measured relative to the energy of the uniform state with a flat interface. Negative energies may therefore exist and mean that the energy of the state with waves is less than that of the undisturbed flow.

It follows from the dispersion relation (2.4) that for linear waves $(h \to 0)$ and given values of ρ_2 and U, there are two solutions corresponding to the two roots of the quadratic equation for C in terms of ρ_2 and U. We denote these two solutions by C_+ and C_-, where

$C_+ > C_-$. For the linear case, steady solutions cease to exist when U exceeds a critical value U_{c0} given by

$$U_{c0} = [(1 + \rho_2)(1 - \rho_B)/\rho_2]^{1/2} \qquad (2.6)$$

for which the two wave speeds are equal with the value $C_+ = C_- = \rho_2 U_{c0}/(1 + \rho_2)$.

The values of C_+ and C_- are

$$\rho_2 U \pm [\rho_2^2 U^2 - (1 + \rho_2)(\rho_2 U^2 - 1 + \rho_B)]^{1/2} . \qquad (2.7)$$

For $U = 0$, the values are equal and opposite. As U increases, the speed of the wave propagating with the current originally increases but eventually decreases. The speed of the wave propagating against the current increases monotonically (in the algebraic sense), becomes zero when $U = [(1 - \rho_B)/\rho_2]^{1/2}$ and then increases to equal C_+ when U is given by (2.6). According to the linear approximation, the energy density E equals $\frac{1}{8} h^2[C^2(1+r) - rCU]$, and it is interesting that the energy becomes negative when the direction of the C_- waves changes.

For finite amplitude waves, the two solutions corresponding to C_+ and C_- waves continue into two families of solutions marked by wave speeds $C_+(h, \rho_2, U)$ and $C_-(h, \rho_2, U)$. For any given value of h and ρ_2, there will again be a critical current U_c beyond which steady solutions no longer exist. For the weakly nonlinear approximation, this value is given by

$$U_c = U_{c0}\left[1 + \frac{h^2}{4}\frac{(1 + \rho_2^2)}{(1 + \rho_2)^2}\right]^{1/2} . \qquad (2.9)$$

It is noteworthy that increasing h increases U_c.

3. NUMERICAL METHODS.

For values of h that are not small, it is necessary to employ numerical methods. Three different techniques were employed. The first was to compute in physical space, i.e. the interface, potentials and stream function were expanded as Fourier series in x with coefficients which are exponential in y. The series were truncated to N modes and the boundary conditions were then satisfied at $N + 1$ equally horizontally spaced points on the interface. This procedure gives $3N + 4$ equations for $3N + 4$ unknowns. These equations were solved by Newton's method, using continuation in either U or h to give the first guesses. Note that this formulation is essentially equivalent to calculating numerically the coefficients of the Stokes expansion as done in [3].

The second method used the potential and stream function as the independent variables and expands the physical coordinates as series in these. The boundary conditions are now satisfied at equally spaced values of the velocity potential and the resulting system of $3N + 3$ equations in $3N + 3$ variables, the expansions being truncated to N modes, was also solved by Newton's method with continuation in U and h employed to give a first guess.

The third method used a vortex sheet representation in which the unknowns are the shape of the interface and the dipole strength of the equivalent double layer. This gives a nonlinear integrodifferential equation, which was solved by discretization and collocation, the resulting system of nonlinear equations again being solved by Newton's method with continuation.

For details, see [4,5]. All methods worked extremely well for small values of h/L, which generally meant $h < 0.6$, with some dependence on ρ_2 and U. (With our scaling, the surface wave of greatest height has $h = 0.89$). The first method was the first to fail as h increased. It is of course clear that this approach of working in physical space must fail when the wave becomes very steep, but the failure, marked by the apparent failure of the Fourier series to converge, seemed to be due to other causes. What actually happened was that the singularities of the analytic continuation of the lower velocity potential, say, into the upper half plane moved down below the crest. In this case, the expansion of the velocity potential would have to diverge near the crest, even though the solution was perfectly well behaved and physically meaningful. This difficulty would not affect the other two methods which were used for values of h up to 1.2 for various values of ρ_2 and U. For large values of h, 100 modes were used in the second method and this seemed adequate except for the largest h. The vortex sheet method with 65 intervals was employed for this case. This method offers in principle the advantage of being able to concentrate points near regions of high curvature, although this was not done.

The accuracy of the calculations was checked by comparing the results of the somewhat different methods with each other in regions of apparent validity and by performing the usual tests of internal consistency by investigating the dependence on number of retained modes. The calculations were carried out on a PRIME 750 and the CRAY-1 at NCAR.

4. A SPECIAL CLASS OF SOLUTIONS.

It is interesting to note that a special class of solutions exist which are simple transformations of the well-known surface permanent wave solutions, which have been extensively studied both numerically and theoretically by many authors. For each value of ρ_2, these solutions describe the shape of the interface for the C_+ case when

$$C_+ = U = (1 - \rho_B)^{1/2} c_s(h) \tag{4.1}$$

where $c_s(h)$ is the wave speed of the surface wave of permanent frm for the given wave height h. Since $C_+ = U$, the upper fluid is stagnant in the wave-fixed coordinates. The dynamic boundary condition for the motion in the lower fluid then becomes that for surface waves with a reduced gravity $g(1 - \rho_B)^{1/2}$. The velocities and wave speed are therefore those of the surface wave multiplied by the factor $(1 - \rho_B)^{1/2}$.

For the C_- branch, special solutions exist with

$$C_- = 0, \quad U = [(1 - \rho_B)/\rho_2]^{1/2} c_s(h) \ . \tag{4.2}$$

In this case, the lower fluid is stagnant and the dynamic boundary condition on the motion of the upper fluid is that with a reduced upside down gravity. The wave profiles are inverted surface waves, with negative gravity multiplied by the factor $[(1 - \rho_B)/\rho_2]^{1/2}$.

These special solutions have geometrical limits when $h = 0.892$, where the waves have a corner at the crest for the C_+ wave, and a corner at the trough for the C_- wave, with an interior angle of 120°. However, these special geometrical limits are only for the case when one of the fluids is moving with the wave. In general, it is expected (see below) that the geometrical limit is associated with the wave surface crossing itself.

5. RESULTS.

The equations have been solved numerically for various values of h, $r = \rho_2/\rho_1$, and U. The existence of the dynamical limit was confirmed, and it was found that the weakly nonlinear approximation (2.9) is a good approximation for values of h up to 0.6. A typical set of results is shown in figure 1. These results are for $r = 0.5$ and $h = 0.6$, and show the wave speeds C, total energies E, and kinetic energies T for both the + and - waves as functions of the current velocity U. The existence of the dynamical limit where $C_+ = C_-$ is clearly demonstrated.

One feature of remarkable interest is the existence of a region of negative energy. This implies that there will be a range of parameters in which the energy of the state with finite amplitude waves is less than that with the same current and a flat surface. Spontaneous generation of such flows is then a definite possibility. The computed results and linear analysis suggest that negative energies appear when C_- is zero and continue for values of U up to that for which the dynamical limit is reached. This aspect of the solutions needs to be explored further in detail.

The geometrical limit or the shape of the wave of greatest height has been addressed for the case of zero current. Solutions were obtained for three values of r (1.0, 0.9, 0.1), using the vortex sheet method as this seemed to provide the best resolution when the waves are large. Values of the wave speed C for + waves are shown in figure 2. For the larger values of r, it was possible to calculate solutions with vertical slope and the shapes

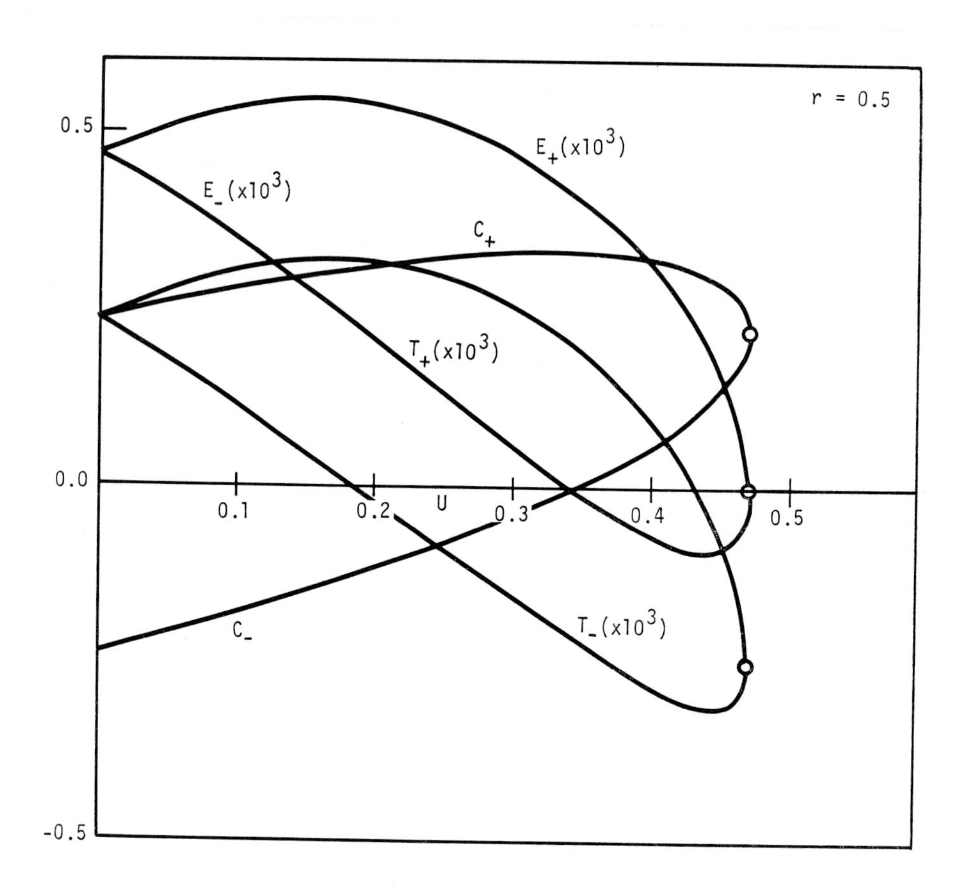

Figure 1. Properties of interfacial waves as functions of current velocity U for a fixed wave height $h = 0.6$ and density ratio $r = 0.5$. Wave speed C, total energy E, and kinetic energy T are shown for both branches.

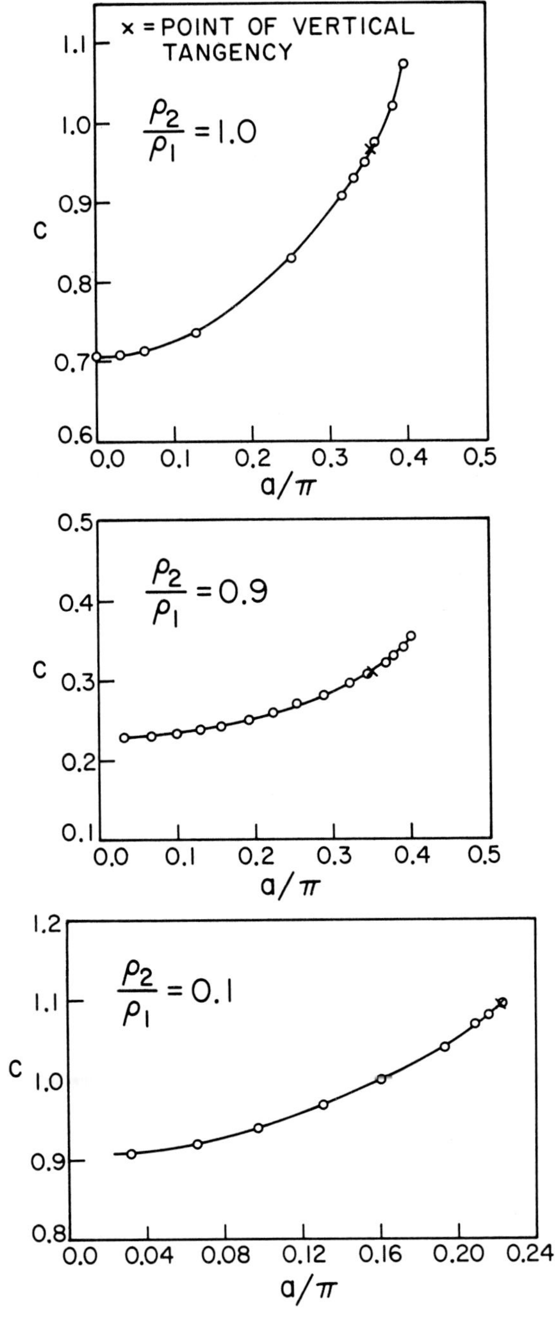

Figure 2. Phase speed C vs wave steepness $a/\pi = h/L$ for $r = 0.1$, 0.9, 1.0. x denotes point of vertical tangency.

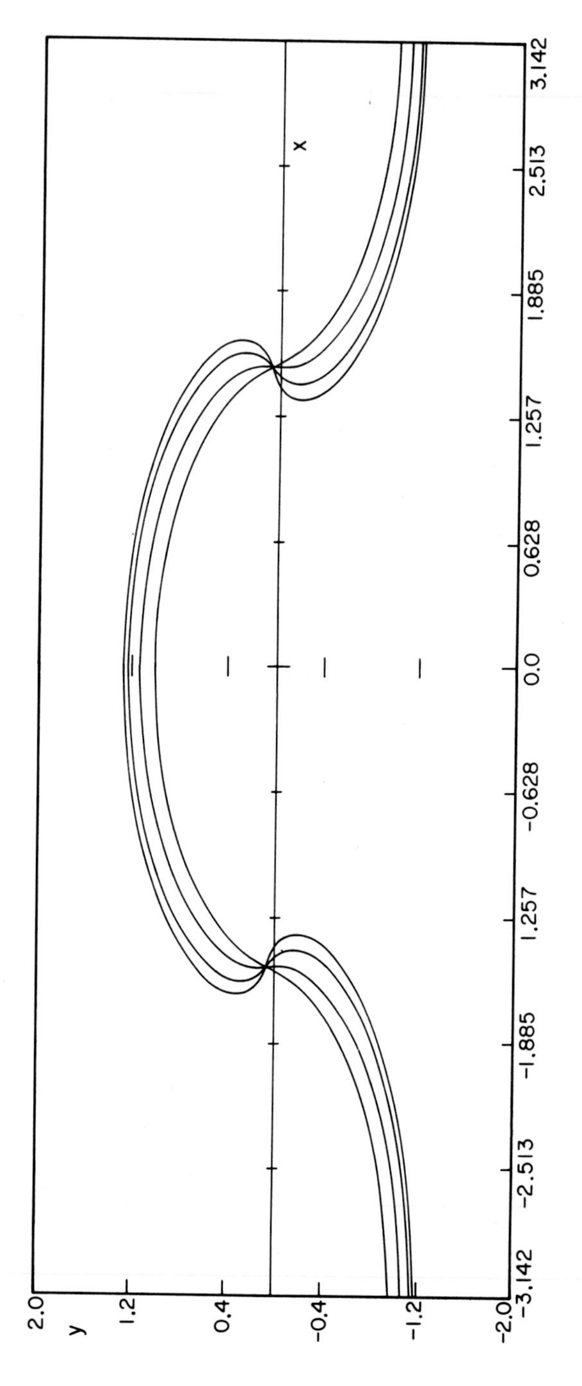

Figure 3. Profiles of steady interfacial waves for r = 0.9. The x-axis is the mean level.

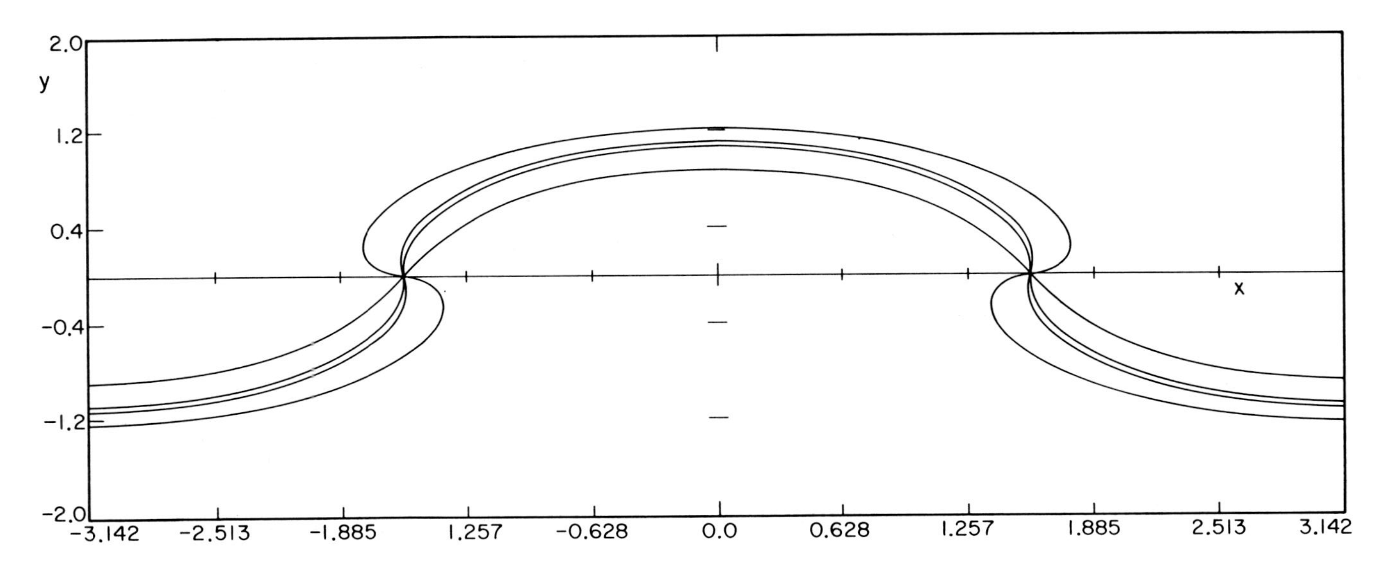

Figure 4. Profiles of steady interfacial waves for $r = 1.0$. The x-axis is the mean level. These waves are symmetrical about $y = 0$.

are shown in figures 3 and 4 for $r = 0.9$ and $r = 1.0$, respectively. These demonstrate clearly the existence of waves of permanent form with a substantial overhang region in which heavy fluid lies on top of light fluid. It is interesting to note that the fluid particles on the interface between the points of vertical tangency in the overhang region are moving faster than the wave. For the smallest value of the density ratio, we did not have sufficient resolution to distinguish the wave shape near the geometrical limit. This difficulty is to be expected, since the smaller the density ratio, the closer the geometrical limit will be to the 120° cusp and the smaller the size of the overhang region.

REFERENCES

1. Chen, B. and P. G. Saffman, Steady capillary-gravity waves on deep water, II, Numerical results for finite amplitude, Stud. App. Math. 61 (1980), 95.
2. Crapper, G. D., An exact solution for progressive capillary waves of arbitrary amplitude, Journ. Fluid Mech. 62 (1957), 532.
3. Holyer, J. Y., Large amplitude progressive interfacial waves, Journ. Fluid Mech. 93 (1979), 433.
4. Meiron, D. I. and P. G. Saffman, Overhanging interfacial gravity waves of large amplitude, Journ. Fluid Mech. (sub judice).
5. Meiron, D. I., P. G. Saffman, and H. C. Yuen, Calculation of steady three-dimensional deep-water waves, Journ. Fluid Mech. 124 (1982), 109.
6. Saffman, P. G. and H. C. Yuen, Finite amplitude interfacial waves in the presence of a current, Journ. Fluid Mech. (1982) to be published.
7. Tsuji, Y. and Y. Nagata, Stokes expansion of internal deep water waves to the fifth order, J. Ocean Soc. Japan 29 (1973), 61.

The work described here was done in collaboration with Dr. H. C. Yuen and Dr. D. I. Meiron. The author was supported partially by the Office of Naval Research and the National Science Foundation Grant OCE 8100517.

Applied Mathematics
California Institute of Technology
Pasadena, CA 91125

Mathematics Research Center
University of Wisconsin-Madison
Madison, WI 53705

INSTABILITY OF FINITE-AMPLITUDE INTERFACIAL WAVES

H. C. Yuen

1. INTRODUCTION

In this paper, we study the stability of steady, finite amplitude interfacial waves in the presence of a current jump to infinitesimal three-dimensional disturbances. The flow is assumed to be inviscid, incompressible, irrotational, and the effect of surface tension is neglected. The steady waves have been calculated and described by Saffman and Yuen [7] and Saffman [5].

It should be noted that the presence of a current jump at the interface causes the flat surface to be unstable to disturbances of sufficiently short wavelength. This short wave instability is the well-known Kelvin-Helmholtz instability, which can be suppressed by the inclusion of surface tension (Lamb [1]). Our main interest here is in instabilities of longer scales which are Kelvin-Helmholtz stable. On such long length scales, the neglect of surface tension is justified.

This study is motivated by the recent discovery of McLean et al [4] (see also McLean [2,3]) of two and three dimensional instabilities of water waves of finite amplitude in which new types of instability behavior were identified. These instabilities were also observed experimentally by Su [8]. The experimental observations of interfacial waves in the presence of current jumps have been sparse, and one purpose of the

ISBN 0-12-493220-7

present study is to establish the presence of long wave instabilities different from the classical Kelvin-Helmholtz type which may be dominant when the unperturbed waves are of finite amplitude.

The method of analysis follows that used by McLean et al [4]. In a frame of reference moving with phase speed C of the steady wave, the unperturbed interface can be expressed as

$$Z(x) = \sum_{n=1}^{\infty} A_n \cos nx. \tag{1.1}$$

The scale is chosen so that the wavelength of the undisturbed wave is 2π. We impose a three-dimensional infinitesimal disturbance η of the form

$$\eta = e^{-i\sigma t} e^{i(px+qy)} \sum_{-\infty}^{\infty} a_n e^{inx}. \tag{1.2}$$

It is a consequence of Floquet theory that the normal modes of the disturbed system are of this form where p and q are arbitrary real numbers. The equations of motion will determine σ as a function of p and q and the amplitude of the undisturbed wave. σ is a complex number. The real part of σ gives the speed of the disturbance relative to the undisturbed wave, and the imaginary part of σ gives the growth rate of an unstable mode. Imaginary part of σ being zero indicates stability. We shall now describe the procedure to calculate σ as a function of p,q and the amplitude of the unperturbed wave as given by (1.1).

2. GOVERNING EQUATIONS

The velocity potentials of the perturbed motion below and above the interface are given respectively by

$$\Phi'(x,y) + \phi'(x,y,z,t) \qquad \text{ABOVE} \tag{2.1}$$

$$\Phi(x,y) + \phi(x,y,z,t). \qquad \text{BELOW} \tag{2.2}$$

We shall denote by U the magnitude of the current jump, and by C the phase speed of the unperturbed wave relative to the fluid at great depths. Without loss of generality, we shall take g = 1, the density of the lower fluid also to be unity, and denote the density of the upper fluid by R (R < 1). The height of the undisturbed wave (i.e., vertical distance between crest and trough) is denoted by H.

We assume that the disturbances are small compared with the unperturbed quantities:

$$|\eta| << Z, \quad |\phi| << |\Phi| \tag{2.3}$$

so that the governing equations can be linearized about the unperturbed state. We then obtain

$$\nabla^2\phi' = 0 \qquad \text{ABOVE} \tag{2.4}$$

$$\nabla^2\phi = 0 \qquad \text{BELOW} \tag{2.5}$$

in the fluids above and below the interface, with the dynamical condition

$$\phi_t + \eta + \nabla\Phi\cdot\nabla\phi + (\nabla\Phi\cdot\nabla\Phi_z)\eta = R\left\{\phi'_t + \eta + \nabla\Phi'\cdot\nabla\phi' + (\nabla\Phi'\cdot\nabla\Phi'_z)\eta\right\} \tag{2.6}$$

and the kinematic conditions

$$\nabla\Phi'\cdot\nabla\eta + \nabla Z\cdot\nabla\phi' + (\nabla\Phi'_z\cdot Z_x - \Phi'_{zz})\eta - \phi'_z = 0, \tag{2.7}$$

$$\nabla\Phi\cdot\nabla\eta + \nabla Z\cdot\nabla\phi + (\nabla\Phi_z\cdot Z_x - \Phi_{zz})\eta - \phi_z = 0, \tag{2.8}$$

to be applied at the unperturbed surface z = Z(x). In addition, the disturbances are assumed to vanish far from the interface, so that we have

$$\phi' \sim 0 \qquad z \to \infty \tag{2.9}$$

$$\phi \sim 0 \qquad z \to -\infty. \tag{2.10}$$

Consistent with the assumed form (1.2) of the perturbation to the surface, the disturbed velocity potentials take the forms

$$\phi' = e^{-i\sigma t}e^{i(px+qy)}\sum_{-\infty}^{\infty} b'_n e^{inx} e^{-\sqrt{(p+n)^2+q^2}(z-z_t)}, \tag{2.11}$$

$$\phi = e^{-i\sigma t}e^{i(px+qy)}\sum_{-\infty}^{\infty} b_n e^{inx} e^{\sqrt{(p+n)^2+q^2}(z-z_c)}. \tag{2.12}$$

where z_c is the height of the crest and z_t is the height of the trough. Substitution of the expressions (1.2), (2.11) and (2.12) into the governing equations (2.6), (2.7) and (2.8) gives the following system of equations for σ, a_n, b_n and b'_n:

$$\left[(1 + \Phi_x\Phi_{xz} + \Phi_z\Phi_{zz}) - R(1 + \Phi'_x\Phi'_{xz} + \Phi'_z\Phi'_{zz})\right]\sum_{-\infty}^{\infty}a_n e^{inx}$$

$$+\sum\left\{i(p+n)\Phi_x + \left[(p+n)^2+q^2\right]^{\frac{1}{2}}\Phi_z\right\}$$

$$\times\ b_n e^{inx} e^{\left[(p+n)^2+q^2\right]^{\frac{1}{2}}\left[Z(x)-z_c\right]}$$

$$-\ R\sum\left\{i(p+n)\Phi'_x - \left[(p+n)^2+q^2\right]^{\frac{1}{2}}\Phi'_z\right\}$$

$$\times\ b'_n e^{inx} e^{-\left[(p+n)^2+q^2\right]^{\frac{1}{2}}\left[Z(x)-z_t\right]}$$

$$= i\sigma\left\{\sum b_n e^{inx} e^{\left[(p+n)^2+q^2\right]^{\frac{1}{2}}\left[Z(x)-z_c\right]}\right.$$

$$\left.+\sum b'_n e^{inx} e^{-\left[(p+n)^2+q^2\right]^{\frac{1}{2}}\left[Z(x)-z_c\right]}\right\} \tag{2.13}$$

$$\sum_{-\infty}^{\infty}\left[\Phi_{xz}Z_x - \Phi_{zz} + i(p+n)\Phi_x\right]a_n e^{inx}$$

$$+\sum_{-\infty}^{\infty}\left\{i(p+n)Z_x - \left[(p+n)^2+q^2\right]^{\frac{1}{2}}\right\}$$

$$\times\ b_n e^{inx} e^{\left[(p+n)^2+q^2\right]^{\frac{1}{2}}\left[Z(x)-z_c\right]} = i\sigma\sum_{-\infty}^{\infty}a_n e^{inx} \tag{2.14}$$

$$\sum_{-\infty}^{\infty}\left[\Phi'_{xz}Z_x - \Phi'_{zz} + i(p+n)\Phi'_x\right]a_n e^{inx}$$

$$+\sum_{-\infty}^{\infty}\left\{i(p+n)Z_x + \left[(p+n)^2+q^2\right]^{\frac{1}{2}}\right\}$$

$$\times\ b'_n e^{inx} e^{-\left[(p+n)^2+q^2\right]^{\frac{1}{2}}\left[Z(x)-z_t\right]} = i\sigma\sum_{-\infty}^{\infty}a_n e^{inx} \tag{2.15}$$

This can be interpreted as an eigenvalue problem for the normal modes of the disturbance with σ as the eigenvalue and a'_n b_n and b'_n as the eigenfunctions. From the structure of the

Euler equations, which are time reversible (since there is no dissipation) and real, we know that whenever σ is an eigenvalue, so are $-\sigma$, σ^* and $-\sigma^*$. Stable eigenvalues appear in pairs, whereas unstable eigenvalues appear in quartets. Consequently, the transition from stability to instability is marked by the coalescence of pairs of eigenvalues which can also be interpreted as a resonance between two modes which have the same frequency in the frame of reference moving with the unperturbed wave.

It is important to note that there is a degeneracy in the dependence of the eigenmodes on p, since the value of p is arbitrary to the addition of an integer as can be seen in the representations (1.2), (2.11) and (2.12). This means that when p is changed to p + n, where n is an integer, we obtain the same eigenmode provided that a_n, b_n and b'_n are changed to a_{n-p}, b_{n-p} and b'_{n-p} accordingly. In principle, there would be no loss of generality in confining ourselves to the range $0 < p < 1$, but it turns out to be convenient not to do this and allow for some arbitrariness in the labeling of the eigenmodes.

3. THE WAVE OF ZERO HEIGHT

For the limiting case where the undisturbed wave has zero wave height (the flat surface), the solutions of the eigenvalue problem give

$$\sigma_{\pm} = \frac{kUR}{1 + R} \pm \sqrt{\frac{(1 - R^2)k - Rk^2U^2}{(1 + R)^2}} - (p + n)C \tag{3.1}$$

$$k = \left[(p + n)^2 + q^2\right]^{\frac{1}{2}} \tag{3.2}$$

and the eigenvectors are such that only those components with suffices p + n are nonzero. Since the flat surface here is regarded as the limit of steady waves with the wave height approaching zero, the eigenvalue contains the wave speed C. It follows from (3.1) that the flat surface is unstable if

$$U^2 > \frac{(1 - R^2)}{Rk} \qquad \text{FOR GIVEN } k \tag{3.3}$$

or

$$k > \frac{(1 - R^2)}{U^2R} \qquad \text{FOR GIVEN } U \tag{3.4}$$

which is the classical Kelvin-Helmholtz instability result. Note that (3.4) confines the Kelvin-Helmholtz instability to sufficiently short waves for a given value of U. We are concerned with properties of waves and disturbances with wavelengths longer than this critical value and which are hence Kelvin-Helmholtz stable.

As pointed out by McLean et al [4], instability modes of finite amplitude waves can be identified by resonance among normal modes associated with the flat surface. This resonance will occur whenever there exist p, q and integers n and n' such that

$$\sigma_+(p + n, q) = \sigma_-(p + n', q). \tag{3.5}$$

Making use of the degeneracy associated with p, we can separate the resonance conditions into two classes

$$n = m, \; n' = -m \qquad \text{CLASS I} \tag{3.6}$$

$$n = m, \; n' = -(m + 1) \qquad \text{CLASS II} \tag{3.7}$$

The results for interfacial waves with current are rather complicated, and it will be helpful to consider first a special case.

4. WATER WAVES: A SPECIAL CASE

An important and interesting special case is water waves. In this case, the density ratio R is assumed to be zero, and the current jump U drops out of the equations. The wave height H then becomes the only parameter describing the unperturbed wave. This case was studied and reported in detail by McLean et al [4] and McLean [2]. The results are summarized here since they permit a clearer insight into the effects of finite amplitude and provide a framework for the discussion of the more complicated effects when density ratio and current strength are considered.

The condition (3.5) gives the two classes of resonance

$$\left[(p + m)^2 + q^2\right]^{\frac{1}{4}} + \left[(p - m)^2 + q^2\right]^{\frac{1}{4}} = 2m \qquad \text{CLASS I} \tag{4.1}$$

$$\left[(p + m)^2 + q^2\right]^{\frac{1}{4}} + \left[(p - m - 1)^2 + q^2\right]^{\frac{1}{4}} = 2m + 1 \qquad \text{CLASS II} \tag{4.2}$$

where m is a positive integer. A plot of the two sets of resonance curves for m = 1 and m = 2 is given in Figure 1.

These curves describe the loci of perturbation wave vectors which have two pairs of eigenvalues that coincide in the limit $H = 0$. The effect of finite H is to merge these pairs of real eigenvalues into a quartet of complex eigenvalues representing instability. The result is a band of instability developing from the resonance curves, with the width of the band proportional to some power of H. The stability diagrams for the two classes of instabilities with $m = 1$ are shown in Figure 2 for different values of H, up to nearly the limiting value of $H = 0.892$. The location and growth rate of the most unstable disturbance in each class are noted. It can be seen that for values of $H < 0.55$, the Class I instability dominates, and the strongest instability is two dimensional ($q = 0$). For larger values of H, the Class II instability overtakes the Class I, with the most unstable mode being three dimensional. The most unstable Class II modes always have $p = 0.5$, which represents modulations in which every other wave crest is enhanced. The value of q for the most unstable Class II mode decreases with increasing H but remains nonzero even for the highest value of H calculated, indicating that the instability is three dimensional. Another interesting property of the Class II waves is that along the line $p = 0.5$, the unstable disturbances (including the most unstable disturbances) have zero real parts, so that they copropagate with the unperturbed wave. It has been pointed out by Saffman and Yuen [6] that this copropagating instability can lead to the bifurcation of the uniform wave train to a steady three-dimensional modulated wave train, a phenomenon which has been observed by Su [8].

5. RESULTS FOR INTERFACIAL WAVES WITH CURRENT JUMP

In this general case, the undisturbed system is described by three parameters, namely, H, U and R. For a given density ratio, it proves convenient to scale U against U_{cr}, which is defined as the value of the current at which the unperturbed wave of wavelength 2π in the zero amplitude limit is Kelvin-Helmholtz unstable,

$$U_{cr} = \left[\frac{(1 - R)(1 + R)}{R}\right]^{\frac{1}{2}} \tag{5.1}$$

We define $U' = U/U_{cr}$.

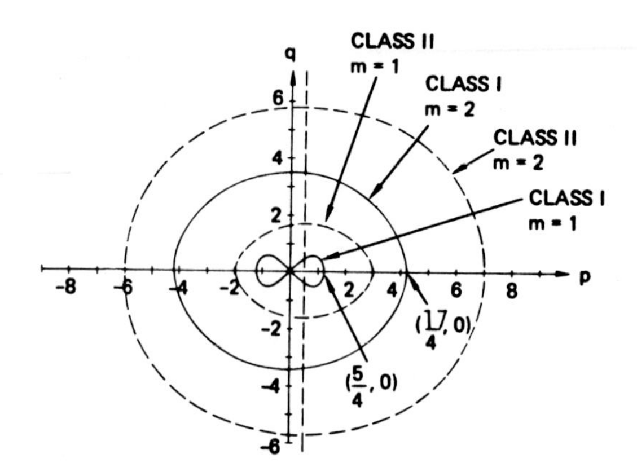

Figure 1. Resonance diagram for water waves (R = 0, U = 0) in p,q plane. Solid lines correspond to Class I resonance (eq. 4.1). Dashed lines correspond to Class II resonance (eq. 4.2).

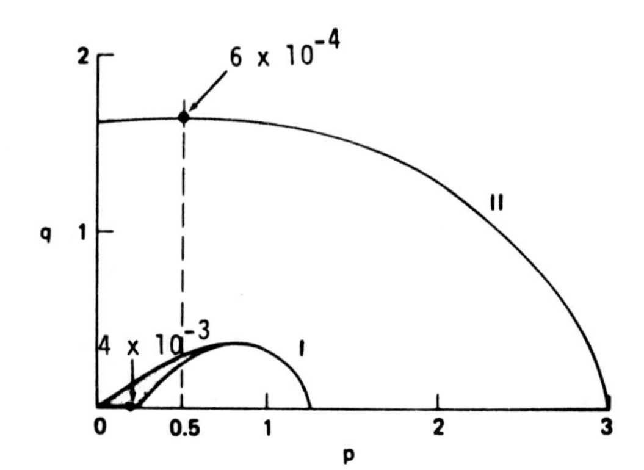

Figure 2. Instability diagram for water waves (R = 0, U = 0) in p,q plane for various values of H. The location and magnitude of the maximum instability for each class of instability are indicated. (a) H = 0.2, b) H = 0.4, (c) H = 0.6, (d) H = 0.7, (e) H = 0.8, f) H = 0.823.

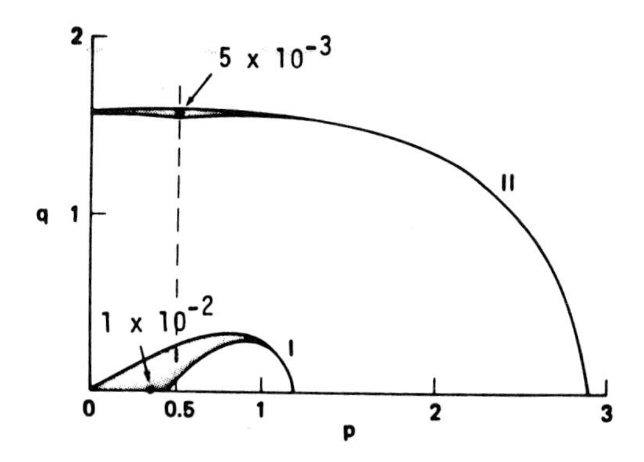

Figure 2a. See caption - Figure 2.

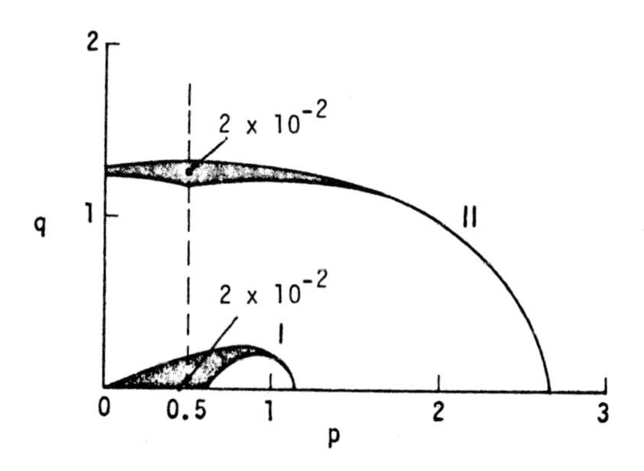

Figure 2b. See caption - Figure 2.

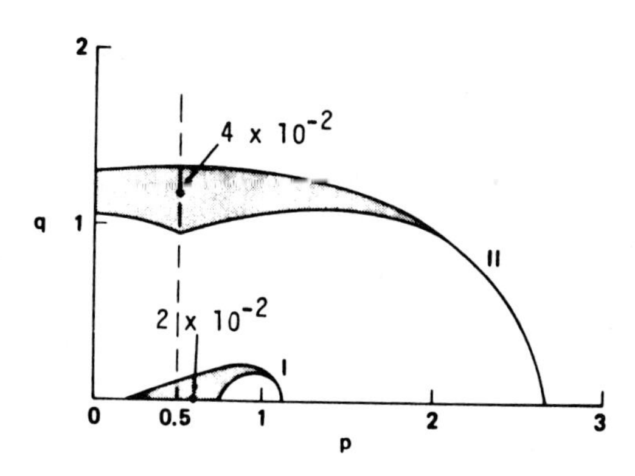

Figure 2c. See caption - Figure 2.

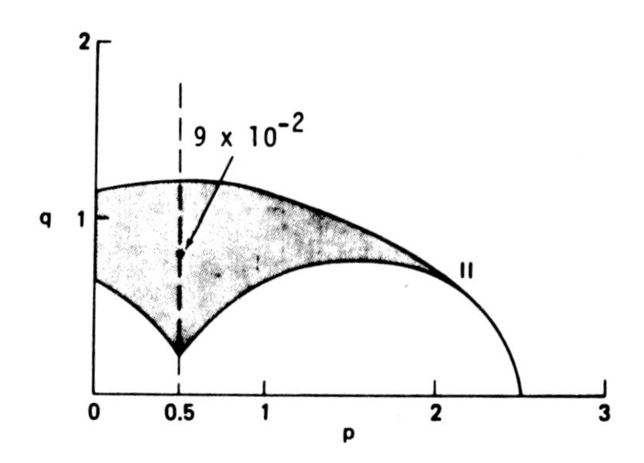

Figure 2e. See caption - Figure 2.

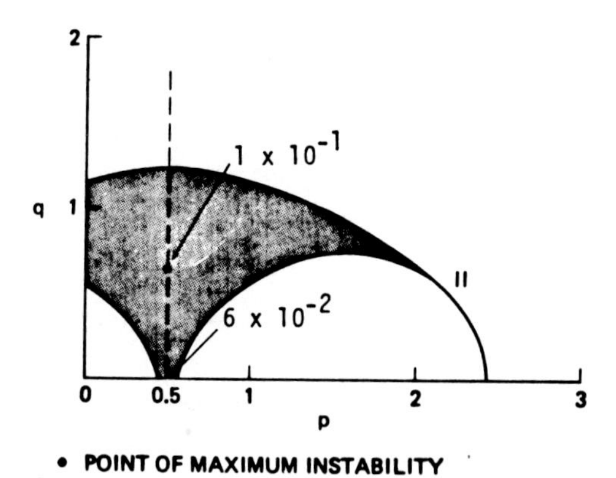

Figure 2f. See caption - Figure 2.

When U' is zero, the resonance curves are not altered from the water wave case since the effect of density ratio is merely to scale the value of gravity. In Figure 3, we show the resonance curves for U' = 0.0, 0.5 and 0.9, for Class I and Class II with m = 1. It can be seen that the U' = 0.0 curves are identical with those in the water wave case. When U' is nonzero, a Kelvin-Helmholtz instability boundary exists, and disturbances lying in the unstable regime are themselves Kelvin-Helmholtz unstable in the limit when the undisturbed waves have zero amplitude. The Class I resonance takes on a new branch which merges into the Kelvin-Helmholtz boundary. The Class II resonance curve, instead of closing on itself as in the U' = 0.0 case, also turns up and merges into the Kelvin-Helmholtz boundary. For values of U' greater than 0.5, the two branches of the Class I resonance curve become one and merge into the Kelvin-Helmholtz boundary as seen in the U' = 0.9 example shown. Based on the water wave results, we expect bands of instability to develop along these resonance curves when the unperturbed wave has finite amplitude.

We present the results of a number of selected cases listed in Table 1. At the present time, cases with larger values of H have not been calculated. This justifies the use of the Fourier series representation (1.1) for the unperturbed wave. As discussed in Saffman [5], large amplitude overhanging waves exist when H is greater than values around one. For such cases, other formulations of the stability analysis would be required. For the values of H in Table 1, it was found that the undisturbed waves can be adequately represented by 20 Fourier modes. We calculated numerically the eigenvalues for finite values of H by truncating the infinite system (2.13)-(2.15) to 5 modes and using library routines for the complex generalized eigenvalue problem. The calculations were carried out on a PRIME 750 minicomputer. Accuracy of the calculations was spot checked by repeating with a larger number of modes, and it was determined that the accuracy of the stability boundary results was within plotting resolution. It is expected that large values of H will require considerably more modes for the description of the unperturbed wave and the perturbation.

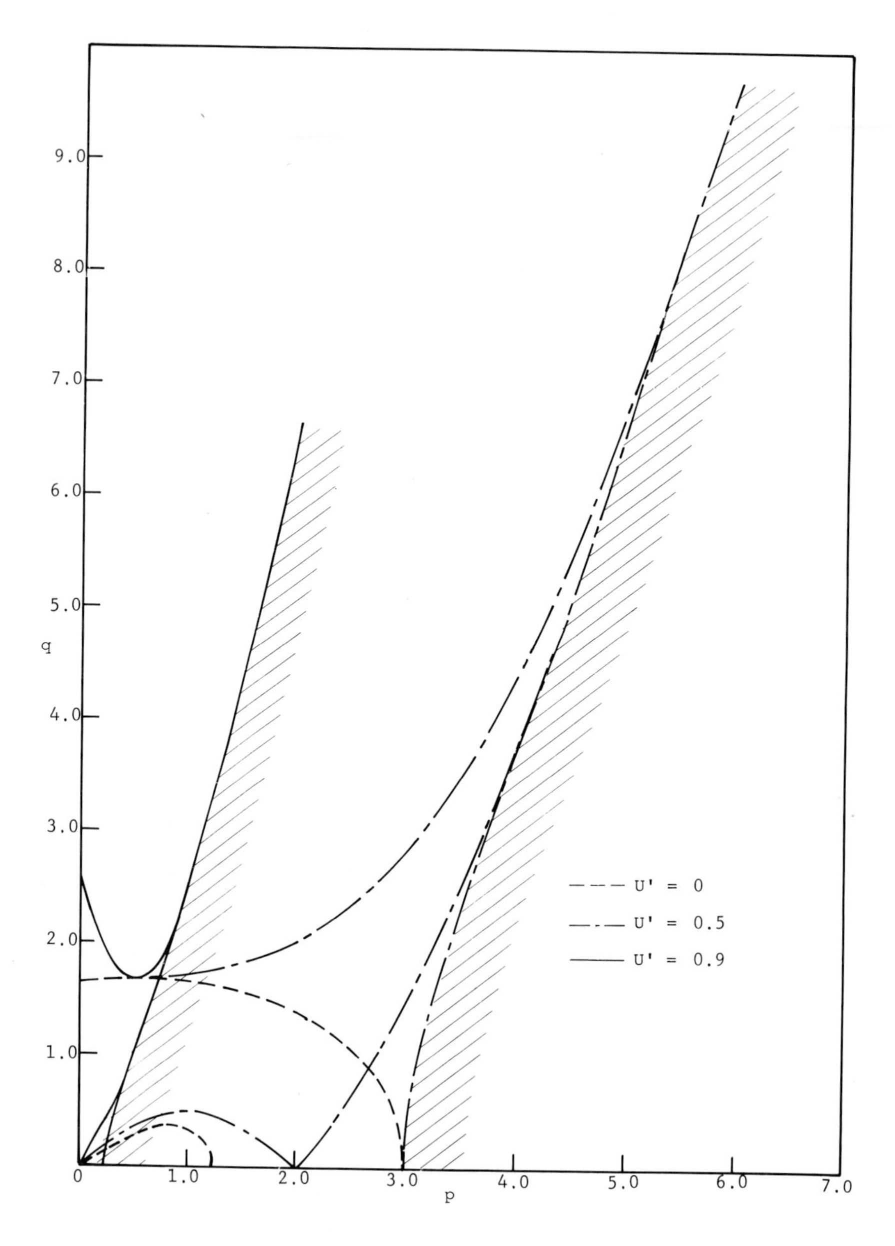

Figure 3. Resonance diagram for interfacial waves with a current jump. Dashed lines correspond to U' = 0 case. Dash-dot lines correspond to U' = 0.5 case. Solid lines correspond to U' = 0.9 case. Shaded regions indicate the boundaries of Kelvin-Helmholtz instability, which is present only for nonzero U'. Note that the U' = 0 case is similar to that for water waves.

R	H	U'	FIGURE
0.1	0.2	0.0	4a
		0.5	4b
		0.9	4c
0.9	0.2	0.0	4d
		0.5	4e
		0.9	4f
0.9	0.5	0.5	4g
		0.9	4h

Table 1: Finite amplitude cases calculated. Results for each case are shown in Figures as indicated.

The stability diagrams for the selected cases are shown in Figure 4.

It can be seen that the cases where $U' = 0$ (Figures 4a and 4d) closely resemble each other and the water wave results. The cases with $H = 0.2$ and $U' = 0.5$ (Figures 4b and 4e) show that the two branches of the Class I resonance curves develop into two distinct regions of instability, but that the general trend is consistent with the expectation that instability bands lie near the resonance curves. The cases with $H = 0.2$ and $U' = 0.9$ (Figures 4c and 4f) show the merging of the Class I instability into the Kelvin-Helmholtz instability. For this value of H, the Class II instability has a very narrow bandwidth.

The large amplitude cases (Figures 4g and 4h) show significantly broadened instability bands for Class II, and a strong increase in the maximum growth rate for the Class II instability, even though it is still smaller than that for the Class I. Figure 4h shows that the maximum growth rate associated with the Class I instability is no longer two dimensional as in the case of $U' = 0$ (and the water wave case).

An interesting phenomenon is encountered in the results shown in Figure 4g. It is seen that a band of instability, not identified with any resonance curves and absent for smaller values of H, appears around $p = 0.7$. Detailed examination of this instability reveals that it is associated with the Kelvin-Helmholtz instability in the following way. We recall that the value of p is arbitrary to the addition of an integer. In the results shown in Figure 4, the value of p was fixed by the requirement that the Fourier component of the perturbation with largest magnitude has wavenumber $p + 1$. In other words, we identify a disturbance by its dominant component. The effect of finite amplitude is usually to increase the relative magnitude of the other components. It can then happen that at some wave height a different component becomes dominant and "nonlinear mode jumping" occurs. This is what has happened in Figure 4g. The effect of finite amplitude on the eigenfunctions of the Kelvin-Helmholtz unstable waves is to introduce a low wavenumber (or long-wave) component into their structures, and the appearance of the apparently new

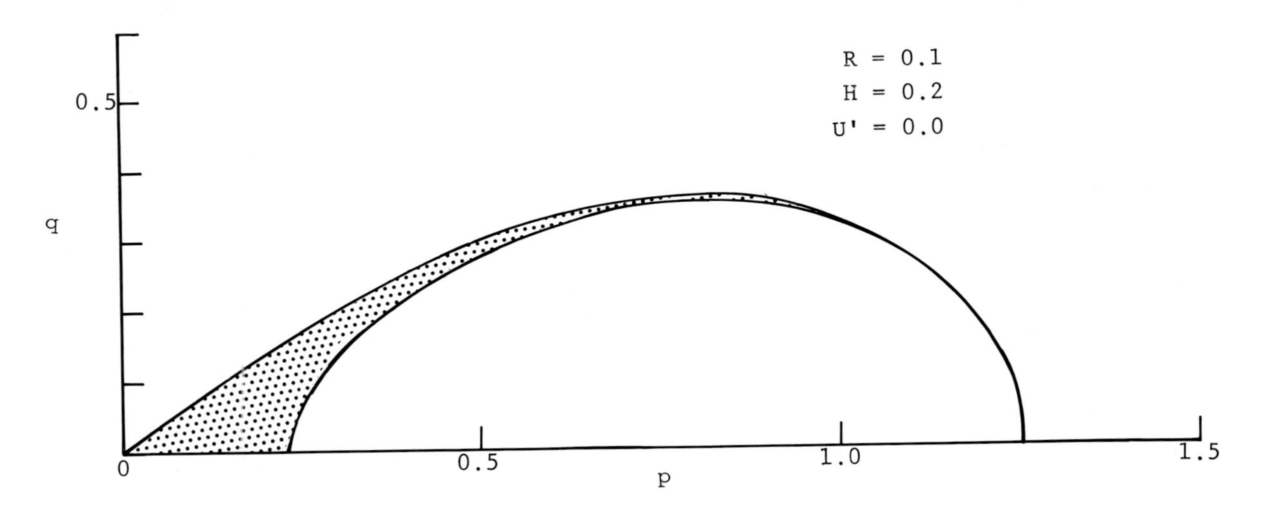

Figure 4a. Instability diagrams for interfacial waves with a current jump. Different values of R, H and U' correspond to cases indicated in Table 1.

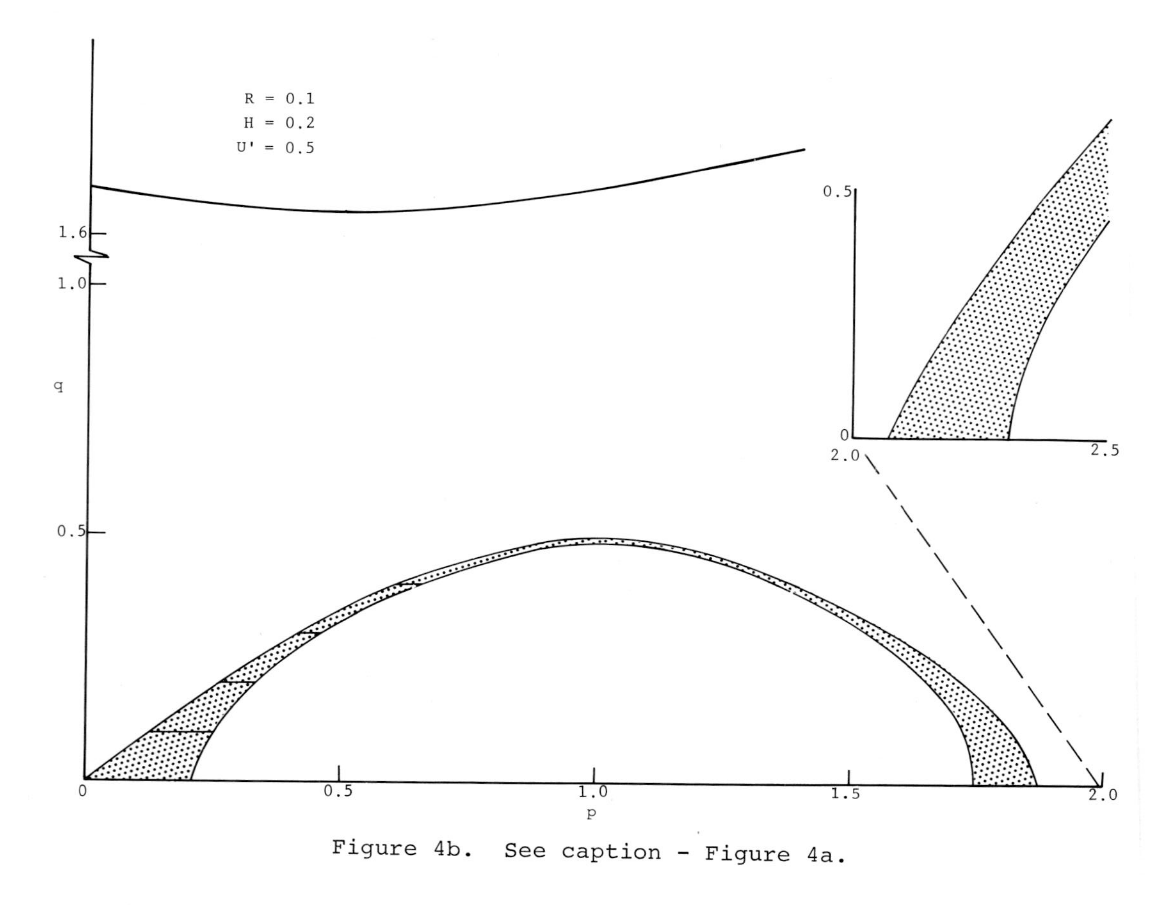

Figure 4b. See caption - Figure 4a.

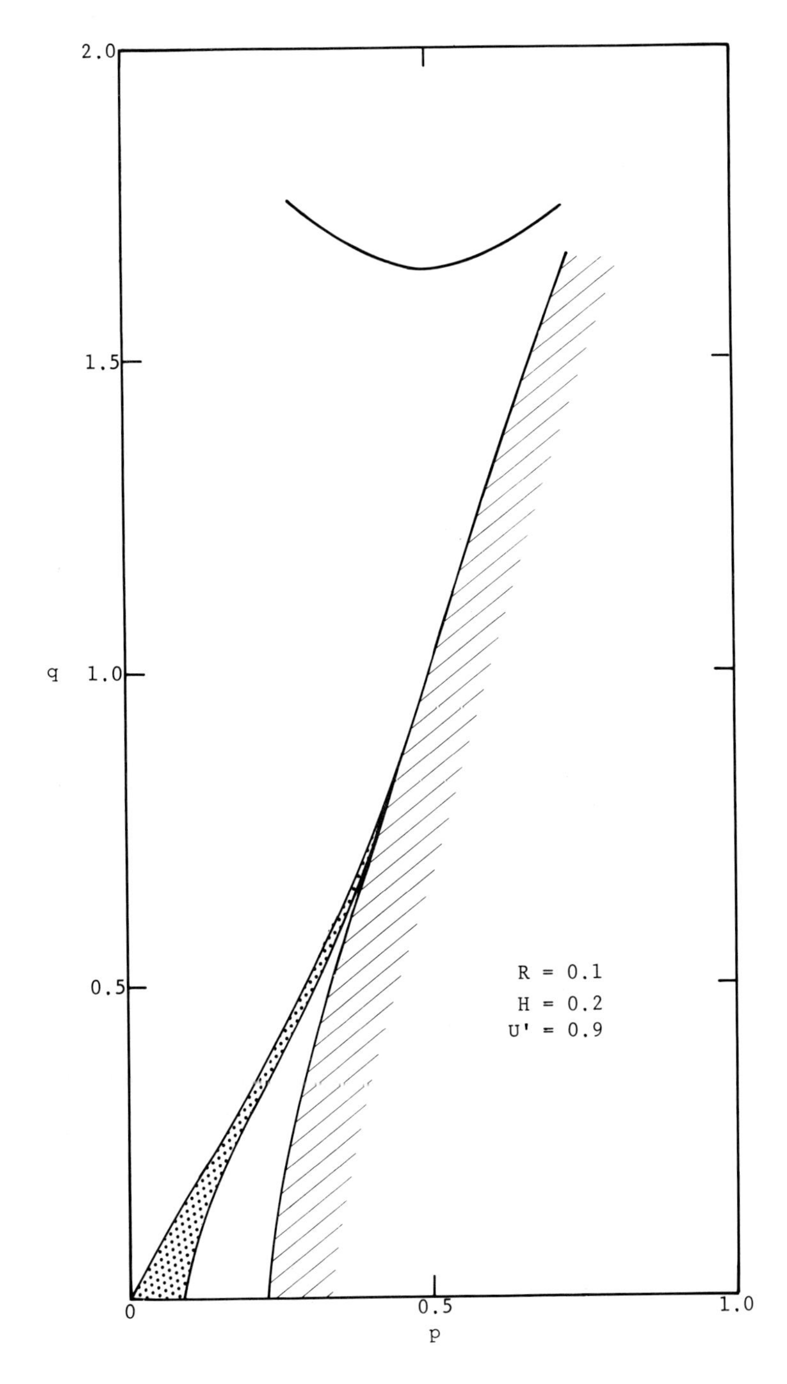

Figure 4c. See caption - Figure 4a.

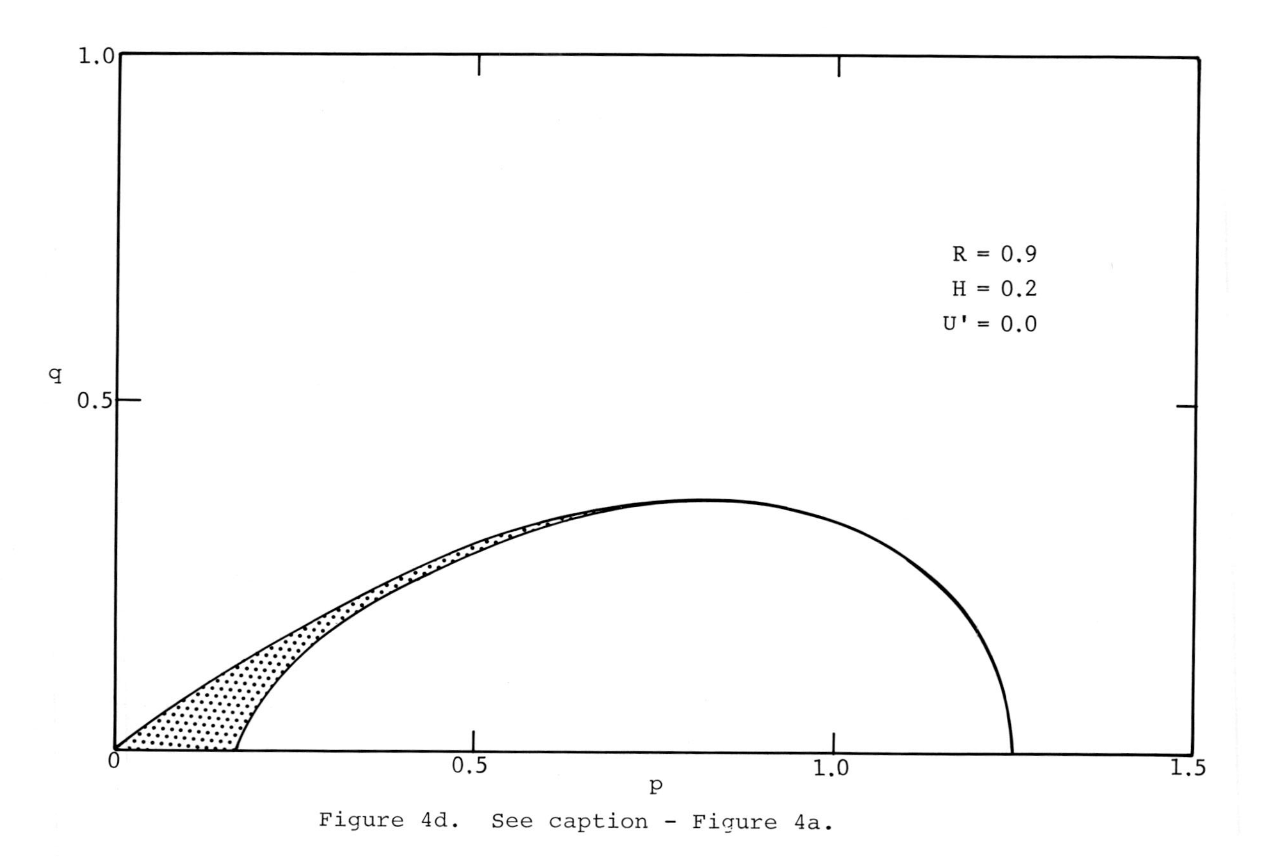

Figure 4d. See caption - Figure 4a.

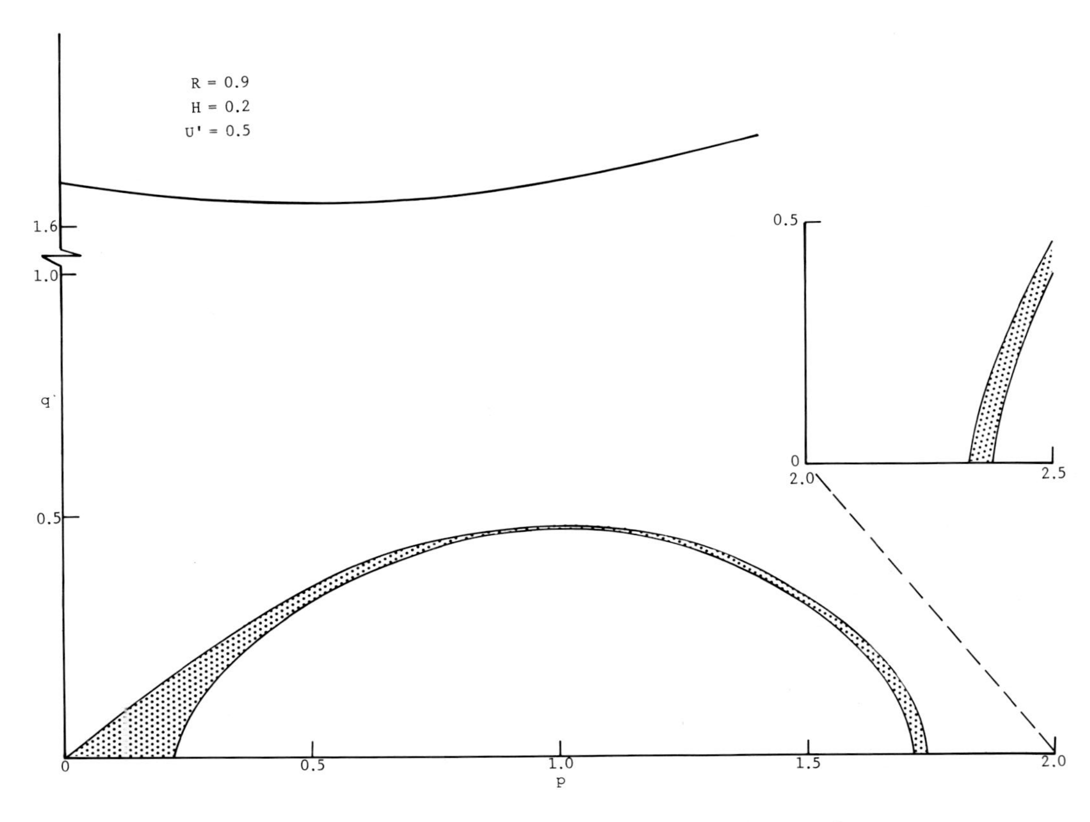

Figure 4e. See caption - Figure 4a.

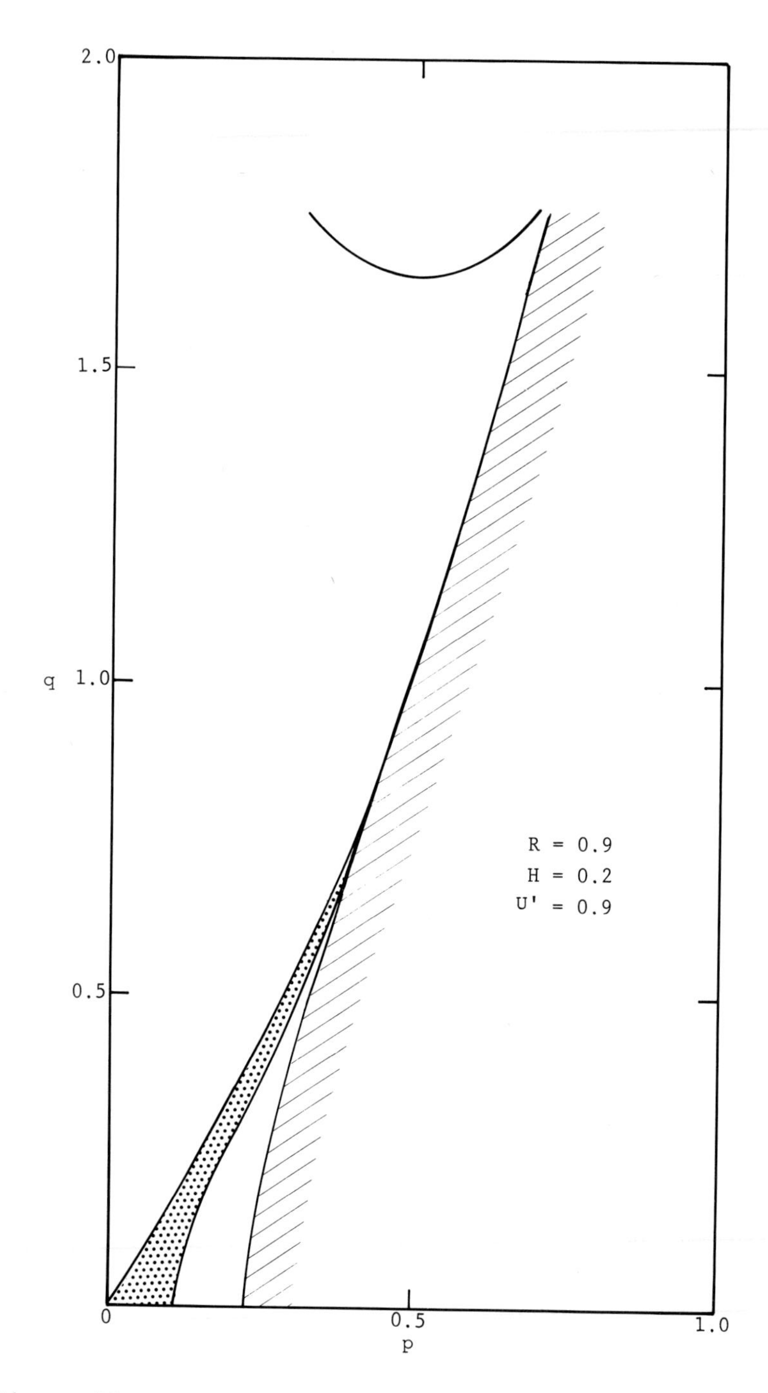

Figure 4f. See caption - Figure 4a.

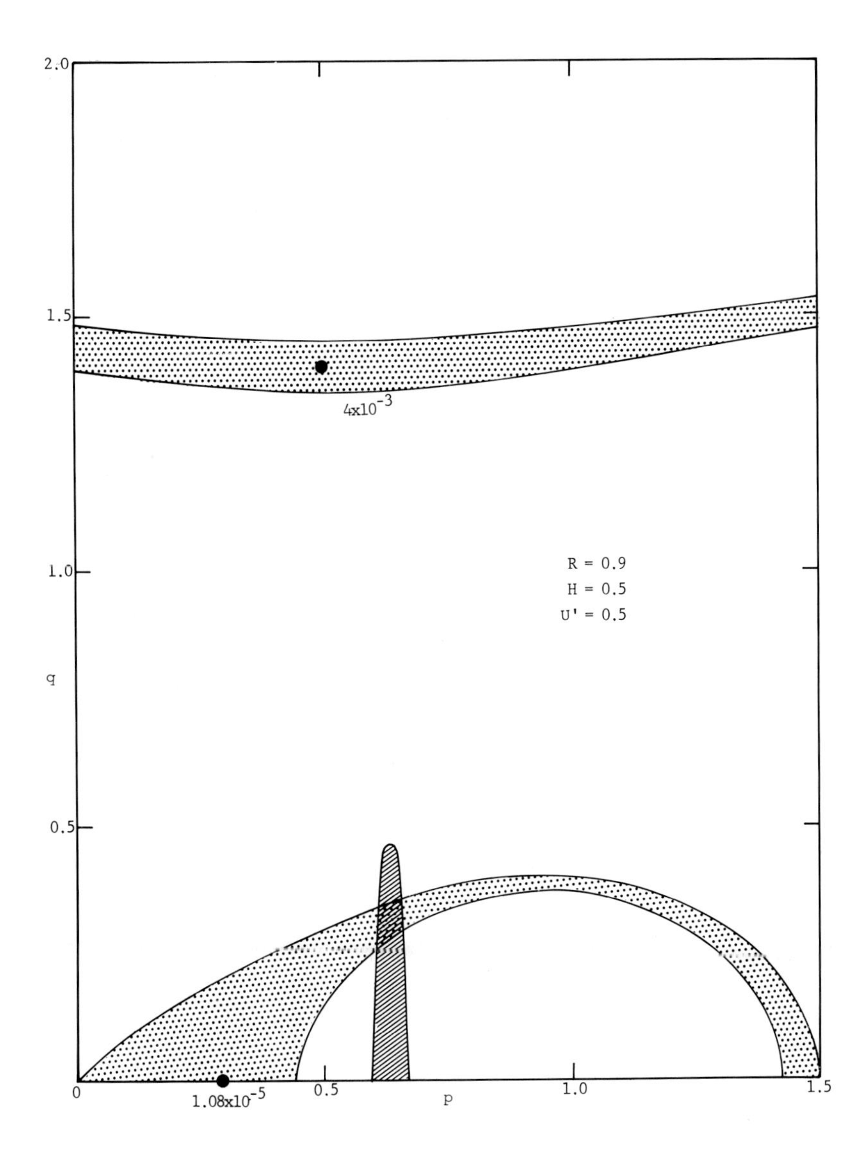

Figure 4g. See caption - Figure 4a.

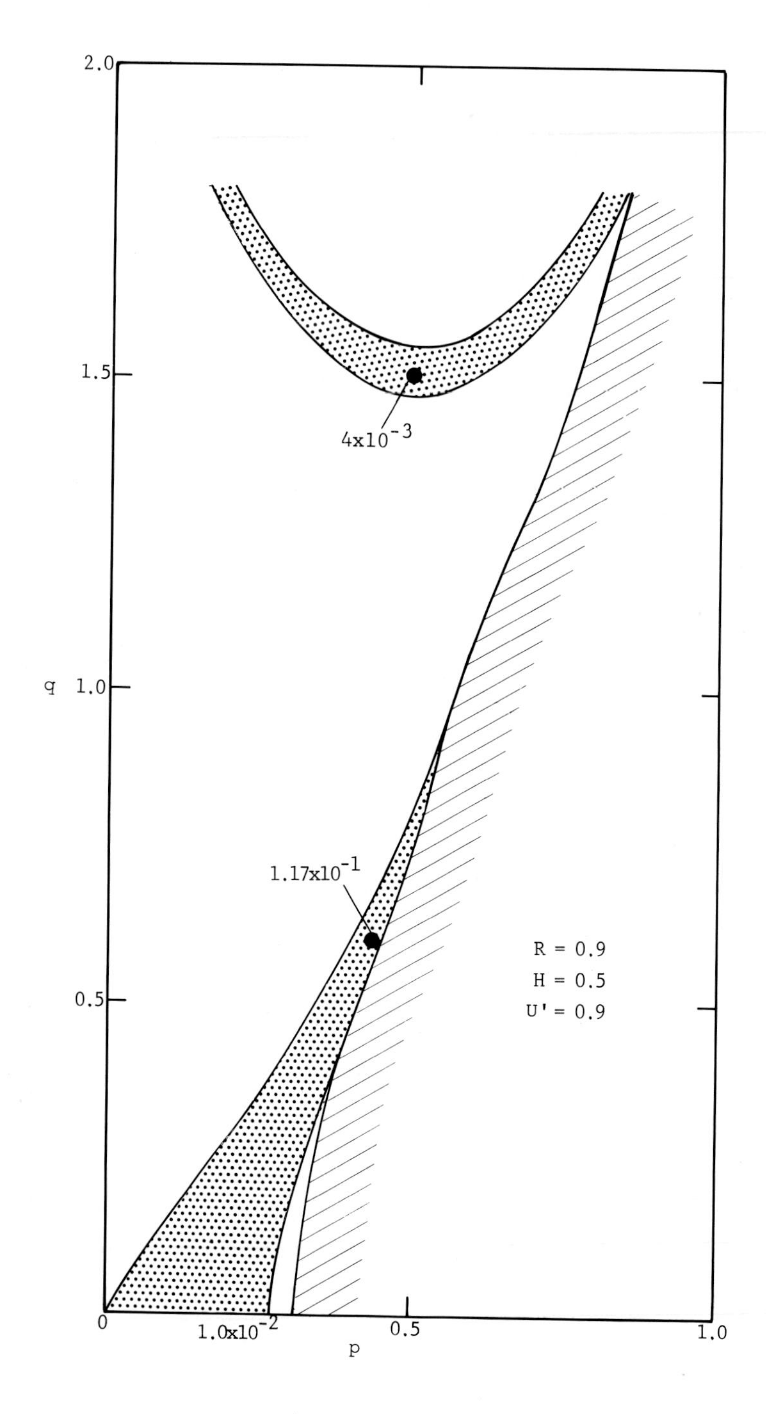

Figure 4h. See caption - Figure 4a.

longer wave instability is when the low wavenumber component becomes dominant. The appearance of this type of instability is abrupt as one increases H. To our knowledge, it is the first reported example of a nonlinear mode jumping phenomenon caused by finite amplitude effects.

6. DISCUSSION

Apart from the mode jumping phenomenon just referred to, the results more or less conform to expectations and resemble the phenomena found in surface waves. Mode resonance found for the flat surface translates into narrow bands of instability, which become wider and are associated with larger growth rates, as the unperturbed wave height increases. In addition to the Kelvin-Helmholtz instability, which is present for short waves in the presence of current, there are long wave instabilities of finite amplitude waves. We have neglected surface tension in our discussion. Since the stabilizing effect of surface tension is more effective the shorter the wave, it is expected that its presence will make the long wave finite amplitude instabilities more important, especially when the Kelvin-Helmholtz is suppressed.

It should be noted that the present results are all for C_+ waves. It is expected that similar results will hold for C_- waves, but this remains to be confirmed. The cases considered are also all for undisturbed waves with positive energy. It was mentioned in the discussion of steady waves of permanent form [5] that such waves exist with negative energy (relative to that of the flat surface). It remains to investigate these solutions, but there is no reason to believe that their stability properties are in any way significantly different. Indeed, it should be noted that there is not necessarily any significance to negative energy, because energy is not Galilean invariant. In the case of waves, it is dangerous to compare the energies of states which do not have the same momentum, and calculation indicates that the interfacial waves of the type considered here always have non-zero momentum compared to the flat surface case when the relative energies are negative. There is clearly a pressing need for good experiments on interfacial waves in the presence of current in a stably stratified situation.

REFERENCES

1. Lamb, Sir Horace, Hydrodynamics (Book), Dover Press (1932).
2. McLean, J. W., Instabilities of finite-amplitude water waves, Journ. Fluid Mech. 114 (1982), 315.
3. McLean, J. W., Instabilities of finite-amplitude gravity waves on water of finite depth, Journ. Fluid Mech. 114 (1982), 331.
4. McLean, J. W., Ma, Y. C., Martin, D. U., Saffman, P. G. and Yuen, H. C., Three-dimensional instability of finite amplitude water waves, Physical Review Letts. 46 (1981), 817.
5. Saffman, P. G., Finite amplitude interfacial waves (this Volume) (1982).
6. Saffman, P. G. and Yuen, H. C., A new type of three-dimensional deep-water wave of permanent form, Journ. Fluid Mech. 101 (1981), 797.
7. Saffman, P. G. and Yuen, H. C., Finite amplitude interfacial waves in presence of a current, Journ. Fluid Mech. (to be published).
8. Su, M. Y., Three-dimensional deep-water waves. Part I. Laboratory experiments on spilling breakers, Journ. Fluid Mech. (to be published).

The work reported here was done in collaboration with Professor P. G. Saffman and was partially supported by the National Science Foundation Grant OCE 8100517.

Fluid Mechanics Department
R1/1008
TRW Space and Technology Group
One Space Park
Redondo Beach, CA 90278

FINITE-AMPLITUDE WATER WAVES WITH SURFACE TENSION

J.-M. Vanden-Broeck

1. INTRODUCTION

This paper deals with the calculation of symmetric finite amplitude waves propagating without change of form on the surface of a liquid above a horizontal flat bottom. The free surface condition is used in its exact nonlinear form, including the effects of capillarity.

The problem is formulated in section 2 and some perturbation solutions for small values of the amplitude and of the depth are reviewed in section 3. In sections 4 and 5 the fully nonlinear problem is solved by discretization of an equivalent integro-differential equation. The numerical results are discussed in section 6.

2. FORMULATION

We consider two-dimensional, periodic waves propagating on the surface of a liquid of density ρ under the combined effects of gravity g and surface tension T over a horizontal bottom. We choose a frame of reference in which the waves are steady, as is the fluid motion, which is assumed to be a potential flow. The phase velocity C is defined as the average fluid velocity at any horizontal level completely within the fluid.

We introduce a potential function ϕ and a stream function Ψ. Let the stream function assume the values zero and $-Q$ on the free surface and on the bottom

ISBN 0-12-493220-7

respectively. The undisturbed fluid depth H is defined by

$$H = \frac{Q}{C} . \tag{2.1}$$

We take the origin of our coordinate system on the undisturbed level of the free surface, so that the bottom is given by $y = -H$. The x-axis is parallel to the bottom and the y-axis is chosen as a line of symmetry of the wave. The equation of the free surface is given by $y = \zeta(x)$.

The exact dimensional equations for ϕ and ζ are

$$\phi_{xx} + \phi_{yy} = 0 \qquad -H < y < \zeta(x) \tag{2.2}$$

$$\phi_y = 0 \quad \text{on} \quad y = -H \tag{2.3}$$

$$\phi_x \zeta_x - \phi_y = 0 \quad \text{on} \quad y = \zeta(x) \tag{2.4}$$

$$\frac{1}{2}(\phi_x^2 + \phi_y^2) + g\zeta - \frac{T}{\rho} \frac{\zeta_{xx}}{(1 + \zeta_x^2)^{3/2}} = \frac{1}{2} b^2$$

$$\text{on} \quad y = \zeta(x) \tag{2.5}$$

Equation (2.2) follows from conservation of mass and absence of vorticity. Equations (2.3) and (2.4) are the conditions that the bottom and the free surface are streamlines and equation (2.5) is the dynamic free surface condition. The Bernoulli constant b on the right hand side of (2.5) is to be found as part of the solution.

We seek solutions for periodic waves of wavelength λ. Solitary waves are the limit of periodic waves as $\lambda/H \to \infty$. In that case $b = C$.

3. PERTURBATION SOLUTIONS

In order to derive perturbation solutions it is convenient to introduce the following dimensionless parameters

$$\alpha = \frac{A}{H}, \qquad \beta = \frac{H^2}{L^2}, \qquad \tau = \frac{T}{\rho g H^2} \tag{3.1}$$

Here L is a length scale of the wave in the x direction and A is a measure of the amplitude.

Stokes (1847, 1880) derived a perturbation solution for pure gravity waves by assuming α small and β of order one. His results were generalized to include the effect of surface tension by Harrison (1909), Wilton (1915), Nayfeh (1970), Schwartz and Vanden-Broeck (1979), Hogan (1980, 1981) and Chen and Saffman (1980). These calculations indicate that many different families of gravity capillary waves exist for $\tau < 1/3$. However this problem of nonuniqueness disappears for $\tau > 1/3$.

Chen and Saffman (1980) showed that gravity waves cannot be obtained as the continuous limit of a capillary gravity wave as $T \to 0$. Although their analysis was restricted to water of infinite depth it seems likely that similar results hold for $0 < \tau < 1/3$. Therefore the nonuniqueness of gravity capillary waves does not imply the nonuniqueness of gravity waves.

Stokes' perturbation calculations become invalid as $\beta \to 0$ because the ratio of successive terms is then unbounded.

The shallow water equations are derived by assuming β small and α of order one. These equations do not have travelling wave solutions because the dispersive effects are neglected (Whitham (1974) p. 457). The inclusion of dispersive effects into the shallow water theory leads to the Korteweg de Vries equation. This equation can be derived by assuming

$$\alpha = \beta = \varepsilon \tag{3.1}$$

and expanding ϕ, ζ and b in powers of ε. (Keller (1948), Vanden-Broeck and Shen (1982)). Korteweg and de Vries (1895) showed that this equation possesses periodic solutions. As the wavelength tends to infinity these waves tend to the solitary wave

$$\frac{\zeta}{H} = A \operatorname{sech}^2\left[\frac{3A}{4(1 - 3\tau)}\right]^{1/2} \frac{x}{H} \tag{3.2}$$

$$c^2 = (1 + A)gH \tag{3.3}$$

When $\tau < 1/3$ these are elevation waves, when $\tau > 1/3$ they are depression waves. Equation (3.2) shows that the slope of the wave profile becomes large near $\tau = 1/3$ and the solution ceases to exist altogether when $\tau = 1/3$. Thus the Korteweg de Vries equation is not valid in the neighborhood of $\tau = 1/3$. This is due to the fact that the dispersive effects disappear as τ approaches 1/3.

A new equation analogous to the Korteweg de Vries equation which is valid in the neighborhood of $\tau = 1/3$ was obtained by Hunter and Vanden-Broeck (1983a). This equation can be derived by assuming

$$\alpha = \varepsilon^2 \qquad \beta = \varepsilon \tag{3.4}$$

and expanding ϕ, ζ and b in powers of ε. Hunter and Vanden-Broeck (1983a) calculated depression solitary waves by integrating this equation numerically. In addition they obtained periodic solutions by a Stokes type expansion. As expected they found that many different families of periodic waves exist for $\tau < 1/3$.

The validity of the Korteweg de Vries equation for τ not close to 1/3 will be discussed in section 6.

4. REFORMULATION AS AN INTEGRO-DIFFERENTIAL EQUATION

It is convenient to reformulate the problem as an integro-differential equation by considering the complex velocity $u - iv$. Here u and v are the horizontal and vertical components of the velocity respectively. The variables are made dimensionless by using H as the unit length and C as the unit velocity. We choose the complex potential

$$f = \phi + i\psi \tag{4.1}$$

as the independent variable.

In order to satisfy the boundary condition (2.3) on the bottom $\psi = -1$, we reflect the flow in the boundary $\psi = -1$. Thus we seek $x + iy$ and $u - iv$ as analytic functions of f in the strip $-2 < \psi < 0$.

We define the dimensionless wavelength $\ell = \lambda/H$ and introduce the function

$$p = \exp[2\pi i f/\ell] , \tag{4.2}$$

writing $|p| = \exp(-2\pi\psi/\ell) = r$ and $\arg p = 2\pi\phi/\ell = \theta$.

This maps the bottom $\psi = -1$, the free surface $\psi = 0$ and its image $\psi = -2$, respectively, onto the circles $r = r_0 = \exp(2\pi/\ell)$, $r = 1$ and $r = r_0^2$.

Next we define the function

$$G(p) = u - iv - 1 .$$

The symmetry of the wave about the crest $\theta = 0$ yields

$$G(\bar{p}) = \bar{G}(p) \tag{4.3}$$

Here the bar denotes complex conjugation. Moreover the real and imaginary parts of the function $G(p)$ on the free surface $r = 1$ and its image $r = r_0^2$ are related by the identities

$$\text{Real}\{G(e^{i\theta})\} = \text{Real}\{G(r_0^2 e^{i\theta})\} , \tag{4.4}$$

$$\text{Im}\{G(e^{i\theta})\} = -\text{Im}\{G(r_0^2 e^{i\theta})\} \tag{4.5}$$

We denote by $u(\phi)$ and $v(\phi)$ the horizontal and vertical components of the velocity on the free surface $\psi = 0$. In order to find a relation between $u(\phi)$ and $v(\phi)$ we apply Cauchy's theorem to the function $G(p)$ in the annulus $r_0^2 \leqslant |p| \leqslant 1$. Using the relations (4.3)-(4.5) we find after some algebra

$$u(\phi) - 1 = -\frac{2}{\ell}\int_0^{\ell/2} [u(s) - 1]ds \tag{4.6}$$

$$+ \frac{1}{\ell}\int_0^{\ell/2} v(s)[\text{cotg}\,\frac{\pi}{\ell}(s - \phi) + \text{cotg}\,\frac{\pi}{\ell}(s + \phi)]ds$$

$$+ \frac{2}{\ell} r_0^2 \int_0^{\ell/2} \frac{[u(s)-1][r_0^2-\cos\frac{2\pi}{\ell}(s-\phi)]+v(s)\sin\frac{2\pi}{\ell}(s-\phi)}{1+r_0^4-2r_0^2\cos\frac{2\pi}{\ell}(s-\phi)}ds$$

$$+ \frac{2}{\ell} r_0^2 \int_0^{\ell/2} \frac{[u(s)-1][r_0^2-\cos\frac{2\pi}{\ell}(s+\phi)]+v(s)\sin\frac{2\pi}{\ell}(s+\phi)}{1+r_0^4-2r_0^2\cos\frac{2\pi}{\ell}(s+\phi)}ds$$

The third integral in (4.6) is of Cauchy principal value form.

The surface condition (2.5) can now be rewritten as

$$\frac{1}{2}F^2[u(\phi)]^2 + \frac{1}{2}F^2[v(\phi)]^2 + \int_0^{\phi} \frac{v(s)}{[u(s)]^2 + [v(s)]^2}\,ds \tag{4.7}$$

$$- \tau \frac{u(\phi)v'(\phi) - v(\phi)u'(\phi)}{\{[u(\phi)]^2 + [v(\phi)]^2\}^{1/2}} = \frac{1}{2}F^2[u(0)]^2 - \frac{\tau v'(0)}{u(0)} .$$

with $F^2 = c^2/gH$.

In the remaining part of the paper we shall choose coordinates $\tilde{x}, \tilde{y}$ with the origin at a crest or a trough of the wave. The shape of the free surface is then defined parametrically by the relations

$$\tilde{x}(\phi) = \int_0^{\phi} u(s)\{[u(s)]^2 + [v(s)]^2\}^{-1} ds , \tag{4.8}$$

$$\tilde{y}(\phi) = \int_0^{\phi} v(s)\{[u(s)]^2 + [v(s)]^2\}^{-1} ds . \tag{4.9}$$

Finally we impose the periodicity condition

$$\tilde{x}(\ell/2) = \ell/2 . \tag{4.10}$$

We shall measure the amplitude of the wave by the parameter

$$u_0 = u(0) . \tag{4.11}$$

For given values of τ, u_0 and ℓ, (4.6)-(4.11) define a system of integro-differential equations for $u(\phi)$, $v(\phi)$, $\tilde{x}(\phi)$, $\tilde{y}(\phi)$ and F.

The equations for solitary waves are obtained by taking the limit $\ell \to \infty$ in (4.6). This leads after some algebra to

$$u(\phi)-1 = \frac{1}{\pi}\int_0^{\infty} v(s)\left[\frac{1}{s-\phi} + \frac{1}{s+\phi}\right]ds + \frac{1}{\pi}\int_0^{\infty} \frac{(s-\phi)v(s)+2[u(s)-1]}{(s-\phi)^2 + 4}ds$$

$$+ \frac{1}{\pi}\int_0^{\infty} \frac{(s+\phi)v(s)+2[u(s)-1]}{(s+\phi)^2 + 4}ds . \tag{4.12}$$

In the next section we describe a numerical scheme to solve these equations.

5. NUMERICAL PROCEDURE

(a) Periodic waves

To solve the system (4.6)-(4.11) we introduce the N mesh points

$$\phi_I = \frac{\ell(I - 1)}{2(N - 1)}, \qquad I = 1,\ldots,N . \qquad (5.1)$$

We also define the corresponding quantities

$$u_I = u(\phi_I), \qquad I = 1,\ldots,N , \qquad (5.2)$$

$$v_I = v(\phi_I), \qquad I = 1,\ldots,N . \qquad (5.3)$$

It follows from symmetry that $v_1 = v_N = 0$, so only $N - 2$ of the v_I are unknown. In addition from (4.11) we have $u_1 = u_0$ so only $N - 1$ of the u_I are unknown. We shall also use the $N - 1$ midpoints $\phi_{I+1/2}$ given by

$$\phi_{I+1/2} = \frac{1}{2}(\phi_I + \phi_{I+1}), \qquad I = 1,\ldots,N - 1 . \qquad (5.6)$$

We evaluate $u_{I+1/2} = u(\phi_{I+1/2})$, $v_{I+1/2} = v(\phi_{I+1/2})$, $u'_{I+1/2} = u'(\phi_{I+1/2})$, and $v'_{I+1/2} = v'(\phi_{I+1/2})$ by four-point interpolation and difference formulas.

We now discretize (4.6) by applying the trapezoidal rule to the integrals on the right-hand side with $s = \phi_I$, $I = 1,\ldots,N$, and $\phi = \phi_{I+1/2}$, $I = 1,\ldots,N - 1$. The symmetry of the quadrature formula and of the discretization enables us to evaluate the Cauchy principal value integral as if it were an ordinary integral. In this way we obtain $N - 1$ algebraic equations.

Next we substitute into (4.7) the expressions for $u_{I+1/2}$, $v_{I+1/2}$, $u'_{I+1/2}$ and $v'_{I+1/2}$ at the points $\phi_{I+1/2}$, $I = 2,\ldots,N - 1$. The integral in (4.7) is evaluated by the trapezoidal rule with the mesh points $s = \phi_I$. The derivative $v'(0)$ in (4.7) is approximated by a four-point difference formula. Thus we obtain another $N - 2$ algebraic equations.

For given values of τ, u_0 and ℓ we have therefore $2N - 3$ algebraic equations for the $2N - 2$ unknowns u_I, v_I and F. The last equation is obtained by discretizing (4.10).

The $2N - 2$ equations are solved by Newton's method. After a solution converges for given values of τ, u_0 and ℓ, the surface profile $\tilde{x}(\phi)$, $\tilde{y}(\phi)$ is obtained by applying the trapezoidal rule to (4.8) and (4.9).

For each calculation presented in section 6 the number of mesh points N was progressively increased up to a value for which the results were independent of N within graphical accuracy. Most of the computations were performed with $N = 60$. A few runs were performed with $N = 100$ as a check on the accuracy of the computations.

In the remaining part of the paper we shall refer to the above numerical scheme for periodic waves as numerical scheme I.

(b) Solitary waves

To solve the system (4.7)-(4.12) we introduce the N mesh points

$$\phi_I = E(I - 1), \qquad I = 1,\ldots,N . \tag{5.5}$$

Here E is the interval of discretization.

The quantities u_I, v_I, $\phi_{I+1/2}$, $u_{I+1/2}$ and $v_{I+1/2}$ are defined as in the previous subsection. It follows from symmetry that $v_1 = 0$. In addition $u_1 = u_0$ so that only $2N - 2$ of the u_I and v_I are unknown.

The discretization of (4.12) is entirely analogous to the procedure used to discretize (4.6). In this way we obtain $N - 1$ algebraic equations. The truncation error due to approximating the infinite integrals by integrals over a finite range, was found to be negligible for NE sufficiently large. As shown in the previous subsection, the discretization of (4.7) leads to another $N - 2$ algebraic equations.

Thus, for given values of τ, u_0 and ℓ we have $2N - 3$ algebraic equations for the $2N - 1$ unknowns u_I, v_I and F. The last two equations are obtained by imposing $u_N = 1$ and $v_N = 0$.

We shall refer to this numerical scheme as numerical scheme II.

6. DISCUSSION OF THE RESULTS

6.1. Gravity Waves

Vanden-Broeck and Schwartz (1979) applied the numerical scheme I to compute pure gravity waves in shallow water. Their numerical results confirm the results of Cokelet (1977). Hunter and Vanden-Broeck (1983b) applied the numericial scheme II to compute pure gravity solitary waves. They found that the ratio of the amplitude of the highest solitary wave versus the depth is 0.83322. This value is about 0.006 higher than the value obtained by most previous investigators.

6.2. Gravity-Capillary waves

Schwartz and Vanden-Broeck (1980) used the scheme I to compute gravity-capillary waves in water of infinite depth. Results for gravity-capillary waves in water of finite depth were obtained by Hunter and Vanden-Broeck (1983a).

The numerical results confirm the existence of many different families of gravity-capillary waves. It was found that all the waves are topologically limited by trapped bubbles. The calculations of Hunter and Vanden-Broeck indicate that each family of solutions only exists for a limited range of values of λ/H. In particular their exists a maximum value of λ/H beyond which solutions in a given family cease to exist. Therefore no solitary waves are obtained as the continuous limit of periodic gravity-capillary waves as the wavelength tends to infinity.

A typical profile of a periodic gravity-capillary wave is shown in Figure 1. This profile does not agree with the solution of the Korteweg de Vries equation which predicts profiles without dimples. This is due to the fact that the length scale of the dimples was not taken into account in the derivation of the Korteweg de Vries equation. Therefore the Korteweg de Vries equation is not valid for $\tau < 1/3$.

On the other hand the numerical calculations of Hunter and Vanden-Broeck (1983b) show the Korteweg de Vries equation is valid for $\tau > 1/3$.

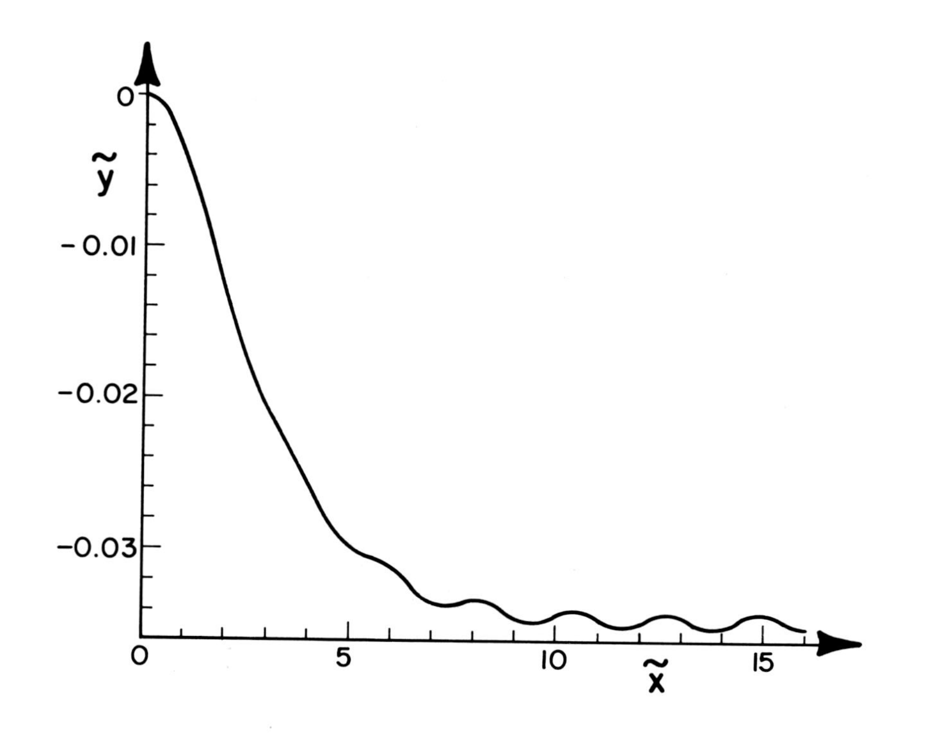

Figure 1. Computed free surface profile of a periodic wave with $\tau = 0.24$ and $u_0 = 0.97$. The wavelength is $\ell = 32$.

REFERENCES

1. Chen, B. and P. G. Saffman, 1980, Studies in Appl. Math. 62, 95.
2. Cokelet, E. D., 1977, Phil. Trans. Roy. Soc. A286, 183.
3. Harrison, W. J., 1909, Proc. Lond. Math. Soc. 7, 107.
4. Hogan, S. J., 1980, J. Fluid Mech. 96, 417.
5. Hogan, S. J., 1981, J. Fluid Mech. 110, 381.
6. Hunter, J. K. and J.-M. Vanden-Broeck, 1983a, (to appear).
7. Hunter, J. K. and J.-M. Vanden-Broeck, 1983b, (to appear).
8. Keller, J. B., 1948, Comm. Pure Appl. Math. 1, 323.
9. Korteweg, D. J. and G. de Vries, 1895, Phil. Mag. 39, 422.
10. Nayfeh, A. H., 1970, J. Fluid Mech. 40, 671.
11. Schwartz, L. W. and J.-M. Vanden-Broeck, 1979, J. Fluid Mech. 95, 119.
12. Stokes, G. G., 1847, Trans. Camb. Phil. Soc. 8, 441.
13. Stokes, G. G., 1880, Mathematical and Physical Papers, vol. 1, p. 314, Cambridge University Press.
14. Vanden-Broeck, J.-M. and M. C. Shen, 1982, (to appear).
15. Vanden-Broeck, J.-M. and L. W. Schwartz, 1979, Phys. Fluids 22, 1868.
16. Whitham, G. B., 1974, Linear and Nonlinear Waves, New York, Wiley.
17. Wilton, J. R., 1915, Phil. Mag. 29, 688.

Mathematics Research Center
University of Wisconsin-Madison
Madison, Wisconsin 53705

GENERALIZED VORTEX METHODS FOR FREE-SURFACE FLOWS

G. R. Baker

1. INTRODUCTION

The unsteady motion of a sharp interface in the irrotational flow of an incompressible, inviscid fluid is about the simplest mathematical model of the motion of a transitional layer between two continuous fluid media. Yet even such a model provides challenging problems to the mathematician and computor among others. The difficulties arise from the nonlinear nature of the equations; the interior flow affects the free surface motion which in turn affects the interior flow. This type of nonlinearity makes analysis extremely difficult but provides the richness of phenomena observed in free surface flow.

In general, numerical techniques must be used to compute free surface motion. There are several different methods that may be adopted and they are thoroughly reviewed by Yeung (1982). In particular, boundary integral techniques are among the most useful since they are naturally adaptive in the sense that only information at the surface is used in the calculations. Their apparent disadvantages are twofold. They are presently limited to those cases where the interior flow is creeping or incompressible, irrotational and inviscid, and they require

ISBN 0-12-493220-7

solutions to integral equations that can be expensive to compute.

The work of Birkhoff (1954) appears to be the first in which boundary integral equations are derived for free surface flow. Unfortunately Birkhoff restricted his numerical calculations to a study of vortex sheet motion, apparently an ill-posed problem. Zaroodny and Greenberg (1973) also derived boundary integral equations but did little to investigate their solution numerically. Encouraged by developments in ship hydrodynamics, there has been a rapid growth in the use of boundary integral equations for free surface flow problems; Longuet-Higgins and Cokelet (1976) used Green's formula to study breaking surface waves, Vinje and Brevig (1981) used Cauchy's formula in their study of ship-wave interactions, and Pullin (1982) studied the Raleigh-Taylor instability in the presence of surface tension. These authors all used costly direct matrix inversion methods to solve their integral equations.

Recently, a generalised vortex method has been developed by Baker et al. (1980, 1982) for incompressible, inviscid, stratified layered flow. This method is a boundary integral method formulated so that solutions to the integral equation may be obtained iteratively and thus expensive direct matrix inversions are avoided. The method has been applied to the study of breaking water waves over finite and infinite bottom topography (Baker et al. 1982), wave generation by external means such as translating, submerged bodies (Baker et al. 1983) and the classical Rayleigh-Taylor instability (Baker et al. 1980).

In their study of the Rayleigh-Taylor instability, Baker et al. (1980) restricted their attention to the two-dimensional, periodic deformations of an unstably stratified interface between incompressible, inviscid fluids of constant, but different densities. They concentrated specifically on the case where fluid falls into vacuum (or very nearly a vacuum). Results show the formation of long spikes falling into vacuum and the rise of a bubble between spikes. In this paper, results will be presented for the long-time development of the instability when the fluid forms a two-dimensional layer of finite thickness. In this case it is possible for the rising bubble to penetrate the fluid layer. These results are important in studies of imploding fluid shells which occur in inertial fusion devices among others.

First, a general discussion of boundary integral methods will be given, followed by a detailed derivation of the generalised vortex method for a single interface. The application to the classical Rayleigh-Taylor instability will be reviewed. Next, a detailed derivation of the generalised vortex method for a fluid layer that is accelerated by an external pressure gradient will be given. The method is then applied to accelerating, thin, periodic, two-dimensional fluid layers.

2. BOUNDARY INTEGRAL TECHNIQUES

Boundary integral equations constitute a set of evolution equations for the location of the free surface and for flow quantities that need be known only at the free surface. The central difficulty in deriving such equations is that normal derivatives at the surface must be computed from information given along the surface. Clearly it is this step that incorporates knowledge of the interior flow in some form. Various boundary integral equations perform this step in different ways. This interpretation of boundary integral equations is best illustrated by a simple, but concrete example.

Consider an incompressible, inviscid fluid below a free surface parametrised by $\vec{x}(\alpha,\beta,t) = (x(\alpha,\beta,t), y(\alpha,\beta,t), z(\alpha,\beta,t))$ where x,z are horizontal coordinates and y is a vertical coordinate pointing upwards (gravity points downwards). The parameters α and β are Lagrangian variables defined by the motion of the free surface;

$$\frac{\partial \vec{x}}{\partial t}(\alpha,\beta,t) = \vec{u}(\alpha,\beta,t) \tag{2.1}$$

where a partial time derivative is used to emphasis that α and β are held fixed, and $\vec{u}$ is the velocity of the fluid at the surface. Since (2.1) specifies that a surface point labelled by α and β moves with the fluid velocity, it is compatible with the kinematic condition. Normally the kinematic condition merely requires that the normal velocity of the surface is that of the fluid or else the surface would rupture. Clearly a surface point following (2.1) has a normal velocity equal to that of the fluid. The tangential velocity of a surface point is physically unconstrained since a surface point need not be a fluid particle. Indeed the usual Eulerian definition of a surface point fixes its motion in the vertical direction. Thus the specification of the tangential velocity of a surface point, together with the required normal velocity, in effect defines the surface coordinates α and β.

So far, no dynamics have been specified. Dynamics will determine $\vec{u}(\alpha,\beta,t)$. The momentum equations for the fluid are

$$\rho \left.\frac{\partial \vec{u}}{\partial t}\right|_E + \rho\vec{u}\cdot\nabla\vec{u} = \nabla p - \rho g\hat{j} \tag{2.2}$$

with

$$\nabla\cdot\vec{u} = 0 , \tag{2.3}$$

where $\vec{u}, \rho$ and p are the fluid velocity, density and pressure respectively, g is the constant of gravity, $\hat{j}$ a unit vector in the y direction and ∇ is the gradient operator. Note that (2.2) has been written in Eulerian form. Finally, in the absence of surface forces, the pressure is continuous at the surface. Since a vacuum is assumed above the free surface, the pressure must vanish at the surface.

If an attempt is made to use (2.2) directly to update $\vec{u}(\alpha,\beta,t)$ at the surface, an immediate difficulty arises; normal derivatives of the velocity components and the pressure must be calculated. Assuming the flow is irrotational; that is

$$\nabla x\vec{u} = 0 , \tag{2.4}$$

the two conditions, (2.3) and (2.4), may be used to express normal derivatives of the velocity components in terms of their tangential derivatives, but there is no simple way to determine the normal derivative of the pressure. The interior flow must be known before the normal derivative of the pressure can be calculated.

There is an alternate way to state this problem. Introduce the velocity potential, ϕ, such that $\vec{u} = \nabla\phi$. Clearly, (2.4) is identically satisfied and (2.3) now gives

$$\nabla^2 \phi = 0 \qquad (2.5)$$

in the interior flow region. Also, (2.2) can be integrated, giving Bernoulli's equation which is evaluated at the free surface as

$$\rho \frac{\partial\phi}{\partial t} - \frac{\rho}{2} (\nabla\phi)^2 + \rho g y = C \qquad (2.6)$$

where C is a constant, and the partial time derivative refers to the change in ϕ following the motion specified by (2.1). Thus, if $\phi = \phi(\alpha,\beta,t)$ is given along the surface, (2.6) describes its change in time. However $\nabla\phi$ must be evaluated at the surface. The tangential components can be evaluated in a straight forward fashion, but the normal component requires knowledge of the interior flow. Because ϕ satisfies (2.5), there are several ways a relationship between the normal derivative of ϕ and its values at the surface can be found. Background information necessary to understand these relationships can be found in detail in Kellogg (1953) or reviewed in Jaswon and Symm (1977).

2.1 Green's Formula

Green's third identity yields the formula,

$$\int_S \frac{\partial\phi}{\partial n}(Q) g(P,Q) dQ = \frac{1}{2}\,\phi(P) + \int_S \phi(Q)\,\frac{\partial g}{\partial n}(P,Q) dQ\ , \qquad (2.7)$$

where $P = \vec{x}(\alpha,\beta)$ and $Q = \vec{x}(\alpha',\beta')$ are points on the surface, the integrals are surface integrals and $\frac{\partial}{\partial n} = \hat{n}(Q)\cdot\nabla$, $\hat{n}(Q)$ being the exterior unit normal at Q. The Green's function $g(P,Q)$ is the free space Green's function for (2.5); $g(P,Q) = \frac{1}{2\pi}\log|P - Q|$ in two-dimensions and $g(P,Q) = \frac{-1}{4\pi|P - Q|}$ in three dimensions. Given that ϕ is known at the surface, (2.7) is a Fredholm integral equation of the first kind for the normal derivative of ϕ. The solution to (2.7), together with the simple calculation of the tangential derivatives of ϕ, gives the velocity at the surface and consequently both the surface location $\vec{x}(\alpha,\beta)$ and $\phi(\alpha,\beta)$ can be advanced in time through (2.1) and (2.6).

2.2 Dipole Representation

A potential is created in the interior region by a distribution of dipoles along the free surface of strength $\mu(\alpha,\beta)$;

$$\phi(\vec{x}) = \int_S \mu(Q) \frac{\partial g}{\partial n} (\vec{x},Q)dQ \ . \tag{2.8}$$

As the free surface is approached along a normal, ϕ has a limiting value,

$$\phi(P) = \oint_S \mu(Q) \frac{\partial g}{\partial n}(P,Q)dQ + \frac{\mu(P)}{2} \ , \tag{2.9}$$

where the principal value of the surface integral must be taken. Given $\phi(P)$, (2.9) is a Fredholm integral equation for μ. Knowing μ, the normal derivative of ϕ can be computed as follows. Introduce the vector potential $\vec{A}$ such that

$$\vec{u} = \nabla \times \vec{A}, \quad \nabla \cdot \vec{A} = 0 \ . \tag{2.10}$$

In terms of the dipole distribution,

$$\vec{A}(P) = \int_S \mu(Q)\hat{n}(Q) \times \nabla g(P,Q)dQ \ . \tag{2.11}$$

Finally, the normal velocity follows from $\hat{n} \cdot \nabla \times \vec{A}$, which involves only tangential derivatives of $\vec{A}$. Thus, μ is determined from (2.9), $\vec{A}$ is computed by (2.11) and then the normal velocity computed from the tangential derivatives of $\vec{A}$. Once again, knowledge of the velocity at the surface allows the temporal advancement of the surface location and ϕ.

There are still further ways to relate the normal derivative of ϕ to its values along a surface such as using conformal mappings in two-dimensional flows, (which can be found by using boundary integral techniques, Gram 1962 and Menikoff and Zemach 1980) or representing the potential by a distribution of sources along the surface. However a discussion of the advantages and disadvantages of the Green's formula and the dipole representation will reveal most of the important features of boundary integral representations.

The essential difference between the two approaches described above is that they involve Fredholm integral equations of different kinds. Much information is known about the mathematical properties of equations of the second kind while less is known about equations of the first kind (Mikhlin 1957). For the dipole representation, it is relatively easy to show that the Fredholm integral of the second kind has a globally convergent Neumann series (proof is available in Baker et al. 1982). That means the equation may be always solved by iteration provided only that the surface is reasonably smooth; let $\mu^{(n)}$ be the nth iterate, then

$$\mu^{(n+1)}(P) = 2\phi(P) - 2 \oint_S \mu^{(n)}(Q) \frac{\partial g}{\partial n}(P,Q)dQ \qquad (2.12)$$

generates a new iterate and the process will converge to μ, the solution to (2.9). On the other hand, (2.7) must be inverted directly.

From a mathematical point of view, whether the Green's formula or the dipole representation is used is immaterial, but when a numerical solution is sought, the ability to solve (2.9) iteratively gives the dipole representation a decided advantage. Suppose N points are used to locate the surface and a quadrature involving values at these points is used to approximate the surface integrals. The integral equations, (2.7) and (2.9), become matrix equations. Direct inversion of a $N \times N$ matrix requires $O(N^2)$ storage locations and $O(N^3)$ arithmetic operations. However, the discretisation of (2.12) leads to an iterative method that requires only $O(N)$ storage locations and $O(N^2)$ arithmetic operations. In general, only several iterations are required for high accuracy since μ at the previous time level makes a good first iterate. By keeping values of μ at several previous time levels, an extrapolated first iterate can be found which will further reduce the number of iterations to typically 1 or 2. More iterations (O(10)) are required when the surface develops regions of high curvature such as in breaking surface waves. If the flow is such that the curvature in some region continues to increase in time, there will be a stage at which the iterations of the numerical approximation to (2.12) will no longer converge. At this point, the resolution of the surface is inadequate; even direct inversion of the matrix will not allow further computation without unphysical behavior.

Currently there are no convergence proofs for the numerical treatment of the various boundary integral equations that have been used. The order of accuracy of the methods is relatively easy to estimate but whether the schemes are numerically stable is extremely difficult to ascertain. Computational experience is mixed. Longuet-Higgins and Cokelet (1976), who used Green's formula, were forced to smooth frequently to suppress rapidly growing small-scale oscillations. Pullin (1982), who used a dipole representation, also found instabilities, but he used direct matrix inversion methods instead of iteration. Baker et al. (1982) found that the iteration method appeared to give relatively stable results. It seems possible that an iterated solution to the matrix equations has errors more favorable than a direct solution for a stable numerical scheme.

It is worth noting that, as far as steady state problems are concerned, there appears to be no advantage nor disadvantage to using dipole representations. Since the surface location is also unknown, the integral equations are nonlinear and recourse must be made to methods like Newton-Raphson iteration for solving systems of nonlinear equations. Of course, an unsteady calculation may be done from some judiciously chosen initial condition until a steady solution is reached, but this approach is only valuable in certain cases. Certainly there are cases where there are no steady solutions or where the transition to a steady-state is of sufficient importance that unsteady calculations must be performed. The classical Rayleigh-Taylor instability is such a case. The rest of this paper will explore the nature of Rayleigh-Taylor instabilities under various flow conditions.

So far the discussion on boundary integral techniques has been of a philosophical nature. As attention is focused on the Rayleigh-Taylor instability, specific consideration will be given to the details of the calculations, for example, the numerical treatment of the singular kernels in the integrals, the range of integration in open geometries and so on.

3. GENERALISED VORTEX METHOD

3.1 Formulation

The approach adopted in the previous section for a free surface between fluid and vacuum can be extended to an interface between two incompressible, inviscid fluids of constant, but different densities. In particular, the dipole strength satisfies an evolution equation that can be identified with the baroclinic generation of vorticity where the dipole strength is the circulation. Hence the method may be termed a generalised vortex method.

Currently the method has been applied to two-dimensional flow problems, where the surface is a curve in two-dimensional space. Only one parameter is needed to describe the curve and so the problem is effectively one-dimensional in spatial terms. Consequently the algebra and numerical computations are simplified, but the method applies equally well to three-dimensional flow problems.

Consider an internal interface, parametrised by $(x(\alpha,t), y(\alpha,t))$, between two fluids of density ρ_1 and ρ_2. The subscripts 1 and 2 will be used to refer to quantities associated with the lower and upper fluids respectively. As before, x and y are horizontal and vertical coordinates respectively and gravity points downwards. Introduce the velocity potential ϕ. It must satisfy Laplaces equation away from the interface. At the interface, the normal derivative of ϕ and the pressure P must be continuous. The potential created by a dipole of strength $\mu(\alpha)$ at $x(\alpha)$, $y(\alpha)$ is given by (2.8),

$$\phi(x,y) = \frac{1}{2\pi} \int \mu(\alpha) \frac{x_\alpha(\alpha)(y-y(\alpha)) - y_\alpha(x-x(\alpha))}{[(x-x(\alpha))^2 + (y-y(\alpha))^2]} \, d\alpha \qquad (3.1)$$

where the subscript α denotes differentiation with respect to α. Introduce the complex field point $z = x + iy$, then (3.1) can be written as

$$\phi(z) = \mathrm{Re}\left\{\frac{1}{2\pi i} \int \frac{\mu(\alpha)\, z_\alpha(\alpha)}{z - z(\alpha)} \, d\alpha\right\} . \qquad (3.2)$$

In addition to the velocity potential, the vector potential $\vec{A}$ must also be calculated. For two-dimensional flow, the vector potential $\vec{A}$ is normal to the x-y plane with magnitude ψ, the stream function. Since ψ is a conjugate harmonic function to ϕ, it follows that the complex potential $\Phi = \phi + i\psi$ is given by

$$\Phi(z) = \frac{1}{2\pi i} \int \frac{\mu(\alpha)\, z_\alpha(\alpha)}{z - z(\alpha)} \, d\alpha \qquad (3.3)$$

Thus both integrals, (2.8) and (2.11), are contained in (3.3). The complex form of the dipole representation also displays very clearly its major properties. As the interface is approached along a normal, the potential takes on limiting values given by the Plemelj formulae,

$$\Phi_1(\alpha) = \frac{1}{2\pi i} \not\oint \frac{\mu(\alpha')\, z_\alpha(\alpha')}{z(\alpha) - z(\alpha')} \, d\alpha' + \frac{\mu}{2}(\alpha) \qquad (3.4a)$$

$$\Phi_2(\alpha) = \frac{1}{2\pi i} \not\oint \frac{\mu(\alpha')\, z_\alpha(\alpha')}{z(\alpha) - z(\alpha')} \, d\alpha' - \frac{\mu}{2}(\alpha) \qquad (3.4b)$$

where the principal value of the integral must be taken. Note first that the complex potential jumps in value across the interface the jump being real and of magnitude $\mu(\alpha)$. By considering a line integral of the velocity that encloses a segment of the interface, the circulation can be related directly to the jump in velocity potential across the interface and hence to μ. Details are available in Jaswon and Symm (1977). Note further that ψ and its tangential derivative, are continuous across the interface and so the Cauchy-Riemann equations imply that the normal derivative of ϕ is continuous in accordance with kinematic requirements. The fact that dipole representations automatically satisfy the kinematic condition at the interface is of great benefit.

The integration in (3.3) is along the interface. If the surface is closed, as in the rise of a buoyant bubble, the integration is over a finite interval and numerical approximation is straightforward. On the other hand, if the interface is open, the integration is infinite and one of two approaches must be adopted. In some cases, it is natural to consider periodic interfaces. Alternatively, the range of integration must be truncated so that outward travelling waves are not reflected at the truncation point. This may be achieved by the imposition of absorption layers at the computational boundaries (Baker et al. 1983). Since attention is being given here to the Rayleigh-Taylor instability, it is natural to consider periodic domains and the long-time evolution of various modes. Let the periodicity interval be 2π and let $0 \leqslant \alpha < 2\pi$ specify the interface in one period. The infinite integral may be written as an infinite sum over a finite interval as follows;

$$\Phi(z) = \frac{1}{2\pi i} \sum_{n=-\infty}^{\infty} \int_{2\pi n}^{2\pi(n+1)} \frac{\mu(\alpha)\, z_\alpha(\alpha)}{z - z(\alpha)}\, d\alpha\,,$$

$$= \frac{1}{2\pi i} \int_0^{2\pi} \mu(\alpha)\, z_\alpha(\alpha) \sum_{n=-\infty}^{\infty} \frac{1}{z - z(\alpha) - n\pi}\, d\alpha\,,$$

$$= \frac{1}{4\pi i} \int_0^{2\pi} \mu(\alpha)\, z_\alpha(\alpha) \cot\{\tfrac{1}{2}(z - z(\alpha))\} d\alpha\,, \tag{3.5}$$

where use has been made of the identity (Carrier, Krook and Pearson 1966, p. 12)

$$\frac{1}{z} + 2z \sum_{n=1}^{\infty} \frac{1}{z^2 - n^2\pi^2} = \frac{1}{2}\cot\left(\frac{z}{2}\right). \tag{3.6}$$

Note that the pole singularity in the integrand of (3.3) has been replaced by a cotangent function, and that results such as (3.4) are similarly changed. Numerical approximations to (3.4) and (3.5) will be discussed in detail later.

The complex velocity, $q = u + iv$, is given by

$$q^*(z) = \frac{\partial\phi}{\partial x} - \frac{i\partial\phi}{\partial y} = \frac{d\Phi}{dz} \tag{3.7}$$

where the asterisk indicates complex conjugation. Actually the velocity needs to be known only at the interface. However, while the normal component is continuous as required by the kinematic condition, the tangential component jumps in value across the interface. In accordance with the discussion in the previous section, a surface point must move with the prescribed normal velocity component, but its tangential component is unconstrained. While previously there was a natural choice, that is not the case here. Instead the velocity of a surface point is defined as a weighted average of the limiting values of velocities on either side of the interface;

$$\frac{\partial z}{\partial t}(\alpha) = \tilde{q}(\alpha) = q(\alpha) + \frac{w}{2}(q_1(\alpha) - q_2(\alpha)), \tag{3.8}$$

$$(-1 \leqslant w \leqslant 1)$$

where

$$q(\alpha) = \frac{q_1(\alpha) + q_2(\alpha)}{2} \tag{3.9}$$

The limiting velocity on either side of the interface can be obtained by evaluating (3.7) at the interface and choosing the direction of differentiation to be along the interface. Thus,

$$q_1^*(\alpha) = \frac{\Phi_{1\alpha}(\alpha)}{z_\alpha(\alpha)} \tag{3.10a}$$

$$q_2^*(\alpha) = \frac{\Phi_{2\alpha}(\alpha)}{z_\alpha(\alpha)} \tag{3.10b}$$

Define the complex potential $\Phi(\alpha)$ at the interface to be the average value,

$$\Phi(\alpha) = \frac{\Phi_1(\alpha) + \Phi_2(\alpha)}{2} = \frac{1}{2\pi i} \text{P}\!\!\oint \frac{\mu(\alpha')z_\alpha(\alpha')}{z(\alpha) - z(\alpha')}\, d\alpha' . \tag{3.11}$$

where (3.4) has been used. Substitute (3.11) and (3.10) into (3.9),

$$q^*(\alpha) = \frac{\Phi_\alpha(\alpha)}{z_\alpha(\alpha)} \; . \tag{3.12}$$

From (3.4),

$$\mu(\alpha) = \Phi_1(\alpha) - \Phi_2(\alpha) \; , \tag{3.13}$$

and so

$$\frac{\mu_\alpha(\alpha)}{z_\alpha(\alpha)} = q_1^*(\alpha) - q_2^*(\alpha) \; . \tag{3.14}$$

Finally, (3.8) can be rewritten as

$$\frac{\partial z^*(\alpha)}{\partial t} = \tilde{q}^*(\alpha) = (\Phi_\alpha(\alpha) + \frac{w}{2}\, \mu_\alpha(\alpha))/z_\alpha(\alpha) \tag{3.15}$$

Recall that the partial derivative in time is used to emphasize that α is kept fixed; (3.15) defines the Lagrangian motion of surface points. When $w = 1(-1)$, the surface points follow the motion of the lower (upper) fluid respectively.

Dynamic considerations lead to an evolution equation for μ. The starting point is the evaluation of Bernoulli's equation on either side of the interface;

$$\rho_1 \left.\frac{\partial \phi_1}{\partial t}\right|_E + \frac{\rho_1}{2}(u_1^2 + v_1^2) + P + \rho_1 g y = 0 \tag{3.16a}$$

$$\rho_2 \left.\frac{\partial \phi_2}{\partial t}\right|_E + \frac{\rho_2}{2}(u_2^2 + v_2^2) + P + \rho_2 g y = 0 \tag{3.16b}$$

Here the time derivatives are Eulerian and the pressure is continuous across the interface. If surface forces, such as surface tension, are present, the pressure jumps by a known amount that is easily incorporated into (3.16) as an extra term. Since the potential evaluated at the interface takes the form,

$$\phi(\alpha,t) = \phi(x(\alpha,t),y(\alpha,t),t) \tag{3.17}$$

the rate of change of ϕ following the motion prescribed by (3.15) is

$$\frac{\partial \phi}{\partial t}(\alpha,t) = \tilde{u}\,\frac{\partial \phi}{\partial x} + \tilde{v}\,\frac{\partial \phi}{\partial y} + \left.\frac{\partial \phi}{\partial t}\right|_E \tag{3.18}$$

Once again, the partial time derivative is used consistently with keeping α fixed, and $\hat{q} = \hat{u} + i\hat{v}$. Consequently, (3.16) can be written as

$$\rho_1 \frac{\partial\phi_1}{\partial t} - \rho_1(\tilde{u}u_1+\tilde{v}v_1) + \frac{\rho_1}{2}(u_1^2+v_1^2) + P + \rho_1 gy = 0 \qquad (3.19a)$$

$$\rho_2 \frac{\Phi\phi_2}{\partial t} - \rho_2(\tilde{u}u_2+\tilde{v}v_2) + \frac{\rho_2}{2}(u_2^2+v_2^2) + P + \rho_2 gy = 0 \qquad (3.19b)$$

The values of the potential and velocity components on either side of the interface can be expressed in terms of μ, $\Phi(\alpha)$ and their derivatives by using (3.9), (3.11), (3.13) and (3.14);

$$\phi_1 = \mathrm{Re}\{\Phi(\alpha)\} + \frac{\mu(\alpha)}{2} \qquad (3.20a)$$

$$\phi_2 = \mathrm{Re}\{\Phi(\alpha)\} - \frac{\mu(\alpha)}{2} \qquad (3.20b)$$

$$q_1^* = q^*(\alpha) + \frac{\mu_\alpha(\alpha)}{z_\alpha(\alpha)} \qquad (3.21a)$$

$$q_2^* = q^*(\alpha) - \frac{\mu_\alpha(\alpha)}{z_\alpha(\alpha)} \qquad (3.21b)$$

Upon substitution into (3.19), subtraction of (3.19b) from (3.19a) to eliminate P and after some straightforward algebra, one obtains

$$\frac{\partial\mu}{\partial t} - \frac{w\mu_\alpha^2}{2z_\alpha z_\alpha^*} =$$

$$= -2A\left[\mathrm{Re}\left\{\frac{\partial\Phi}{\partial t}\right\} - \frac{1}{2}q^*q - \frac{w}{2}\mu_\alpha \mathrm{Re}\left\{\frac{q}{z_\alpha}\right\} + \frac{\mu_\alpha^2}{8z_\alpha z_\alpha^*} + gy\right] \qquad (3.22)$$

where

$$A = \frac{\rho_1 - \rho_2}{\rho_1 + \rho_2} \qquad (3.23)$$

The quantity $\frac{\partial\Phi}{\partial t}$ in (3.22) is obtained by Lagrangian time differentiation of (3.11)

$$\frac{\partial\Phi(\alpha)}{\partial t} = \frac{1}{2\pi i} \oint \frac{\partial\mu(\alpha')}{\partial t} \frac{z_\alpha(\alpha')}{z(\alpha) - z(\alpha')} d\alpha' \tag{3.24}$$

$$+ \frac{1}{2\pi i} \oint \frac{\mu(\alpha')}{z(\alpha)-z(\alpha')} [\tilde{q}_\alpha(\alpha') - \frac{\tilde{q}(\alpha)-\tilde{q}(\alpha')}{z(\alpha)-z(\alpha')} \cdot z_\alpha(\alpha')] d\alpha'$$

Of course, for periodic domains, the cotangent function would be used instead of the pole singularity in (3.11) and (3.24).

Physically, (3.22) describes the baroclinic generation of vorticity. When $\rho_1 = \rho_2$, (3.22) states that the circulation stays constant along trajectories whose motion is determined by the average fluid velocity $(w = 0)$, a result well-known in fluid dynamics. Since $\frac{\partial\Phi}{\partial t}$ is related to an integral of $\frac{\partial\mu}{\partial t}$, (3.22) is a Fredholm integral equation of the second kind; the form of the integral equation is

$$\frac{\partial\mu(\alpha)}{\partial t} = -2A \; \mathrm{Re}\left\{\frac{1}{2\pi i} \int \frac{\partial\mu(\alpha')}{\partial t} \frac{z_\alpha(\alpha')}{z(\alpha)-z(\alpha')} d\alpha'\right\} + R(\alpha) \tag{3.25}$$

where R represents the remaining terms that depend on z, μ, q and their derivatives. The above equation is directly related to (2.9) where $\frac{\partial\mu}{\partial t}$ takes the place of μ. Hence $\frac{\partial\mu}{\partial t}$ may be found by the iteration scheme (2.12).

Equations (3.15) and (3.22) constitute a set of evolution equations for the location of the interface and μ. Suppose that $\mu(\alpha)$ and the surface location $z(\alpha)$ are known at some time. The complex potential $\Phi(\alpha)$ is calculated from (3.11) and consequently the velocity $q(\alpha)$ follows from (3.12). Thus the right hand side of (3.15) may be evalauated. Then $R(\alpha)$ in (3.25) is computed and the integral equation solved by iteration to give $\frac{\partial\mu(\alpha)}{\partial t}$. When $A = 1$, the generalised vortex method is a variation of the method using a dipole representation as described in section 2.2.

The generalised vortex equations may be used to study steady and unsteady interfacial flow problems, although care must be used in formulating the steady problem because the Lagrangian time derivatives do not necessarily vanish in steady flow. Baker et al. (1982), Meiron and Saffman (1983) have used this method to calculate steadily advecting waves of permanent form. For unsteady problems, initial values of z and μ must be given. If the flow is initially at rest, $\mu = 0$, but if not, μ must be determined from (3.13) or (2.9).

The derivation of the generalised vortex method above has been for a single interface in irrotational, incompressible, inviscid stratified-layered flow. Generalisations to several interfaces and solid boundaries are straightforward (Baker et al. 1982, Verdon et al. 1982). The assumption that the flow is irrotational and incompressible may be modified. In particular, source and vorticity distributions away from the interface are easily included. Normally the distribution must be restricted in some way or else major computational difficulties may arise in determining the motion of the source or vorticity distribution. For example, point sources and vortices are easily included into the formulation. Generalised vortex methods may also be applied to two-dimensional flow containing regions of constant vorticity across whose boundaries the density may jump but is otherwise constant. These equations are an extension of the method of contour dynamics promoted by Zabusky et al. (1979). Details will be presented elsewhere. Finally, generalised vortex methods are applicable to creeping flow and porous media flow, although difficulties discussed elsewhere in these proceedings are present for moving contact lines. Work is also in progress on hybrid schemes to solve Poisson equations in complex geometries.

3.2 Numerical Approximation

Numerical approximations to (3.15) and (3.22) must deal with the singular nature of the integrands in the principal-value integrals, (3.11) and (3.24). The singularity can always be removed in two-dimensional problems or weakened in three-dimensional problems by using

$$\oint_S \mu(Q) \frac{\partial g}{\partial n}(P,Q)dQ = 0 \qquad (S \text{ an open surface})$$
$$= \frac{\mu(P)}{2} \qquad (S \text{ a closed surface}) \tag{3.26}$$

Specifically for two-dimensional problems, (3.11) may be rewritten as

$$\Phi(\alpha) = \frac{1}{2\pi i} \int \frac{\mu(\alpha') - \mu(\alpha)}{z(\alpha) - z(\alpha')} z_\alpha(\alpha')d\alpha' + \frac{\mu(\alpha)}{2} \tag{3.27}$$

for a closed interface, and

$$\Phi(\alpha) = \frac{1}{2\pi i} \int_0^{2\pi} (\mu(\alpha') - \mu(\alpha))z_\alpha(\alpha') \times$$
$$\times \cot\{\tfrac{1}{2} (z(\alpha) - z(\alpha'))\}d\alpha' \tag{3.28}$$

for a periodic, open interface. Furthermore, (3.24) may be regularised by differentiating (3.27) (or (3.28)) instead of (3.11).

The most convenient numerical approximation to (3.28) is

$$\Phi_j = \frac{h}{2\pi i} \sum_{\substack{k=1 \\ k+j=\text{odd}}}^{N} (\mu_k - u_j) \frac{\partial z_k}{\partial \alpha} \cot\{\tfrac{1}{2} (z_j - z_k)\} \tag{3.29}$$

where the subscripts j and k corresponds to evaluation at jh and kh respectively, $h = \frac{2\pi}{N}$. The summation corresponds to the composite trapezoidal rule using alternate points, thus avoiding the evaluation of the integrand at $j = k$. If $\frac{\partial z}{\partial \alpha}$ were known analytically, the approximation would be spectrally accurate (Isaacson and Keller 1966, p. 340), but in general the derivative must be computed numerically. Derivatives of all quantities with respect to α were usually computed using cubic spline approximations, and consequently the error in (3.29) is of order h^4. The integrals in the periodic version of (3.24) can be similarly approximated. Besides the integrals, only derivatives and algebraic terms have to be evaluated to compute the rates of change of z and μ. When the difference between successive iterates was less than some fixed tolerance, the iteration process indicated by (2.12) was considered converged and the best value used as the solution. Finally, z and μ were updated with a fourth-order Adams-Moulton predictor-corrector scheme, where the initial starting values are generated by a fourth-order Runge-Kutta scheme.

Several tests were conducted to study the behavior of the numerical method. The accuracy of the numerical integration was tested by considering a circular interface with various dipole representations for which exact values of the potential are known. The average potential at the interface was calculated numerically using the quadrature indicated in (3.29). As h decreases, one observes spectral improvement in the error until the error associated with the spline approximations used to compute z_α becomes noticeable. Since that error behaves as Ch^4 where C is small for periodic functions, good accuracy can be obtained with relatively moderate values for N. This behavior was also observed when the numerical method was used to compute the long time behavior of the Rayleigh-Taylor instability.

3.3 Classical Rayleigh-Taylor Instability: Results

The generalised vortex method has been applied to a study of the classical Rayleigh-Taylor instability by Baker et al (1980). Their results are reviewed here as useful background information to the subsequent study of accelerating fluid layers.

Imagine a semi-infinite fluid initially suspended above a vacuum with its surface at $y = 0$. Of course the fluid may subsequently fall without distortion under the influence of gravity, but the physically interesting case is when the fluid is held in place by an infinite pressure difference between the pressure at the surface and the pressure of the fluid at infinity. Any initial perturbation of the surface is unstable and results in spikes of fluid falling into the vacuum but without any motion of the fluid at infinity; that is, there is no mean external flow. Consequently, the vacuum will penetrate the fluid as rising bubbles between the spikes. In this way the fluid and vacuum attempt to change places. This behavior is easily observed in the evolution of the interface, initially perturbed by

$$y = \varepsilon \cos(2\pi x/\lambda) \ . \tag{3.30}$$

In Figure 1, the interface is plotted at various times for the case where $\varepsilon = 0.5$, $\lambda = 2\pi$, $g = 1$ and $A = 1$. The choice $w = -1$ ensures that the surface points flow with the fluid into the spike region, maintaining good resolution as the spike curvature increases. The spike's downward acceleration is shown in Figure 2 for several choices of ε. In all cases, the acceleration overshoots before tending to the steady value $g = 1$. On the other hand, the bubble rises with a uniform speed after a sufficient time has elapsed; see Figure 3.

Since the fluid is semi-infinite, the long-time behaviour described above will persist until other physical effects become important. For instance, surface tension may be expected to cause drops to form at the head of the spikes that eventually detach and fall away. In the next section, attention will be given to the case of a finite layer falling under gravity.

4. ACCELERATING FLUID LAYERS

There are several instances where accelerating fluid layers are of practical concern, notably in inertial confinement devices. There is a close relationship between accelerating fluid layers and fluid layers falling under the influence of gravity since the fictitious force induced by an accelerating frame resembles a gravity force. In this section, the modifications to the generalised vortex method necessary to describe the motion of accelerating fluid layers will be given first, followed by some specific numerical results for layers in two-dimensional, periodic planar geometry.

4.1 Formulation

Consider an incompressible, inviscid fluid lying between two non-intersecting surfaces $\vec{x}_1$ and $\vec{x}_2$. Adjacent to the fluid layer, there lies fluid so low in density that the region it occupies may be considered effectively a vacuum, yet it is capable of supporting an external pressure. Alternatively the region is a vacuum and

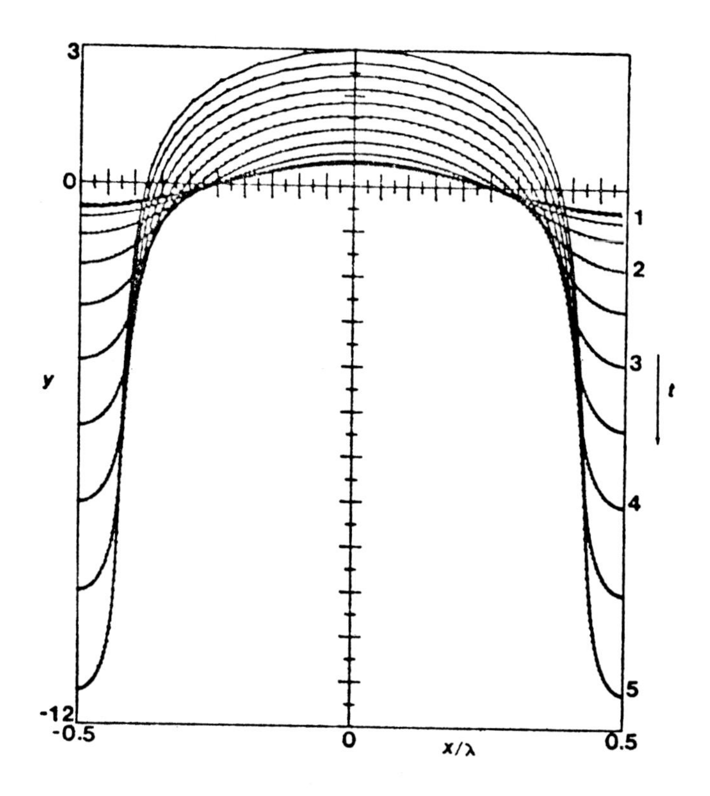

Figure 1. Plot of the interface $y(x,t)$ at times $t = 0.5$ (0.5). The dots on the interface indicate the position of markers moving with the Lagrangian velocity of the upper fluid. The fluid is initially at rest with the initial interface given by Eq. (3.30) with $\varepsilon = 0.5$, $\lambda = 2\pi$. Here $g = 1$ and the Atwood ratio $A = 1$.

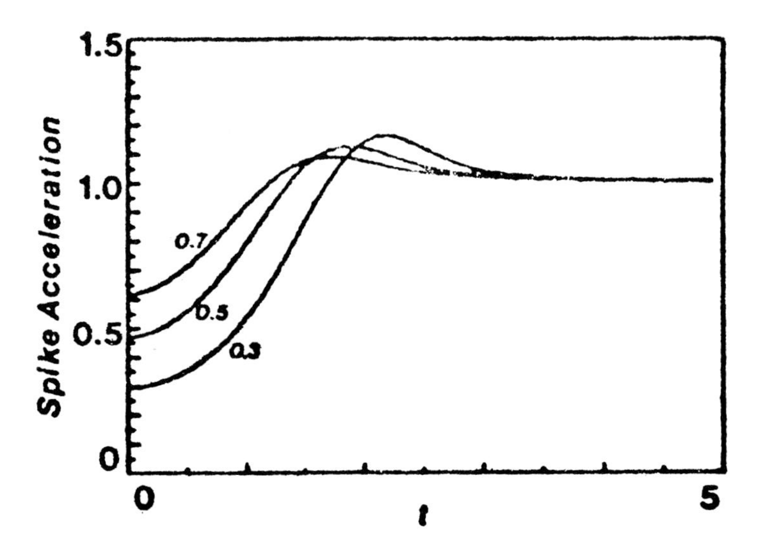

Figure 2. A plot of normalized acceleration a/g vs. time at the tip of the spike for Rayleigh-Taylor instability runs with the fluid initially at rest and the initial interface given by Eq. (3.30) with $\lambda = 2\pi$, $\varepsilon = 0.3$, 0.5, and 0.7. Here $g = 1$ and $\Lambda = 1$.

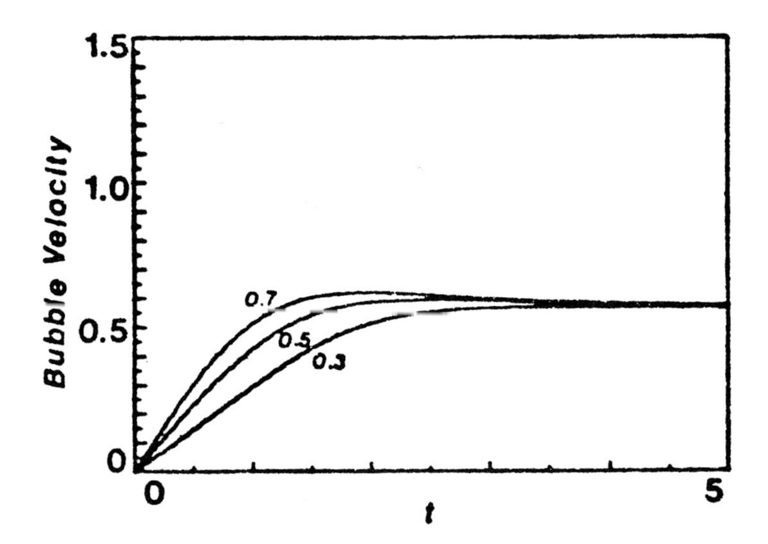

Figure 3. A plot of the velocity vs. time at the peak of the bubble for the same three instability runs as in Figure 2.

there are external surface pressures present. In either case, assume that the external pressure is constant, but with different values P_0 and P_1 in each external region adjacent to $\vec{x}_1$ and $\vec{x}_2$ respectively. Clearly the pressure difference $P_0 - P_1$ will accelerate the fluid layer. Since there is a mean flow present, the velocity potential can no longer be expressed solely in terms of dipole distributions.

Once again, for simplicity, the flow is assumed two-dimensional but the approach adopted below applies equally well to three-dimensional flow. The complex velocity potential may be written as

$$\Phi = \Phi_E + \Phi_D \tag{4.1}$$

where Φ_E accounts for the external flow and Φ_D may be expressed in terms of dipole representations along each surface;

$$\Phi_D(z) = \sum_{n=1}^{2} \frac{1}{2\pi i} \int \frac{\mu_n(\alpha) z_{n\alpha}(\alpha)}{z - z_n(\alpha)} d\alpha \tag{4.2}$$

where $z_1(\alpha)$ and $z_2(\alpha)$ are the complex locations of the two surfaces and $\mu_1(\alpha)$ and $\mu_2(\alpha)$ are the corresponding dipole strengths. The choice for Φ_E depends on the geometrical configuration of the layer. For a periodic layer, parallel to the x-axis,

$$\Phi_E = -iV(t)z \ . \tag{4.3}$$

The Lagrangian motion of each interface is given by (3.15) as long as the full potential Φ from (4.1) is used and the principal values of the integrals in (4.2) are taken when necessary. Thus,

$$\frac{\partial z_j^*}{\partial t}(\alpha) = \tilde{q}_j^*(\alpha) = (\Phi_{j\alpha}(\alpha) + \frac{w_j}{2} \mu_{j\alpha}(\alpha))/z_\alpha(\alpha) - iV(t) \tag{4.4}$$

where $\Phi_j(\alpha)$ is the average value of Φ_D at the surface $z_j(\alpha)$, and w_j is the weighting parameter introduced in (3.8).

The equations for the dipole strengths μ_1 and μ_2 are obtained using the methods described in section III;

$$\frac{\partial \mu_1}{\partial t}(\alpha) - \mathrm{Re}\{\frac{1}{\pi i} \oint \frac{\partial \mu_1}{\partial t}(\alpha') \frac{z_{1\alpha}(\alpha')}{z_1(\alpha)-z_1(\alpha')} d\alpha'$$

$$+ \frac{1}{\pi i} \int \frac{\partial \mu_2}{\partial t}(\alpha') \frac{z_{2\alpha}(\alpha')}{z_1(\alpha)-z_2(\alpha')} d\alpha'\} = g_1 + 2\frac{dV}{dt} y_1 \quad (4.5a)$$

$$\frac{\partial \mu_2}{\partial t}(\alpha) + \mathrm{Re}\{\frac{1}{\pi i} \int \frac{\partial \mu_1}{\partial t}(\alpha') \frac{z_{1\alpha}(\alpha')}{z_2(\alpha)-z_1(\alpha')} d\alpha'$$

$$+ \frac{1}{\pi i} \oint \frac{\partial \mu_2}{\partial t}(\alpha') \frac{z_{2\alpha}(\alpha')}{z_2(\alpha)-z_2(\alpha')} d\alpha'\} = g_2 - 2\frac{dV}{dt} y_2 \quad (4.5b)$$

where

$$g_1 = \mathrm{Re}\{\frac{1}{\pi i} \oint \frac{\mu_1(\alpha')}{z_1(\alpha)-z_1(\alpha')}\{q_{1\alpha}(\alpha') - \frac{q_1(\alpha)-q_1(\alpha')}{z_1(\alpha)-z_1(\alpha')} z_{1\alpha}(\alpha')\}d\alpha'$$

$$+ \frac{1}{\pi i} \int \frac{\mu_2(\alpha')}{z_1(\alpha)-z_2(\alpha')}\{q_{2\alpha}(\alpha') - \frac{q_1(\alpha)-q_2(\alpha')}{z_1(\alpha)-z_2(\alpha')} z_{2\alpha}(\alpha')\}d\alpha'$$

$$+ (\frac{W}{2} + \frac{1}{4})\frac{\mu_{1\alpha}^2}{z_{1\alpha}z_{1\alpha}^*} - q_1 q_1^* - w\mu_{i\alpha}\mathrm{Re}\{\frac{q_1}{z_{1\alpha}}\} + 2P_0 \quad (4.6a)$$

$$g_2 = -\mathrm{Re}\{\frac{1}{\pi i} \int \frac{\mu_1(\alpha')}{z_2(\alpha)-z_1(\alpha')}\{q_{1\alpha}(\alpha') - \frac{q_2(\alpha)-q_1(\alpha')}{z_2(\alpha)-z_1(\alpha')} z_{1\alpha}(\alpha')\}d\alpha'$$

$$+ \frac{1}{\pi i} \oint \frac{\mu_2(\alpha')}{z_2(\alpha)-z_2(\alpha')}\{q_{2\alpha}(\alpha') - \frac{q_2(\alpha)-q_2(\alpha')}{z_2(\alpha)-z_2(\alpha')} z_{2\alpha}(\alpha')\}d\alpha'$$

$$+ (\frac{W}{2} - \frac{1}{4}) \frac{\mu_{2\alpha}^2}{z_{2\alpha}z_{2\alpha}^*} + q_2 q_2^* + w\mu_{2\alpha}\mathrm{Re}\{\frac{q_2}{z_{2\alpha}}\} - 2P_i \quad (4.6b)$$

and

$$q_j = \frac{\Phi_{j\alpha}}{z_\alpha} \quad (4.7)$$

Terms that are constant in Bernoulli's equation throughout the layer have been ignored since they contribute nothing to the kinematics or dynamics but merely reflect the fact that the potential is not unique to any constant.

Upon close examination, one may recognise that (4.5) has the same form as (3.23) with $\frac{dV}{dt}$ taking the role of g, but with an additional term present which includes the effect of the external pressure. The surface labelled by the subscript 1 lies below that labelled by 2, so the values of A must be -1 and 1 respectively.

The dipole equations, (4.5), are coupled Fredholm integral equations of the second kind. They possess non-trivial homogeneous solutions $\frac{\partial\mu_1}{\partial t} = \frac{\partial\mu_2}{\partial t} = $ constant, and so, according to the Fredholm Alternative, the inhomogeneous equations have solutions if and only if

$$2\,\frac{dV}{dt}\,[\int y_2\tau_2 d\alpha + \int y_1\tau_1 d\alpha] = \int g_2\tau_2 d\alpha - \int g_1\tau_1 d\alpha \qquad (4.8)$$

where τ_1 and τ_2 satisfy the homogeneous adjoint system,

$$\tau_1(\alpha) + \mathrm{Re}\{\frac{1}{\pi i} \oint \tau_1(\alpha')\,\frac{z_{1\alpha}(\alpha)}{z_1(\alpha)-z_1(\alpha')}\,d\alpha'$$

$$+ \frac{1}{\pi i} \int \tau_2(\alpha')\,\frac{z_{2\alpha}(\alpha)}{z_1(\alpha)-z_2(\alpha')}\,d\alpha'\} = 0 \qquad (4.9a)$$

$$\tau_2(\alpha) - \mathrm{Re}\{\frac{1}{\pi i} \int \tau_1(\alpha')\,\frac{z_{1\alpha}(\alpha)}{z_2(\alpha)-z_1(\alpha')}\,d\alpha'$$

$$+ \frac{1}{\pi i} \oint \tau_2(\alpha')\,\frac{z_{2\alpha}(\alpha)}{z_2(\alpha)-z_2(\alpha')}\,d\alpha'\} = 0 \qquad (4.9b)$$

The condition (4.8) determines $\frac{dV}{dt}(t)$. In effect, it guarantees that the decomposition of the potential into an external and dipole contributions is legitimate. The solutions to (4.5) are arbitrary to any $\frac{\partial\mu_1}{\partial t} = \frac{\partial\mu_2}{\partial t} =$ constant. Thus it is necessary to impose a further restriction to ensure a unique solution for $\frac{\partial\mu_1}{\partial t}$ and $\frac{\partial\mu_2}{\partial t}$. Then (4.5) may be solved iteratively following a slight modification of the scheme suggested by (2.12), details will be given elsewhere. Similarly (4.9) may be solved iteratively.

The process of computing the evolution equations for z_1, z_2, μ_1, μ_2 and V, given their values at any time t, is as follows. The velocities $\tilde{q}$ are evaluated directly via (4.4). Then τ_1 and τ_2 are solved by iteration from (4.9); g_1 and g_2 are evaluated and consequently $\frac{dV}{dt}$ is determined by (4.8). Finally, $\frac{\partial \mu_1}{\partial t}$ and $\frac{\partial \mu_2}{\partial t}$ are solved by iteration from (4.5). Consequently the values of z_1, z_2, μ_1, μ_2 and V may be updated and the process continued.

4.2 Numerical Results

While the procedure outlined above is straightforward for computing the motion of a finite fluid layer driven by an external pressure gradient, an even simpler problem can be studied by these methods. Suppose that $\frac{dV}{dt}$ is given. The pressure difference $P_0 - P_1$ now varies in time such that (4.8) is satisfied, but one need not know what that variation is in order to compute $\frac{\partial \mu_1}{\partial t}$ and $\frac{\partial \mu_2}{\partial t}$. In fact, externally driven flow in a frame moving with the fluid layer is entirely equivalent to that of a fluid layer falling under the influence of a downward gravity $g = \frac{dV}{dt}$ without any mean flow.

A preliminary study has been made of the fall of a fluid layer, initially lying between

$$y_1 = \varepsilon \cos x \tag{4.10a}$$

and

$$y_2 = H , \tag{4.10b}$$

and subject to a downward gravity $g = 1$. In Figure 4, the location of the surfaces are plotted at various times for the case when $\varepsilon = 0.2$, $H = 1$, $w_1 = -1$ and $w_2 = 1$ so that the markers follow the fluid motion at the surfaces. Several features are easily observable. The fluid falls rapidly into long spikes while a bubble rises between them. A small reverse jet forms behind the spike. The results raise a natural question. Does the bubble burst through the layer in a finite time? The layer thickness at the bubble peak $(x = 0)$ is plotted in Figure 5 in time.

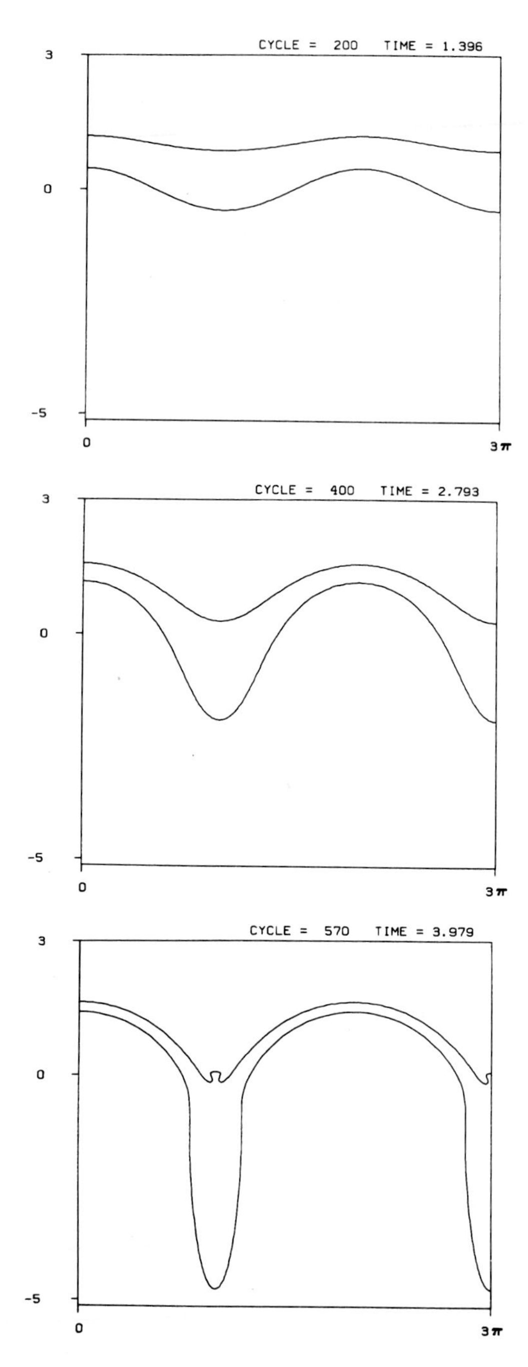

Figure 4. Plot of the surface of the fluid layer of various times.

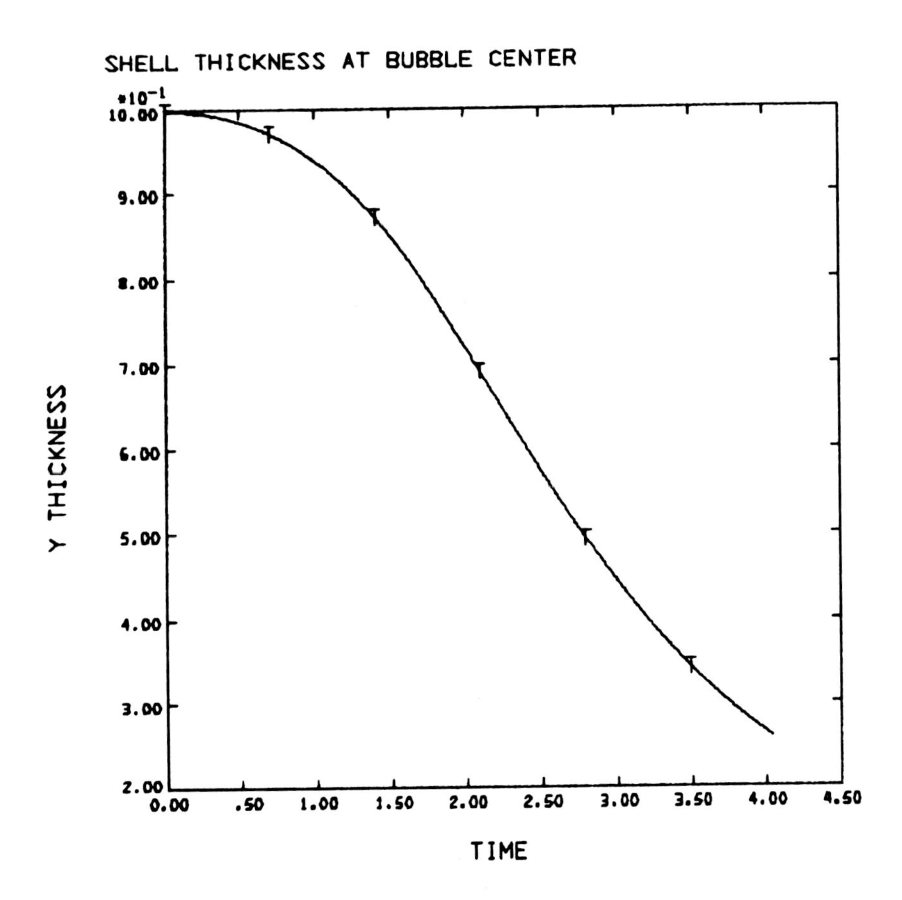

Figure 5. The ratio of the thickness of the fluid layer at the center of the bubble to its initial value as a function of time

The results are not yet conclusive. The layer appears to thin exponentially, but the code stops before the question can be laid to rest. The difficulty is that as the surfaces approach one another, large errors appear in the numerical quadrature because of the strong variation in the integrands. This error has been discussed before for vortex sheets by Baker (1980) and Maskew (1977). Clearly an improved quadrature or another suitable subtraction to remove the dominant behavior of the integrand is necessary. Work is continuing and more detailed results are expected to be published at some later time.

REFERENCES

1. Baker, G. R., 1980, J. Fluid Mech. 100, p . 209.
2. Baker, G. R., D. I. Meiron and S. A. Orszag, 1980, Phys. Fluids 23, p. 1485.
3. Baker, G. R., D. I. Meiron and S. A. Orszag, 1982, J. Fluid Mech. 123, p. 477.
4. Baker, G. R., D. I. Meiron and S. A. Orszag, 1983, submitted to J. Fluid Mech.
5. Birkhoff, G., 1954, Los Alamos Sci. Lab. Report LA-1862.
6. Carrier, G., M. Crook and C. Pearson, 1966, Functions of a Complex Variable, McGraw-Hill, New York.
7. Gram, C., 1962, Selected Numerical Methods, Regnecentralen, Copenhagen
8. Isaacson, E. and H. B. Keller, 1966, Analysis of Numerical Methods, Wiley and Sons, New York.
9 Jaswon, M. A. and G. T. Symm, 1977, Integral Equation Methods in Potential Theory and Elastostatics, Academic, New York.
10. Kellogg, O. D., 1953, Foundations of Potential Theory, Dover.
11. Longuet-Higgins, M. S. and E. D. Cokelet, 1976, Proc. Roy. Soc. London Ser. A364, p. 1.
12. Maskew, B., 1977, J. Aircraft 14, p. 188.
13. Meiron, D. I. and P. G. Saffman, 1983, submitted to J. Fluid Mech.
14. Menikoff, R. and C. Zemach, 1980, J. Comp. Phys. 36, p. 366.

15. Mikhlin, S. G., 1957, Integral Equations and Their Applications to Certain Problems in Mechanics, Mathematical Physics and Technology, Macmillan, New York.

16 Pullin, D. I., 1982,J. Fluid Mech. 119, p. 507.

17. Verdon, C. P., R. L. McCrory, R. L. Morse, G. R. Baker, D. I. Meiron and S. A. Orszag, 1982, to appear in Phys. Fluids.

18. Vinje, T. and P. Brevig, 1981, Adv. Water Resources 4, p. 77.

19. Yeung, R. W., 1982, Ann. Rev. Fluid Mech. 14, p. 395.

20. Zabusky, N. J., M. H. Hughes and K. V. Roberts, 1979, J. Comp. Phys. 30, p. 96.

21. Zaroodny, S. J. and M. D. Greenberg, 1973, J. Comp. Phys. 11, p. 440.

Much of this work has been done in collaboration with D. I. Meiron, S. A. Orszag, C. Verdon and R. McCrory.

Mathematics Research Center
University of Wisconsin-Madison
Madison, Wisconsin 53705

A REVIEW OF SOLUTION METHODS FOR VISCOUS FLOW IN THE PRESENCE OF DEFORMABLE BOUNDARIES

L. G. Leal

A. INTRODUCTION

The motion of a viscous fluid which is bounded by a deformable interface constitutes a difficult and classical problem of fluid mechanics. In general, the shape of this interface is not known in advance, but must be determined as part of the solution. The goal is to develop methods of solution which are *not* restricted to small deformations from some obvious shape. In general, however, viscous flow problems involving two fluids are exceedingly difficult to solve even when the shapes of the interface are assumed to be *known*. Among the few solutions of this latter type the most commonly quoted are those involving a spherical drop in some linear flow (for example, simple shear) or in a uniform stream. Although a number of solutions have been published for creeping motions of particles near a solid wall, almost nothing has been done (until recently) on the corresponding problems of particle motions near an initially flat fluid interface.

Recent years have seen a number of important advances in solution techniques for viscous flow problems involving fluid interfaces, both with a known shape, and of the free (deformable) boundary type. Probably the most common approach for the latter problems has been numerical analysis via finite-element approximations. However, finite-element methods have been the subject of numerous recent books, reviews and papers [cf. 32,40], and we shall thus focus on

ISBN 0-12-493220-7

alternate techniques in this paper. These include, particularly, the so-called singularity methods of low Reynolds number hydrodynamics, and numerical methods based upon finite-difference approximations which can be used for free or fixed boundary problems at finite Reynolds numbers.

Although many problems could be used as a vehicle for illustration of these solution methods, we have concentrated in our research on the motions of particles or drops/bubbles near a fluid interface, and we shall thus focus on that class as a convenient framework for our discussion. The archetypal problem is translation (and/or rotation for rigid particles) in the vicinity of an interface that would be *flat* in the absence of the fluid motions induced by the particles. In the case of a bubble or drop, even the limiting problem of translation far from the interface (i.e. in an "unbounded" fluid) is of considerable interest, and, for finite Reynolds numbers, unsolved except for cases of infinitesimal deformation [4,29,42].

It may be noted at the outset that the particle motion problems considered here have no connection with wave motions at the interface. This is not to say that wave motions could never be important for this class of problems. However, the parameter domain of dominant interest in many of the related technological applications is highly viscous flows at low Reynolds number where wave motions play, at most, a secondary role.

B. THE BASIC PROBLEM

The problem which will form the basis for most of the discussion in this paper is the motion of a particle or drop through a quiescent viscous fluid near a fluid-fluid interface. The particle is assumed to be wholly immersed in one of the two fluids; and in the problem which we have investigated most extensively, the interface is assumed to be flat prior to the existence of any fluid motion. When the particle or drop moves, the motions induced in the fluids cause the interface to deform. The problem, in general, is to determine the velocity and pressure fields in the two fluids (as well as inside the drop if the particle is a drop), and the interface (and drop) shape, all as a function of time starting from a known initial state. The most common technological application where this problem plays a fundamental role is the coalescence (or capture) of a small drop (or particle) at an interface as occurs in a variety of industrially important phase separation processes.

For simplicity, we focus on the case of a single rigid particle which translates and/or rotates with either a prescribed velocity, or under the action of a prescribed force and/or torque in the presence of a pair of incompressible, Newtonian fluids. The governing equations in the two fluids are thus simply the Navier-Stokes equations

$$\begin{aligned} \nabla^2 \underline{u}_1 - \nabla p_1 &= Re(\lambda)^{-1} \left\{ \frac{\partial \underline{u}_2}{\partial t} + \underline{u}_1 \cdot \nabla \underline{u}_1 \right\} \\ \nabla \cdot \underline{u}_1 &= 0 \end{aligned} \tag{1a}$$

$$\begin{aligned} \nabla^2 \underline{u}_2 - \nabla p_2 &= Re \left\{ \frac{\partial \underline{u}_2}{\partial t} + \underline{u}_2 \cdot \nabla \underline{u}_2 \right\} \\ \nabla \cdot \underline{u}_2 &= 0 \end{aligned} \tag{1b}$$

where the two fluids are designated I and II, and the particle is assumed to be in the lower fluid II. These equations (1) are to be solved subject to the boundary conditions

$$\underline{u}_1, \underline{u}_2 \to 0 \qquad \text{as} \qquad |\underline{x}| \to \infty \tag{2}$$

$$\underline{u}_1 = \underline{u}_2 \tag{3a}$$

$$\underline{n} \cdot \underline{u}_1 = \underline{n} \cdot \underline{u}_2 = \kappa \frac{\partial H}{\partial t} \quad \text{on the interface} \tag{3b}$$

$$\underline{n} \cdot \underline{\underline{T}}_1 - \underline{n} \cdot \underline{\underline{T}}_2 = \frac{1}{Ca} (\nabla \cdot \underline{n}) \underline{n} + \frac{1}{Cg} H \underline{n} \tag{3c}$$

and either

$$\underline{u}_2 = \underline{e} \text{ on the sphere surface when the velocity is prescribed} \tag{4a}$$

or

$$1 = \frac{1}{6\pi} \underline{e} \cdot \int_{S_p} \underline{\underline{T}}_2 \cdot \underline{n} dS \text{ when the force is prescribed.} \tag{4b}$$

Appropriate modifications can be made if the angular velocity or the torque is given.

The Reynolds number which appears in (1) is defined as

$$\mathrm{Re} \equiv \frac{U_p \ell_p}{\nu_2}$$

where U_p and ℓ_p are the characteristic velocity and length scales of the particle, and ν_2 is the kinematic viscosity of fluid II. The viscosity ratio μ_1/μ_2 is denoted as λ. The dimensionless parameters appearing in the normal stress balance are

$$\mathrm{Ca} \equiv \frac{\mu_2 U_p}{\sigma} \quad ; \quad \mathrm{Cg} \equiv \frac{\mu_2 U_p}{g \ell_p^2 (\rho_2 - \rho_1)} .$$

The stress $\underline{\underline{T}}_i$ which appears in (3c) is defined in terms of the dynamic pressure and is non-dimensionalized with respect to $T_c = \mu_2 U_p/\ell_p$, so that

$$\begin{aligned} \underline{\underline{T}}_1 &= -p_1 \underline{\underline{I}} + \lambda \underline{\underline{\tau}}_1 \\ \underline{\underline{T}}_2 &= -p_2 \underline{\underline{I}} + \underline{\underline{\tau}}_2 \end{aligned} \tag{5}$$

and the body force term appears explicitly on the right-hand side of (3c).

The interface shape is defined as

$$H(z,r,\phi,t) = 0 \tag{6}$$

where $z = 0$ is the plane of the undeformed interface. In terms of this function

$$\underline{n} = \frac{\nabla H}{|\nabla H|} \tag{7a}$$

and

$$\kappa = \frac{1}{|\nabla H|} . \tag{7b}$$

As noted in the Introduction, the mathematical problem represented by Eqs. (1-7) is highly nonlinear and also fully time-dependent. In part, this is a consequence of the governing Navier-Stokes equations, as is true in all Newtonian fluid mechanics problems at non-zero Reynolds number.

Additionally, however, the dependence of the solution on the unknown boundary shape, i.e. the function H, introduces further nonlinearities and time-dependence through the boundary conditions (3a-c).

In the present paper, we first consider solution methods at low Reynolds number where the creeping motion limit of Eqs. (1) can be used. We then describe a general numerical approach for the finite Reynolds number case — but applied here only to the limiting problem of a single gas bubble rising due to buoyancy in an unbounded, viscous fluid.

C. SOLUTION METHODS FOR VANISHINGLY SMALL REYNOLDS NUMBERS

We begin by considering the limit, Re → 0, of the general problem defined in the preceding section. In this case, the equations of motion, (1), reduce to the well-known creeping motion equations

$$\left.\begin{aligned} \nabla^2 \underline{u}_i - \nabla p_i &= 0 \\ \nabla \cdot \underline{u}_i &= 0 \end{aligned}\right\} \quad i = I, II \tag{8}$$

which are linear, and "quasi-static" in form. As noted in the preceding section, however, the resulting problem of solving (8) subject to the conditions (2-7) is still highly nonlinear and time-dependent, as a consequence of the unknown interface shape which enters the boundary conditions (3a-c). The usual solution methods for low Reynolds hydrodynamics, which rely on linearity of both the differential equations and boundary conditions, can thus be applied only as a first approximation in those limiting cases (e.g. Ca → 0, Cg → 0 or particle far from the interface) where the interface remains very nearly flat. The general case (arbitrary Ca, Cg or particle position) when interface deformation is *not* small is still simplified compared to Re ≠ 0, however, because the linearity of (8) allows a completely general solution to be obtained, which need then only be specialized for a particular case by application of boundary conditions.

We shall discuss the general case shortly. First, however, we consider briefly the solution of particle motion problems in the fixed shape, flat interface approximation. Although the case of a flat wall ($\lambda \to \infty$) has received considerable attention [16,31], relatively few solutions have yet been obtained even for a *flat* interface with arbitrary λ.

(1) Special Case – Flat Interface

When Ca or Cg is small enough, or the particle is sufficiently far from the interface, its shape will depart very little from the initial, flat form. Thus, to a first approximation, the interface can be treated as flat and the problem is linear. A variety of solution techniques can then be used to obtain a solution, mirroring the many methods of solution that have been developed for creeping motions in unbounded domains.

Most common among the unbounded domain techniques have been straightforward, though sometimes tedious, eigenfunction expansions [16,30,31], and this approach can also be used in the presence of a flat wall or interface. However, exact creeping motion solutions via eigenfunction expansions can be achieved only when *all* boundaries of the flow domain are coordinate surfaces for some known coordinate system and this restricts use in the presence of a flat interface to the special cases of either a spherical particle where a bipolar spherical coordinate system has both the sphere surface and the interface as coordinate surfaces, or possibly ellipsoids/spheroids with special (i.e. parallel or normal) orientations relative to the interface.

Methods designed specifically to treat more general particle geometries have been developed for *unbounded* fluids, based upon superposition of fundamental solutions of Stokes' equations for distributions of point force (or higher order) singularities. In the unbounded fluid case, these singularities may be distributed over the boundaries of the particle, in which case only point force singularities are required, or inside the particle (usually on a line or, in the case of a sphere, at the center point) where higher order singularities are also necessary. *Boundary* distributions of point forces lead to integral equations for the "weighting functions" which can generally be solved quite easily by numerical methods. The resulting "boundary integral" technique is particularly powerful, as we shall discuss in the next subsection, for problems with deformable boundaries or for problems with highly complex boundary shapes, both at zero Reynolds number. Here we consider the use of *internal* distributions of point force and higher order singularities, a technique which has been developed extensively for unbounded flows by Wu and co-workers [6-9,22]. In unbounded fluids, this method leads to exact solutions for relatively simple flows and particles such as spheres and spheroids, and to approximate solutions for slender-bodies with only a line distribution of point force singularities required in this latter case.

The basic development underlying both types of application are the fundamental solutions of Stokes' equations which satisfy continuity of velocity, continuity of tangential stress and the kinematic condition, $\underline{n}\cdot\underline{u}_I - \underline{n}\cdot\underline{u}_{II} = 0$, at the flat interface. (The normal stress balance is not applicable for this case, where the boundary shape is assumed known at first order of approximation.) With these fundamental solutions available, arbitrary distributions of singularities will automatically satisfy boundary conditions at the interface and thus the problems reduce, as in the *unbounded* fluid case, to satisfying boundary conditions at the particle surface. The necessary fundamental solutions are obtained easily using a generalization by Lee, Chadwick and Leal [25] of a method due to Lorentz, who utilized the so-called reciprocal theorem of low Reynolds number hydrodynamics, to obtain a solution of Stokes' equations for the motion generated by a point force in the presence of a plane solid wall.

Lee, Chadwick and Leal [25] gave, as a lemma to Lorentz' theorem, a general formula for obtaining solutions of Stokes' equations that satisfy all of the required boundary conditions at a flat interface, starting with any arbitrary solution of Stokes' equations for the corresponding motion in an unbounded domain with no interface. One use of Lee's lemma is to generate fundamental solutions for a point-force (or higher order singularities) in the presence of a flat interface from the corresponding fundamental solutions (such as the Stokeslet, force dipole, potential dipole, etc.) in an unbounded fluid (cf. Chwang and Wu [7]). For convenience, we can represent the fundamental solution for a point force (Stokeslet) located at an arbitrary point, $\underline{x}_s$ as

$$\begin{aligned} \underline{u}(\underline{x},\underline{x}_s;\underline{\alpha}) &= \underline{\alpha}(\underline{x}_s)\cdot\underline{\underline{\psi}}(\underline{x},\underline{x}_s) \\ p(\underline{x},\underline{x}_s;\underline{\alpha}) &= \underline{\alpha}(\underline{x}_s)\cdot\underline{\pi}(\underline{x},\underline{x}_s) \end{aligned} \tag{9}$$

where the point force is

$$\underline{f}_s(\underline{x}_s) = 8\pi\mu_2\underline{\alpha}(x_s)$$

and $\underline{\underline{\psi}}$ and $\underline{\pi}$ are the tensorial (and vector) Green's function for Stokes' equations. As noted earlier, these fundamental solutions can be used in two ways to solve problems of particle motion near a flat fluid interface: first, via internal distributions of point force and higher order singularities for bodies of general, but non-slender, shape; and via a line distribution of point forces only for a "slender body" of arbitrary cross section.

(a) Non-Slender Bodies near a Flat Interface

The case of non-slender bodies, exemplified by the problem of a sphere in simple translation and rotation, follows the analogous studies of Wu and co-workers [6-9,22] for unbounded fluid domains, but differs in the sense that exact solutions cannot be obtained with any finite number of interior force multipoles, even for the simplest case of a spherical body. Only asymptotic solutions can be obtained, valid when the sphere is far enough from the interface. This is a consequence of the fact that the solution generated by applying Lee's lemma to the solution for an unbounded domain satisfies boundary conditions only at the interface — in particular, it no longer satisfies boundary conditions at the particle surface. The form of the "disturbance flow" produced at the sphere surface simplifies to a form which can be "cancelled" by a small number of additional internal singularities only when the particle is far from the surface.

Solutions are outlined in detail for a sphere by Lee et al. [25]. Here we simply summarize the method of solution starting with the exact solution of Stokes' equations for translation of a sphere in an *unbounded* fluid. Wu et al. [7] (and others, earlier) have shown that the latter solution can be represented as the motion generated by a point force (Stokeslet) and potential dipole located at the center of the sphere. We thus apply the lemma of Lee et al. to this solution, obtaining a new solution corresponding to a point force and potential dipole of the same strength in the presence of an interface at some prescribed position. As guaranteed by the lemma, this new solution, call it $(\underline{u}_1^+, \underline{u}_2^+)$, satisfies boundary conditions at the interface (here, as before, the subscripts denote the velocity fields in fluids 1 and 2; we assume the sphere to be located in fluid 2). However, it does not now satisfy boundary conditions on the sphere surface. In particular, even though the unbounded solution $\underline{u}_2^{(o)}$ satisfies exactly the no-slip conditions at the sphere, the velocity field correction

$$\underline{u}_2^{(1)} = \underline{u}_2^+ - \underline{u}_2^{(o)}$$

is nonzero at the sphere surface. Further, since $\underline{u}_2^{(1)}$ evaluated at the sphere surface is highly complicated, it is not possible to cancel $\underline{u}_2^{(1)}$ precisely (so as to satisfy the no-slip and zero normal velocity boundary conditions) with any small number of additional internal singularities. Indeed, Lee et al. [25] found it necessary to consider the asymptotic limit $\varepsilon = \ell^{-1} \ll 1$,* and then choose singularities to cancel only the first few terms of $\underline{u}_2^{(1)}$ at the sphere surface, with $\underline{u}_2^{(1)}$ expressed in powers of ε. In general, when these new singularities are forced to satisfy boundary conditions at the interface, by application of Lee's lemma, they generate new higher-order corrections in ε at the sphere surface, which,

*ℓ = distance from sphere center to the plane of the undeformed interface, non-dimensionalized by the sphere radius, a.

when combined with higher-order terms from $\underline{u}_2^{(1)}$ require still more (though asymptotically weaker) higher-order singularities at the sphere center.

In spite of the asymptotic nature of the solutions generated, comparison for this case of a sphere with exact solutions obtained using eigenfunction expansions in bipolar spherical coordinates [26], showed good qualitative and even quantitative agreement down to a separation distance of two radii between the sphere center and the flat interface. Thus, the method based on internal distributions of singularities constitutes a potentially powerful tool for obtaining solutions for more complicated problems including many where formally exact eigenfunction expansions are impossible. Among the latter problems presently being studied at Caltech is motion of spherical particles in general linear flows, such as simple shear flow, and also the translation or rotation of ellipsoidal particles near an interface (or wall).

(b) Slender Bodies near a Flat Interface

The problem of motion of *slender bodies* near a fluid interface can also be treated using the fundamental solutions of Stokes' equations that were generated by application of Lee's lemma. In this case, the analysis again follows closely the well-developed theory of slender body theory for Stokes' flow in an unbounded fluid (cf. [2,10]). In a paper which has just been submitted for publication, Yang and Leal [45] have considered translation and rotation of an arbitrarily oriented slender body both in the case of quiescent fluids, and in the case of linear flows such as simple shear flow and "hyperbolic" stagnation point flows. The problem of translation with the particle axis parallel or perpendicular to the interface was considered earlier by Fulford and Blake [14].

In the slender-body theory, the velocity field (at lowest order of approximation) is represented by a line distribution of Stokeslets along the particle centerline with line density $\underline{\alpha}(\underline{x}_s)$, i.e.

$$\underline{u}(\underline{x}_s) = \int_{-\ell}^{\ell} \underline{\alpha}(\underline{x}_s) \cdot \underline{\underline{\psi}}(\underline{x}, \underline{x}_s) d\zeta \qquad (11a)$$

$$p(\underline{x}) = \int_{-\ell}^{\ell} \underline{\alpha}(\underline{x}_s)\cdot\underline{\pi}(\underline{x},\underline{x}_s)\,d\zeta \qquad (11b)$$

The boundary conditions at the flat interface are satisfied for any distribution $\underline{\alpha}(x_s)$ which may thus be chosen to satisfy (at least approximately) the boundary conditions on the particles surface.

If the particle translates and rotates in an otherwise quiescent fluid system, application of the boundary conditions to (11a) gives

$$\underline{e} + \left(\frac{\underline{\Omega}_p}{U_p}\right) \wedge \underline{x}_o = \int_{-\ell}^{\ell} \underline{\alpha}(\underline{x}_s)\cdot\underline{\underline{\psi}}(\underline{x}_B,\underline{x}_s)\,d\zeta \qquad (12)$$

in which $\underline{\Omega}$ is the angular velocity of the particle, and $\underline{x}_o$ is the position vector from particle center to the surface point $\underline{x}_B$. Equation (12) is a linear integral equation of the second kind which can be solved for the function $\underline{\alpha}(\underline{x}_s)$ either by numerical or asymptotic analysis. In the latter case, which was carried out in detail in Ref. [45], it is assumed

$$\underline{\alpha} = \varepsilon\underline{\alpha}_o + \varepsilon^2\underline{\alpha}_1 + \varepsilon^3\underline{\alpha}_2 + \ldots$$

and solutions are sought sequentially for the terms at each order in ε, where $\varepsilon \equiv \left\{\ell n \frac{2\ell}{R_o}\right\}^{-1}$. Although self-consistent solutions can be obtained at $0(\varepsilon)$ and $0(\varepsilon)^2$ with only the Stokeslet distribution, satisfaction of the boundary conditions on the surface of a cylindrical body at $0(\varepsilon^3)$ requires, in addition, a line distribution of potential dipoles of strength

$$\underline{\beta}(x) = -\frac{1}{2}\left[r_o^2(x)\underline{\alpha}(x)\right]$$

where $r_o(x)$ is the local effective radius of the cylinder.

The results of Yang and Leal [45] for translation and rotation parallel and perpendicular to the interface along three mutually perpendicular axes, completely specify all components of the translational and rotational resistance and coupling tensors relating the force and torque on the body to its translational and angular velocities. Thus, general trajectory equations can be obtained to determine the particle motions for an arbitrary applied force and/or torque. The interested reader is referred to our paper [45] for details of these results.

(c) Extension of the Flat Interface Solutions to Small Deformations via Domain Perturbations

The flat interface solutions are, of course, only approximations valid whenever the deformation of the interface is asymptotically small. In particular, these solutions assume that the normal velocity is zero at the interface. This condition is more restrictive than the condition of continuous normal velocity and, together with continuity of tangential velocity and stress, allows a unique solution for the velocity and pressure fields without consideration of the normal stress balance at the interface. Because the normal stress condition is not applied, however, the solution does not generally satisfy it, meaning that it can be accepted (as already noted) only as an approximation (with the possible exception of some special flow which allows a flat interface). In reality the interface deforms, by an amount determined by

$$Ca, \; Cg \; \text{and} \; \ell \quad .$$

When these parameters are small, interface deformation remains small (provided the particle does not penetrate the plane of the undeformed interface), and the flat interface solutions with $\underline{u}\cdot\underline{n} = 0$ can be demonstrated rigorously as the correct zeroth order approximation in an asymptotic expansion for small deformation, i.e. small Ca, Cg or ℓ [3]. Following the framework of the considerable body of work on small deformations of bubbles or drops in flow, the first approximation to interface deformation can be obtained from (3c) using the zeroth-order velocity and pressure fields to approximate the normal stress difference. Knowing the first-order approximation for the interface shape, corresponding corrections can be obtained for the velocity and pressure fields and thus also for the motion of the body. However, provided that it is the force or torque on the body that is required (for given velocities), rather than the detailed velocity and pressure fields in the fluids, the first correction to the flat interface results can be obtained using the reciprocal theorem, requiring only the flat interface velocity and pressure fields and the first interface shape correction as inputs. The same general approach has been used previously to obtain the first corrections to creeping motion of a particle or drop when the drop is slightly deformed [5], or when the suspending fluid is slightly non-Newtonian [5], or when there are weak inertia effects present [19]. Its utilization in the case of a rigid sphere moving in the presence of a slightly deformed interface has been given in detail by Berdan and Leal [3], and we will not repeat the arguments here.

(2) Boundary-Integral Methods for Large Interface Deformations at Low Reynolds Number

In spite of the fact that the general problem of particle motion near an interface remains both time-dependent and nonlinear in the creeping motion limit ($Re \to 0$), it is significant that the nonlinearity resides entirely in the boundary conditions where it is a consequence of the appearance of the function H representing the (unknown and usually time-dependent) interface shape. Only in the limiting case (just discussed) of a flat interface where the shape is known does the low Reynolds number problem become completely linear and quasi-steady. Nevertheless, the linearity of the governing Stokes' equations affords a very considerable simplification in the analysis required to obtain solutions in the general case. Specifically, a general solution of Stokes' equations can be obtained in a straightforward fashion, and the problem is reduced to determining the specific form of this solution which satisfies the boundary conditions. This class of nonlinear boundary value problems is therefore very much easier to solve than problems involving the full Navier-Stokes equations, where both the boundary conditions and differential equations are highly nonlinear, and full finite-difference or finite-element solutions are generally necessary.

In order to solve the problem at low Reynolds number, we apply the general solutions of Stokes' equations due to Ladyzhenskaya [23], in which the velocity and pressure fields are expressed in terms of "single-layer" and "double-layer" singularity distributions at the boundaries of the flow domain. In the class of problems considered here, the boundary of fluid domain I is simply the fluid interface while the boundaries of domain II are the interface and the particle surface. If the particle were a drop, the third fluid domain inside the drop would also have the drop surface as its boundary. Utilization of the Ladyzhenskaya solution in the context of numerical calculations to determine the strength of the single and double layer potentials (as a function of position on the flow boundaries) was first reported by Youngren and Acrivos [46] for flow of an unbounded fluid past a single solid particle. Problems of drop motion (and shape) in extensional flows were later considered by Youngren and Acrivos [47] and Rallison and Acrivos [34], and in shear flow by Rallison [35]. Published papers describing initial calculations for the present problem include Lee and Leal [27], Leal and Lee [24] and a forthcoming paper by Geller, Lee and Leal [15].

Let us consider the method in more detail for the case of a solid particle (which could be quite arbitrary in shape) moving with an arbitrary translational or angular velocity in fluid II (see section II). By using the jump condition for the double layer potential at the boundaries (cf. Ladyzhenskaya [23]) and applying the boundary conditions of section II at the interface and particle surface, it can be shown

$$\frac{1}{2}\,\underline{u}^I(\underline{x}) = -\frac{3}{4\pi}\int \frac{\underline{r}\,\underline{r}\,\underline{r}}{R^5}\,\underline{u}^I\cdot\underline{n}dS^I + \frac{1}{8\pi}\int\left(\frac{\underline{\underline{I}}}{R}+\frac{\underline{r}\,\underline{r}}{R^3}\right)\cdot\underline{\underline{T}}^I_2\cdot\underline{n}dS^I + \frac{1}{8\pi}\int\left(\frac{\underline{\underline{I}}}{R}+\frac{\underline{r}\,\underline{r}}{R^3}\right)\cdot\underline{\underline{T}}^P\cdot\underline{n}dS^P, \quad \underline{x}\varepsilon S^I \tag{13}$$

$$\underline{u}^P = -\frac{3}{4\pi}\int \frac{\underline{r}\,\underline{r}\,\underline{r}}{R^5}\,\underline{u}^I\cdot\underline{n}dS^I + \frac{1}{8\pi}\int\left(\frac{\underline{\underline{I}}}{R}+\frac{\underline{r}\,\underline{r}}{R^3}\right)\cdot\underline{\underline{T}}^I_2\cdot\underline{n}dS^I + \frac{1}{8\pi}\int\left(\frac{\underline{\underline{I}}}{R}+\frac{\underline{r}\,\underline{r}}{R^3}\right)\cdot\underline{\underline{T}}^P\cdot\underline{n}dS^P, \quad \underline{x}\varepsilon S^P \tag{14}$$

$$\frac{1}{2}(\lambda+1)\underline{u}^I(\underline{x}) = \frac{3}{4\pi}(\lambda-1)\int \frac{\underline{r}\,\underline{r}\,\underline{r}}{R^5}\,\underline{u}^I\cdot\underline{n}dS^I + \frac{1}{8\pi}\int\left(\frac{\underline{\underline{I}}}{R}+\frac{\underline{r}\,\underline{r}\,\underline{r}}{R^3}\right)\cdot\underline{\underline{T}}^P\cdot\underline{n}dS^P - \frac{1}{8\pi}\int\left(\frac{\underline{\underline{I}}}{R}+\frac{\underline{r}\,\underline{r}}{R^3}\right)\cdot\underline{F}(\eta)dS^I \tag{15}$$

Here the subscripts I and P indicate variables at the interface and particle surface, respectively, while $\underline{r}$ is the position vector from point $\underline{x}$ to a point $\underline{\eta}$ on the boundary where the singularity is applied (with $R \equiv |\underline{r}|$). Integrations indicated are for all possible $\underline{\eta}$ on the particular boundary. The vector function $\underline{F}$ represents the right-hand side of the boundary condition (3c). Although other forms could be chosen to represent the general solution of Stokes' equations, the Ladyzhenskaya formulation is particularly advantageous because the weighting functions for the "potential" distributions are precisely the velocity and stress components at the boundaries. If we consider a particle moving with specified velocity, Eqs. (13-15) can be solved numerically to obtain $\underline{u}^I$, $\underline{\underline{T}}_2^I$ and $\underline{\underline{T}}^P$ for a given boundary shape (i.e. with $\underline{F}$ specified). Thus, starting from some initial shape, we can calculate the instantaneous velocity of points on the interface and use this information together with the particle velocity to obtain the particle position and interface shape at some incremental time Δt later. With the new interface shape, updated values of $\underline{u}^I$, $\underline{\underline{T}}_2^I$ and $\underline{\underline{T}}^P$ can be obtained from (13-15) and so on. For the case where the force (and/or torque) on the particle is specified (rather than its velocity), it is necessary to include an additional integral constraint on the values of the stress components, $\underline{\underline{T}}^P$, at the particle surface. This equation (or equations) is then simply included with (13-15) and the whole set solved simultaneously using a collocation method, with the resulting set of quasi-linear algebraic equations being solved, for example, by Gaussian elimination.

Examples of the type of results obtained using this method for the case of a spherical particle are contained in the papers cited earlier, as well as additional details of the solution technique including treatment of the singular points of the integrands in (13-15).

D. SOLUTION METHODS FOR FINITE REYNOLDS NUMBER

When the Reynolds is not asymptotically small, the complete Navier Stokes' equations must be solved and this means that a full numerical solution technique must be used. The majority of free boundary calculations, to date, have utilized a finite-element formulation, and this approach has been reviewed extensively in recent years. Here we consider briefly an alternative approach based upon a finite-difference formulation of the problem. The technique which we describe below has not yet been developed to the extent, for example, of the boundary-integral techniques that were described in the preceding section, and in its present state is restricted to two-dimensional or axisymmetric problems. Thus, though the method is applicable, in principle, for the general class of free boundary problems represented in section II, it has so far been used extensively for only two problems; buoyancy driven motions of a gas bubble in an unbounded fluid, and motion of a gas bubble in an axisymmetric straining flow. Papers describing these two problems in detail are currently being prepared (Ryskin and Leal [37]). Only a brief description of the method has been published previously (Ryskin and Leal [38]). The description presented here is identical in most respects to that given in Ref. [38] which is, however, not available easily. In view of the fact that the buoyancy driven motion of a bubble in an unbounded fluid can be considered as a limiting case of the approach normal to a fluid interface, we use that problem for illustrative purposes in the present communication.

The difficulty, generally well known, when finite-difference methods are applied in domains with boundaries that are not coincident with coordinate lines or surfaces is a considerable loss of accuracy associated with the need for interpolation between nodal points in the application of boundary conditions. Thus, if one considers only the classical orthogonal coordinates such as cylindrical, spherical, etc., the use of finite-difference methods is generally unacceptable for free boundary problems. The method described here is based on a numerical method of constructing a system of *orthogonal, boundary-fitted coordinates* when the solution domain does not conform to a simple shape.

(a) Orthogonal Mapping

The idea which we pursue is thus to obtain orthogonal boundary-fitted coordinates for the domain exterior to a bubble whose shape is unknown, though smooth and generally nonspherical. In the present development (a detailed description of which will appear soon in the *J. of Comp. Physics* [39]), the unknown shape is generated via an iterative procedure starting from some initial guess. At each step, with the boundary shape specified, mapping functions for the boundary-fitted coordinates are generated numerically, the equations of motion are then solved in the transform domain and the normal stress balance at the bubble surface is used to generate an improved shape. We restrict ourselves to steady, axisymmetric configurations and discuss the mapping problem in 2D, with the axisymmetric boundary shape generated by rotation about the axis of symmetry which is thus required to be a coordinate line for the transform coordinates.

In the remainder of this section, we outline a method of obtaining the desired coordinate transformation. From a mathematical point of view, we require a pair of functions $x(\xi,\eta)$ and $y(\xi,\eta)$ which map points of the physical domain onto a unit square, $0 \leq \xi,\eta \leq 1$, in the transform domain, with lines of constant ξ and η being orthogonal. For convenience, we designate the bubble surface as $\xi = 1$, and the upstream and downstream axes of symmetry as $\eta = 1$ and $\eta = 0$, respectively, with $\xi \to 0$ corresponding to infinity. There has, of course, been a great deal of recent research aimed at the problem of obtaining numerically generated coordinate mappings. Included in this work are methods based on the solution of a pair of elliptic equations for the mapping functions [43], conformal mapping [13], direct integration of "Cauch-Riemann"-type equations as an initial value problem starting from a boundary [11,41], and other methods of orthogonal mapping which are equivalent to conformal mapping with less restrictive constraints on the ratio of the diagonal components of the metric tensor (the latter are, in fact, most closely related to the present approach) [20,28,33]. Limitations of space prevent a detailed review of this prior work.

However, in general, the resulting coordinate systems are either nonorthogonal [43], ill-conditioned in the sense of extreme sensitivity to boundary shape and/or (the possibility of) highly nonuniform spacing of coordinate lines (conformal mapping [13], and some types of "orthogonal mapping" [20,33]) or only suitable in some local subdomain (integrations of Cauchy-Riemann equations [11,41]).

The present objective is a numerically generated mapping which is applicable in the whole domain, automatically orthogonal and free of the usual sensitivity problems of conformal mapping. Our basis is conventional tensor analysis, yielding equations for $x(\xi,\eta)$ and $y(\xi,\eta)$ which are coordinate invariant. These equations follow almost trivially from the observation that a Cartesian coordinate x (or y) is a linear scalar function of position, so that grad x (or grad y) is constant and

$$\text{div grad } x \;=\; 0 \tag{16}$$

The latter is nothing more than the covariant Laplace equation for x. When expressed in terms of the desired (but as yet unspecified) ξ,η coordinates, the equations governing the transformation (mapping) functions become

$$\frac{\partial}{\partial\xi}\left(f\,\frac{\partial x}{\partial\xi}\right) + \frac{\partial}{\partial\eta}\left(\frac{1}{f}\,\frac{\partial x}{\partial\eta}\right) = 0 \tag{17a}$$

$$\frac{\partial}{\partial\xi}\left(f\,\frac{\partial y}{\partial\xi}\right) + \frac{\partial}{\partial\eta}\left(\frac{1}{f}\,\frac{\partial y}{\partial\eta}\right) = 0 \tag{17b}$$

with

$$f(\xi,\eta) \equiv h_\eta/h_\xi, \text{ and } h_\eta \text{ and } h_\xi \text{ the scale factors of the } (\xi,\eta) \text{ coordinates.} \tag{18}$$

The solution of these equations, subject to appropriate boundary conditions (which we shall discuss in the next section), will yield orthogonal coordinates for any f, which can thus be chosen freely. It may be noted that conformal mapping corresponds to the restrictive choice $f(\xi,\eta) = 1$, for all ξ,η.

In general, the "most appropriate" choice of f depends on the type of mapping required. The problem of direct mapping with fixed boundary shape and a specified distribution of coordinate nodes along the boundary is discussed elsewhere

[39]. Here, we consider only the mapping problem in which the boundary shape is unknown and required as part of the solution of the overall problem. In this case, $f(\xi,\eta)$ can be specified directly as a function of ξ,η, with the form for f chosen so as to yield desired properties of the transform coordinates (e.g. nonuniform spacing of coordinate lines in some region of the domain).

(b) The Fluid Dynamics Problem - Basic Formulation

Let us now return to the problem of uniform streaming flow past a bubble. In this case, we adopt the very simple form for f, $f(\xi,\eta) \equiv \pi\xi(1 - \frac{1}{2}\cos\pi\eta)$. In addition, we introduce a relatively simple modification of the mapping procedure outlined above to take care of the fact that infinite values of the mapping functions x and y, corresponding directly to an infinite domain, cannot be generated numerically. To avoid this difficulty, we simply calculate the mapping from the unit square in the ξ,η plane to an auxiliary finite domain, which is then transformed to the physical domain by a conformal inversion.

Now, one great advantage of orthogonal coordinates, in addition to avoiding inaccuracy of numerical approximation in nonorthogonal coordinates, is that physical components of vectors and tensors can be used instead of covariant or contravariant ones. The governing Navier-Stokes equations, plus boundary conditions, can thus be expressed in a straightforward manner in terms of the resulting boundary-fitted coordinates ξ,η,ϕ, obtained by rotation of the two-dimensional coordinates given by $x(\xi,\eta)$ and $\sigma(\xi,\eta)$ (where x is parallel to the axis of symmetry and σ is the distance to this axis along a normal through the point of interest). If we introduce the streamfunction ψ, and use standard expressions for the invariant differential operators in general orthogonal curvilinear coordinates, the Navier-Stokes equations are

$$\frac{1}{h_\xi h_\eta}\frac{\partial\psi}{\partial\eta}\frac{\partial}{\partial\xi}\left(\frac{\zeta}{\sigma}\right) - \frac{1}{h_\xi h_\eta}\frac{\partial\psi}{\partial\xi}\frac{\partial}{\partial\eta}\left(\frac{\zeta}{\sigma}\right) = \frac{2}{Re}\mathscr{L}^2(\zeta\sigma) \tag{19}$$

$$\mathscr{L}^2\psi + \zeta = 0 \tag{20}$$

where ζ is the vorticity, $Re = \frac{d\rho v_\infty}{\mu}$, d is the equivalent diameter of the bubble and

$$\mathscr{L}^2 \equiv \frac{1}{h_\xi h_\eta}\left\{\frac{\partial}{\partial\xi}\left(\frac{f}{\sigma}\frac{\partial}{\partial\xi}\right) + \frac{\partial}{\partial\eta}\left(\frac{1}{f\sigma}\frac{\partial}{\partial\eta}\right)\right\} \quad (21)$$

The streamfunction at infinity, for a uniform streaming flow, takes the form

$$\psi_\infty = \frac{1}{2}\sigma^2 \quad (22)$$

Thus, to avoid dealing with large (or infinite) numbers, we actually solve for

$$\overset{*}{\psi} = \psi - \psi_a \quad (23)$$

where ψ_a is the potential flow solution for flow past a spherical bubble with the given form ψ_∞ at infinity, i.e.

$$\psi_a = \frac{1}{2}\sigma^2(1 - \xi^3) \quad (24)$$

Now, Eqs. (19) and (20), rewritten in terms of ψ^*, are to be solved for ψ^*, ζ and the bubble shape subject to the boundary conditions

$$\psi^* \text{ is bounded, } \zeta = 0 \text{ ; at infinity (i.e. } \xi = 0) \quad (25)$$

$$\psi^* = 0 \text{ , } \zeta = 0 \text{ ; at } \eta = 0,\ \eta = 1 \text{ (symmetry axis)} \quad (26)$$

and, at the bubble surface,

$$\left.\begin{aligned} &\psi^* = 0 \quad \text{(zero normal velocity)} && (27)\\ &\zeta + 2\kappa_{(\eta)} u_s = 0 \quad \text{(zero tang. stress)} && (28)\\ &-\frac{3}{4}C_D x - p_d + \tau_{nn} + \frac{4}{We}\left(\kappa_{(\eta)} + \kappa_{(\phi)}\right) = 0 && (29)\\ &\qquad\text{(normal stress balance)} \end{aligned}\right\} \text{ at } \xi = 1$$

The first term in (29) is the hydrostatic pressure; C_D is the drag coefficient; p_d is the dynamic pressure

$$p_d = -u_s^2 - \frac{4}{Re}\int \frac{1}{\sigma h_\xi}\frac{\partial}{\partial\xi}(\sigma\zeta)h_\eta d\eta \quad (30)$$

u_s is the surface velocity; τ_{nn} is the normal component of viscous stress at the surface, We $= \frac{d\rho v_\infty^2}{\gamma}$; γ is the surface tension, and $\kappa_{(\eta)}$ and $\kappa_{(\phi)}$ are normal curvatures in two perpendicular directions.

(c) Numerical Scheme

In order to solve Eqs. (19) and (20) of the preceding section, together with Eqs. (17a) and (17b) for the mapping functions $x(\xi,\eta)$ and $\sigma(\xi,\eta)$, we used a uniform 41x41 grid in the domain, $0 \le \xi,\eta \le 1$. The computations were carried out using single precision arithmetic on a VAX-11/780 computing system, which has a round-off error of $O(10^{-6})$. Thus, with an $O(h^2)$ finite-difference scheme, this mesh size represents the practical limit of resolution in order that truncation error be comparable to this round-off error divided by h^2 (when computing second derivatives).

The numerical scheme itself must be fast, highly stable and applicable to elliptic equations of quite general form. In the work reported here, we adopt the ADI scheme of Peaceman and Rachford and treat all equations of the problem (i.e. the equations of motion for ζ and ψ^*, and the two mapping equations for x and σ) as "quasi-time-dependent", by writing them in the standard form

$$\frac{\partial w}{\partial t} = q_1 \frac{\partial^2 w}{\partial \xi^2} + q_2 \frac{\partial^2 w}{\partial \eta^2} + q_3 \frac{\partial w}{\partial \xi} + q_4 \frac{\partial w}{\partial \eta} + q_5 w = q_6 \qquad (31)$$

with $\partial/\partial t$ representing a "fictitious" (or artificial) time derivative as required by ADI. An optimal value of the iteration parameter (i.e. time step) was determined [1] to be $O(h)$.

Boundary conditions for Eqs. (17), (19) and (20) are straightforward (see Eqs. (25)-(28) plus [39]), with the exception of conditions at the bubble surface. Here, the necessary boundary values of vorticity are calculated indirectly from the boundary condition (28) on tangential stress using a natural extension of the method for a solid boundary suggested by Dorodnitsyn and Meller [12] and Israeli [21] and utilized previously for a spherical drop [36]. At each new iteration, say n, the new value of the boundary vorticity ζ^n is determined from its previous value and the previous value of the tangential stress, as

$$\zeta^n = \zeta^{n-1} + \beta\left(-2\kappa_{(\eta)} u_s^{n-1} - \zeta^{n-1}\right) \tag{32}$$

where the optimal β was found (by trial and error) to be approximately 0.2. When the solution has converged, of course, the tangential stress will be zero. Boundary conditions for $x(\xi,\eta)$ and $\sigma(\xi,\eta)$ at $\xi = 1$ must also be discussed briefly. Both x and σ cannot be specified directly at $\xi = 1$ if the condition $g_{12} = 0$ is satisfied (i.e. the coordinates are to be orthogonal) as the problems for x and σ are then overdetermined. We would, on the other hand, like to approach the final solution for bubble shape iteratively starting from some initial guess. This involves incrementing the bubble boundary to create a new shape at each iteration, based upon the normal stress imbalance at the interface at the preceding iteration. However, in view of the restriction on simultaneous specification of x and σ, the necessary small displacement of the bubble boundary must be carried out indirectly rather than specifying increments in $x(1,\eta)$ and $\sigma(1,\eta)$ directly. This is accomplished by changing the mapping itself (rather than the position of the bubble surface) via incremental changes in the scale factor h_ξ of the mapping, i.e.

$$h_\xi^{(n+1)}\Big|_{\xi=1} = h_\xi^{(n)}\Big|_{\xi=1}\left(\frac{4\pi}{(\text{vol})^n}\right)^{1/3} + \beta_G\cdot\Delta^n \tag{33}$$

where Δ^n is the normal stress imbalance at iteration n,

$$\Delta = -\frac{3}{4}C_D x - p_d + \tau_{nn} + \frac{4}{We}\left(\kappa_{(\eta)} + \kappa_{(\phi)}\right). \tag{34}$$

The incremented $h_\xi\Big|_{\xi=1}$ is then used to generate "equivalent" boundary conditions for

$$\frac{\partial x}{\partial \xi}\Bigg|_{\xi=1} \quad \text{and} \quad \frac{\partial \sigma}{\partial \xi}\Bigg|_{\xi=1} \tag{35}$$

The normal stress difference, Δ, has to be normalized before it is used in (33) for changing the bubble shape because of the indeterminacy due to incompressibility (Δ contains an integration constant); this indeterminacy is removed by requiring that the volume of the bubble remain constant.

The overall solution algorithm may thus be schematically represented as follows:

1. Start with an initial guess of the shape. Here we choose a spherical shape, i.e. a circle in a plane through the axis of symmetry.

2. For the given bubble shape and coordinate mapping, compute a new approximation for the dynamic fields (ψ and ζ) by advancing the solution of the Navier-Stokes equations one iteration (i.e. one ADI time step).

3. Calculate the normal stress terms at the bubble surface, and if the condition (29) is not satisfied, increment the bubble shape by a small amount by incrementing $h_\xi(1,\eta)$ using Eq. (33), and obtaining corresponding boundary conditions for $\partial x/\partial\xi|_{\xi=1}$ and $\partial y/\partial\xi|_{\xi=1}$.

4. Calculate a new orthogonal mapping fitting the new bubble shape by solving Eqs. (17a,b) with appropriate boundary conditions (in practice we do only one ADI iteration on the mapping equations).

5. Repeat this process starting with step (2) until convergence is achieved.

(d) Results

This solution scheme has been used to obtain solutions of the problem of buoyancy driven motion of a gas bubble for $2 \leq$ Re ≤ 200 and Weber ranging from 2 up to a maximum of 20 for Re ≤ 20 and 10 for Re ≤ 200. These solutions were generated sequentially by holding the Reynolds number fixed and incrementing We, using the solution at the next lowest We as the initial condition in each case. The computation times to obtain these solutions depended, of course, on Re and We, but generally fell in the range 15-45 minutes on a VAX-11/780 computer.

Examples of two solutions with a modest level of deformation (relative to the spherical rest state) were given in the earlier publication, Ryskin and Leal [38]. A streamline plot for an additional example showing greater deformation and a direct comparison with experimental observations of Hnat and Buckmaster [18] is shown in Fig. 1. The bubble shape in this case, Re = 20, We = 15, is close to the well-known spherical cap that is characteristic of experimental results for large Re and We. The calculated indentation at the rear of the bubble is, of course, not visible in the photograph, presumably because it is shielded by the rim but the comparison between both shape and streamlines is otherwise very good. A feature of the flow pictured in Fig. 1, which we found to be characteristic of all solutions for large We, even for Re as small as 2, is the recirculating wake at the rear. We consider this feature to be a consequence of the accumulation of the vorticity generated in regions of strong curvature at the bubble surface (see Eq. 28). This aspect of our solutions is discussed in considerable detail in our forthcoming paper (Ryskin and Leal [37]). Another result worth mentioning here is the existence of steady, axisymmetric solutions for all values of Re $\leq$ 200 with We considerably in excess of 3.2, the value beyond which steady axisymmetric solutions based on a potential flow analysis [44] fail to exist. Although Re = 200 is not, perhaps, large enough to reveal the true nature of the large Re, asymptotic form of any steady, axisymmetric solutions of the bubble rise problem, it is evident from our solutions that the complete neglect of vorticity in the potential flow analysis renders any conclusions relevant only for Re $\gg$ 200 – if, indeed, they are relevant to the full problem at all. The fact that We close to 3.2 was also observed by Hartunian and Sears [17] to delineate the critical condition for onset of unsteady motion (of low M drops) at finite Reynolds, can be viewed only as a coincidence. The criterion We $\approx$ 3.2 for breakdown of rectilinear bubble motion is, we believe,

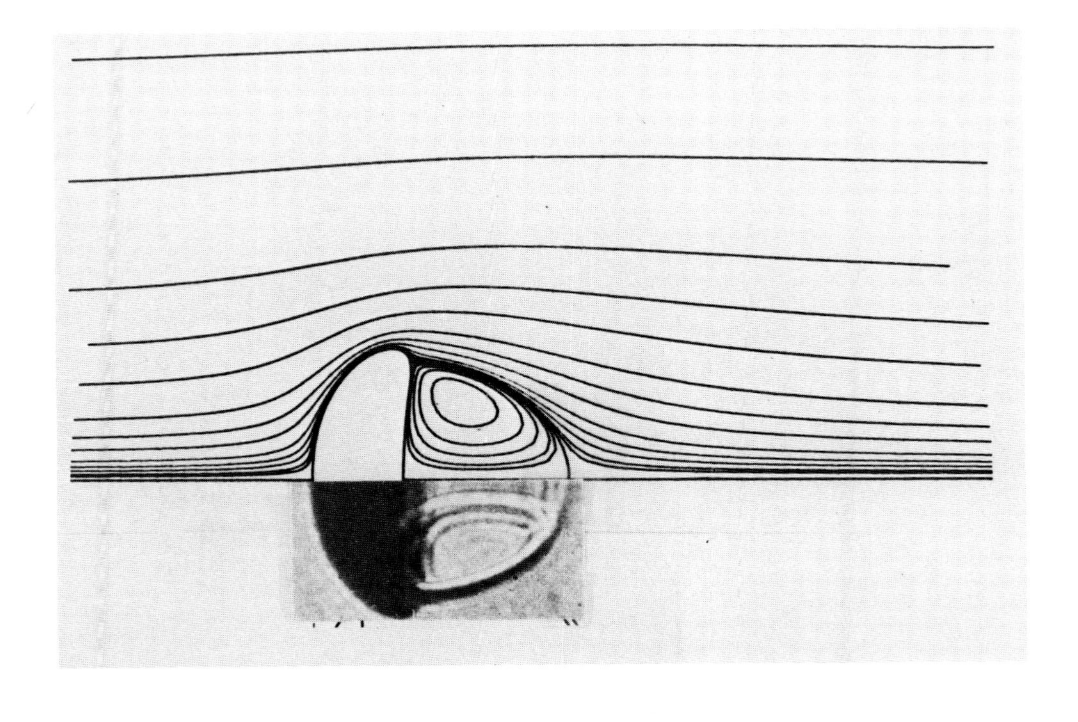

Comparison of Solution with Shadograph Photo from Hnat and Buckmaster (1976): Re = 20, We = 15 (theory); Re = 19.4, We = 15.3 (expt).

$$C_{D_{theory}} = 3.55; \; C_{D_{expt}} = 3.44$$

actually a critical value for onset of unsteady motion in the recirculating wake, rather than indicative of the nonexistence of steady solutions for any physically meaningful problem. It may be noted in this regard that a value of We between 3 and 4 was also found numerically to coincide with the existence of sufficient deformation (i.e. boundary curvature) for generation (and some development) of a recirculating wake.

REFERENCES

1. Ames, W. F., "Numerical Methods for Partial Differential Equations", Barnes and Noble, New York (1969).
2. Batchelor, G. K., Slender-body theory for particles of arbitrary cross section in Stokes' flow, J. Fluid Mech. 44 (1970) 419.
3. Berdan, C. and Leal, L. G., Motion of a sphere in the presence of a deformable interface. Part I. The effects of small deformations, J. Colloid Interface Sci. 87 (1982) 62.
4. Brignell, A. S., The deformation of a liquid drop at small Re, Q. Jour. Mech. Appl. Math. 26 (1973) 99.
5. Chan, P. and Leal, L. G., The motion of a drop in a second-order fluid, J. Fluid Mech. 92 (1979) 131.
6. Chwang, A. T. and Wu, T. Y., Hydromechanics of low-Reynolds number flow. Part 1. Rotation of axisymmetric prolate bodies, J. Fluid Mech. 63 (1974) 607.
7. Chwang, A. T. and Wu, T. Y., Hydromechanics of low-Reynolds number flow. Part 2. Singularity method for Stokes' flows, J. Fluid Mech. 67 (1975) 787.
8. Chwang, A. T., Hydromechanics of low-Reynolds number flow. Part 3. Motion of a spheroidal particle in quadratic flows, J. Fluid Mech. 72 (1975) 17.
9. Chwang, A. T. and Wu, T. Y., Hydromechanics of low-Reynolds number flow. Part 4. Translation of spheroids, J. Fluid Mech. 75 (1976) 677.
10. Cox. R. G., The motion of long slender bodies in a viscous fluid. Part 1. General theory, J. Fluid Mech. 44 (1970) 791.

11. Davies, C. W., An initial value approach to the production of discrete orthogonal coordinates, J. Comp. Phys. 39 (1981) 164.
12. Dorodnitsin, A. A. and Meller, N. A., Approaches to the solution of stationary Navier-Stokes equations, U.S.S.R. Comput. & Math. Phys. 8(2) (1968) 205.
13. Fornberg, B., A numerical method for conformal mapping, SIAM J. Sci. Stat. Comput. 1 (1980) 386.
14. Fulford, G. R. and Blake, J. R., On the motion of a slender body near an interface between two immiscible liquids at very low Reynolds number, J. Fluid Mech. (1983), to appear.
15. Geller, A. S., Lee, S. H. and Leal, L. G., The creeping motion of a spherical particle normal to a deforming fluid interface, J. Fluid Mech., in preparation.
16. Happel, J. and Brenner, H., "Low Reynolds Number Hydrodynamics", Noordhoff (1973), especially Ch. 4 and references therein.
17. Hartunian, R. A. and Sears, W. R., On the stability of small gas bubbles moving uniformly in various liquids, J. Fluid Mech. 3 (1957) 27.
18. Hnat, J. G. and Buckmaster, J. D., Spherical cap bubbles and skirt formation, Phys. Fluids 19 (1976) 182.
19. Ho. B. P. and Leal, L. G., Inertial migration of rigid spheres in two-dimensional shear flows, J. Fluid Mech. 65 (1974) 365.
20. Hung, T.-K. and Brown, T. D., An implicit finite-difference method for solving the Navier-Stokes equation using orthogonal curvilinear coordinates, J. Comp. Phys. 23 (1977) 343.
21. Israeli, M., A fast implicit numerical method for time dependent viscous flows, Stud. Appl. Math. 49 (1970) 327.
22. Johnson, R. E. and Wu, T. Y., Hydromechanics of low-Reynolds number flow. Part 5. Motion of a slender torus, J. Fluid Mech 95 (1979) 263,
23. Ladyzhenskaya, O. A., "The Mathematical Theory of Viscous Incompressible Flow", Gordon-Breach, New York (1963).

24. Leal, L. G. and Lee, S. H., Particle motion near a deformable fluid interface, Adv. in Colloid and Int. Sci. 17 (1982) 61.
25. Lee, S. H., Chadwick, R. S. and Leal, L. G., Motion of a sphere in the presence of a plane interface. Part 1. An approximate solution by generalization of the method of Lorentz, J. Fluid Mech. 93 (1979) 705.
26. Lee, S. H. and Leal, L. G., Motion of a sphere in the presence of a plane interface. Part 2. Exact solutions using bipolar coordinates, J. Fluid Mech. 98 (1980) 193.
27. Lee, S. H. and Leal, L. G., Motion of a sphere in the presence of a deformable interface. Part 2. Large deformations for motion normal to the interface, J. Coll. Int. Sci. 87 (1982) 81.
28. Mobley, C. D. and Stewart, R. J., On the numerical generation of boundary-fitted orthogonal curvilinear coordinate systems, J. Comp. Phys. 34 (1980) 124.
29. Moore, D. W., The velocity of rise of distorted gas bubbles in a liquid of small viscosity, J. Fluid Mech. 23 (1965) 749.
30. Nir, A. and Acrivos, A., On the creeping motion of two arbitrary-sized touching spheres in a linear shear field, J. Fluid Mech. 59 (1973) 209.
31. O'Neill, M. E. and Stewartson, K., On the slow motion of a sphere parallel to a nearby plane wall, J. Fluid Mech. 27 (1967) 705.
32. Orr, F. M. and Scriven, L. E., Rimming flow: numerical simulation of steady, viscous, free-surface flow with surface tension, J. Fluid Mech. 84 (1978) 145.
33. Pope, S. B., The calculation of turbulent recirculating flows in general orthogonal coordinates, J. Comp. Phys. 26 (1978) 197.
34. Rallison, J. M. and Acrivos, A., A numerical study of the deformation and burst of a viscous drop in an extensional flow, J. Fluid Mech. 89 (1978) 191.
35. Rallison, J. M., A numerical study of the deformation and burst of a viscous drop in general shear flows, J. Fluid Mech. 109 (1981) 465.

36. Rivkind, V. Ya. and Ryskin, G., Flow structure in motion of a spherical drop in a fluid medium of intermediate Reynolds numbers", Fluid Dyn. 11 (1976) 5.

37. Ryskin, G. and Leal, L. G., Large deformations of a bubble or drop in axisymmetric flow fields. Part 1, Numerical technique. Part 2, Rising bubble. Part 3, Bubble in extensional flow, J. Fluid Mech., submitted.

38. Ryskin, G. and Leal, L. G., Bubble shapes in steady axisymmetric flows at intermediate Reynolds number, Proc. 2nd Int. Colloq. on Bubbles and Drops, NASA-JPL Publ. 82-7 (1982) 151.

39. Ryskin, G. and Leal, L. G., Orthogonal mapping, J. Comp. Phys. (1983), to appear.

40. Saito, H. and Scriven, L. E., Study of coating flow by the finite element method, J. Comp. Phys. 42 (1981) 53.

41. Starius, G., Constructing orthogonal curvilinear meshes by solving initial value problems, Numer. Math. 28 (1977) 25.

42. Taylor, T. D. and Acrivos, A., On the deformation and drag of a falling viscous drop at low Reynolds number, J. Fluid Mech. 18 (1964).

43. Thames, F. C., Thompson, J. F., Mastin, C. W. and Walker, R. L., Numerical solutions for viscous and potential flow about arbitrary two-dimensional bodies using body-fitted coordinate systems, J. Comp. Phys. 24 (1977) 245.

44. Vanden-Broeck, J.-M. and Keller, J. B., Deformation of a bubble or drop in a uniform flow, J. Fluid Mech. 101 (1980) 673.

45. Yang, S.-M. and Leal, L. G., Slender-body theory in Stokes' flow near a plane fluid-fluid interface. Part 1. Arbitrary motion in a quiescent fluid, J. Fluid Mech. (1983), to appear.

46. Youngren, G. K. and Acrivos A., Stokes' flow past a particle of arbitrary shape: a numerical method of solution, J. Fluid Mech. 69 (1975) 377.

47. Youngren, G. K. and Acrivos, A., On the shape of a gas bubble in a viscous extensional flow, J. Fluid Mech. 76 (1976) 433.

Supported by a grant from the Fluid Dynamics Program of the National Science Foundation.

Department of Chemical Engineering
California Institute of Technology
Pasadena, CA 91125

ON EXISTENCE CRITERIA FOR FLUID INTERFACES IN THE ABSENCE OF GRAVITY

P. Concus

1. INTRODUCTION

We consider here equilibrium interfaces between two fluids, such as the liquid-gas free surface of a liquid that partly fills its container. The shape of the equilibrium free surface is determined by the interaction of surface and gravitational forces, in such a manner that the mechanical energy of the configuration is stationary with respect to displacements that satisfy prescribed constraints. Of principal interest are stable equilibria, for which the energy is minimized.

Attention is focussed on cylindrical containers of general section. A number of interesting — in some cases striking — results have been uncovered in recent years, most pronounced for the situation in which gravity is absent.

We consider a cylindrical container with axis oriented vertically and assume that the equilibrium free surface of a liquid partly filling the container projects simply onto a section Ω of the cylinder (Fig. 1). There is assumed to be sufficient liquid in the container to cover the base entirely. We denote by $u(x,y)$ the height of the free surface above a horizontal reference plane and by Σ the boundary of Ω. The gravitational field, if present, is taken to be uniform and is positive when directed vertically downward.

The condition, that the surface plus gravitational energy be stationary subject to the constraint of fixed liquid volume, leads to the Laplace-Young equation

$$\operatorname{div} Tu = \kappa u + 2H \quad \text{in } \Omega , \tag{1}$$

where

$$Tu = \frac{\nabla u}{\sqrt{1+|\nabla u|^2}} ,$$

and

ISBN 0-12-493220-7

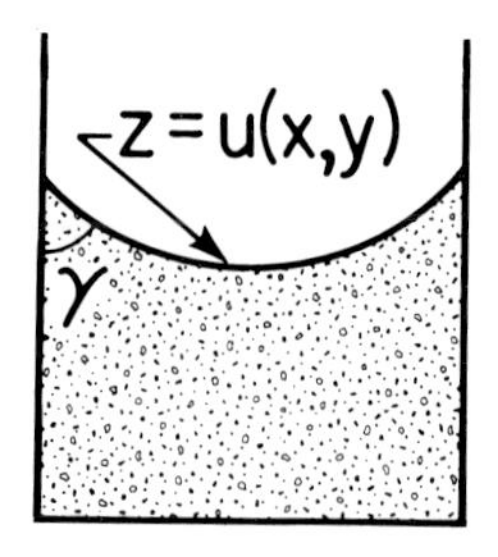

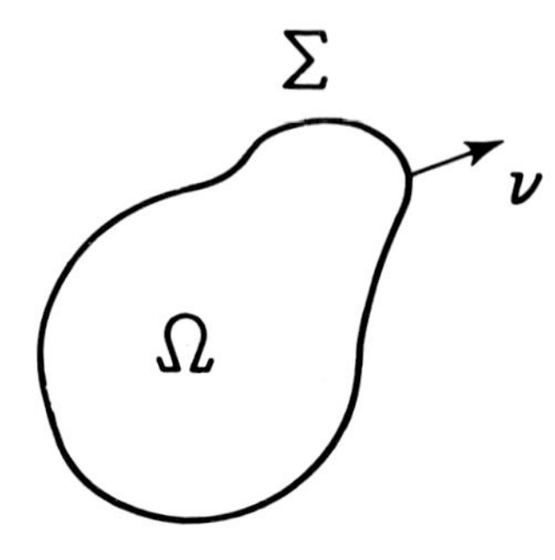

Fig. 1. *A partly filled cylinder and its section Ω with boundary Σ and exterior normal ν.*

$$\nu \cdot Tu = \cos\gamma \qquad \text{on } \Sigma . \tag{2}$$

These equations are based on surface energy being proportional to area. In (1), $\kappa = \rho g / \sigma$ is the capillary constant, with ρ the difference in densities between the gas and liquid phases, g the acceleration due to gravity, and σ the gas-liquid surface tension; $2H$ is the Lagrange multiplier for the volume constraint and is determined in general by the shape of the cylinder section, the volume of liquid present, and the contact angle γ at which the free surface meets the cylinder wall. The contact angle, which is taken to be constant, is a physically determined quantity that depends on the material properties of the liquid, gas, and container.

The boundary Σ is assumed to be smooth except possibly at a finite number of corners. The contact-angle boundary condition (2), where ν is the unit exterior normal, need not be prescribed at the corners.

Since the quantity on the left of (1) is twice the mean curvature of the free surface, a solution of (1),(2) represents a surface whose mean curvature varies linearly with height and that intersects the cylinder with prescribed angle γ. If $\kappa = 0$ then a solution surface has constant mean curvature, a situation that occurs in the absence of gravity and is taken often as an idealization of those terrestrial situations for which $\kappa\Omega \ll 1$, i.e., for which surface forces dominate gravitational ones. (In this paper Ω, Σ, ... are used interchangeably to denote both a set and its measure). We focus attention in what follows on the case $\kappa = 0$. For simplicity only wetting liquids $0 \le \gamma < \pi/2$ are discussed, as the supplementary case $\pi/2 < \gamma \le \pi$ can be obtained immediately by means of a simple transformation. The case $\gamma = \pi/2$ corresponds to the trivial solution $u \equiv$ const.

In this paper we highlight some of the advances made on this problem during approximately the past decade and indicate some current investigations. The bulk of the author's work on capillary surfaces has been carried out jointly with Robert Finn, whose mathematical and physical insight have played a crucial role in obtaining the results discussed here.

2. SPECIALIZED DOMAINS — NECESSARY CONDITIONS

If $\kappa = 0$, then the constant mean curvature H can be determined from the geometry and the boundary data by integrating (1) over Ω. One obtains, after integration by parts and using (2), the relationship $H = \dfrac{\Sigma}{2\,\Omega\cos\gamma}$. Thus (1) becomes

$$\operatorname{div} Tu = \frac{\Sigma}{\Omega}\cos\gamma \qquad \text{in } \Omega\,. \tag{3}$$

The problem then is to solve (2),(3) in the given domain for the prescribed value of γ. The additive constant, to which a solution of (2),(3) is determined, can be fixed by specifying, for example, the volume of liquid in the cylinder.

2.1 *A Closed-Form Solution*

If Ω is the disc $x^2 + y^2 \le a^2$, then it is well known that the unique solution of (2),(3) (up to an additive constant) is the portion of the lower spherical surface

$$u = -[a^2 - (x^2+y^2)\cos^2\gamma]^{1/2}\sec\gamma\,. \tag{4}$$

Less well known is that (4) also is a solution if Σ is a polygon circumscribing the circle $x^2 + y^2 = a^2$, so long as no vertex lies exterior to the concentric circle of radius $a/\cos\gamma$.

What if the polygon does extend beyond this circle? The answer can be obtained from the following result.

2.2 *A Necessary Condition for Existence*

Let a subregion Ω^* and sub-boundary Σ^* be cut from Ω and Σ, respectively, by a curve (or systems of curves) Γ in Ω (Fig. 2). There holds [1]:

A necessary condition for existence of a solution of (2),(3) in Ω is that for every Γ

$$\varphi(\Gamma) \equiv \Gamma - \left(\Sigma^* - \frac{\Sigma}{\Omega}\Omega^*\right)\cos\gamma > 0\,. \tag{5}$$

This result can be obtained by integrating (3) over the subdomain Ω^*, integrating by parts, and using (2) and the property that $|Tu| < 1$ for any differentiable u.

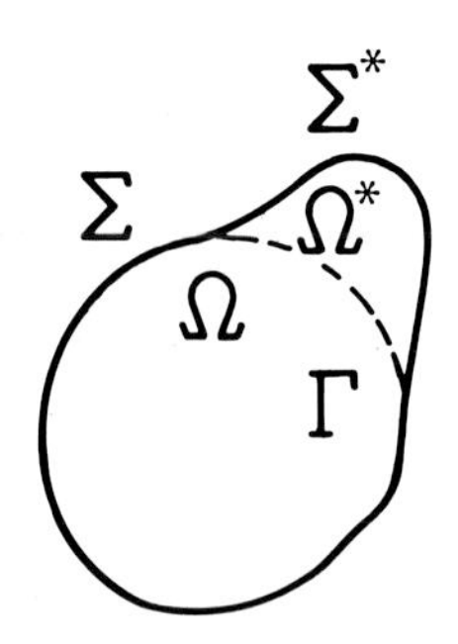

Fig. 2. *The domain Ω subdivided by the curve Γ.*

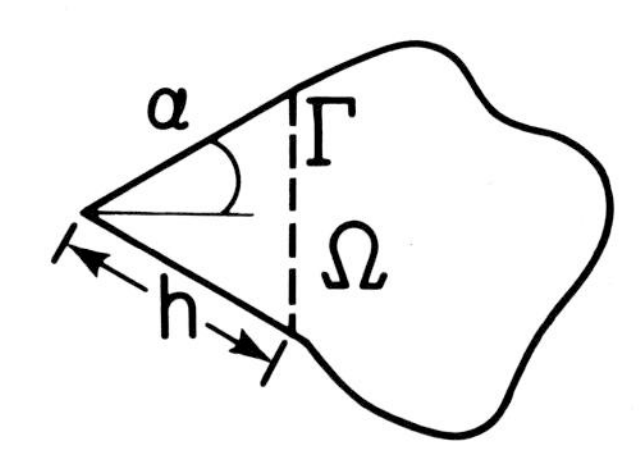

Fig. 3. *Domain with a corner.*

2.3 Corner Phenomenon

By applying the above necessary condition to the domain containing a corner shown in Fig. 3, one can obtain the answer to the question of Sec. 2.1. If the corner has interior angle 2α, and Γ is the straight line segment shown in Fig. 3, then one obtains for this configuration $\varphi(\Gamma) = 2h(\sin\alpha - \cos\gamma) + O(h^2)$. Letting $h \to 0$ gives the following result [1].

If

$$\alpha + \gamma < \pi/2 \tag{6}$$

there is no solution of (2),(3) in Ω.

This non-existence result is sharp, since for polygons circumscribing the disc, the lower spherical cap '4) gives a solution for $\alpha + \gamma \geq \pi/2$, where 2α is the smallest interior angle of the polygon. Thus the solution depends discontinuously on the contact angle γ. As γ decreases from $\pi/2$ to $(\pi/2) - \alpha$, the solution exists and is uniformly bounded and analytic in Ω, but if γ is less than $(\pi/2) - \alpha$ no solution exists.

This remarkable behavior was tested experimentally by W. Masica at the NASA Lewis Research Center Zero-Gravity Facility. Liquid interior to a regular hexagonal cylinder, measuring approximately 4 cm. on a side and made of acrylic plastic, was photographed during free fall. The liquids shown in Fig. 4 make a contact angle $\gamma \approx 48°$ (20% ethanol solution) for case (a) and $\gamma \approx 25°$ (30% ethanol solution) for case (b). Each liquid was initially at rest, and the configurations are shown in Fig. 4 after slightly more than 5 seconds of free fall, at which time free-surface equilibrium had substantially been achieved.

In Fig. 4a the free surface is essentially that given by (4), as would be expected, since $\gamma \geq 30°$. In Fig. 4b, for which $\gamma < 30°$, the fluid has flowed up into the container corners and, after hitting the lid, filled in the corners between the walls and the lid. One would expect basically the same result no matter how tall the cylinder.

The above mathematical results are not spurious ones that depend on the corner being perfectly sharp. They can be obtained as limiting cases of the following result for smooth containers [1].

Suppose there is a point on Σ at which the curvature exceeds Σ/Ω. Then there exists γ_{cr}, $0 < \gamma_{cr} \leq \pi/2$ such that there is no solution whenever $0 \leq \gamma < \gamma_{cr}$.

This result for smooth domains is sharp, as well, since explicit solutions can be given for $\gamma = 0$ for certain configurations in which the boundary curvature achieves the value Σ/Ω.

Discontinuous behavior can occur for domains with corners even if gravity is not zero [2]. If $\kappa > 0$, the solution height at a corner with interior angle 2α can change abruptly from a bounded value to infinity as γ traverses the critical value $(\pi/2) - \alpha$.

3. SUFFICIENT CONDITIONS FOR EXISTENCE — MORE GENERAL DOMAINS

An important result was obtained by Giusti, who gave sufficient conditions for there to exist a solution of (2),(3). With Γ, Ω^*, Σ^*, and $\varphi(\Gamma)$ as in Sec. 2.2, he proved in effect [7]:

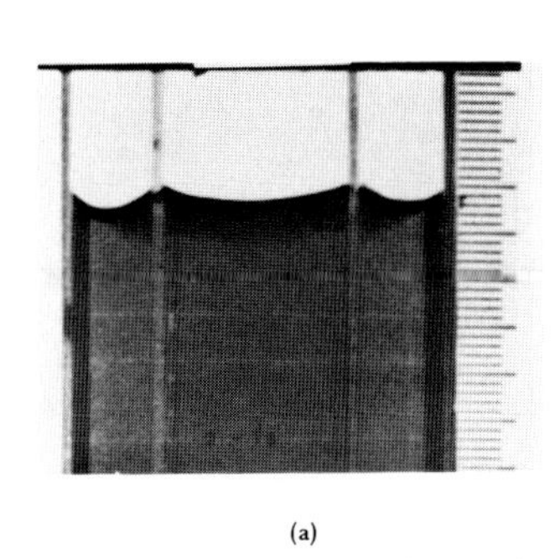

(a)

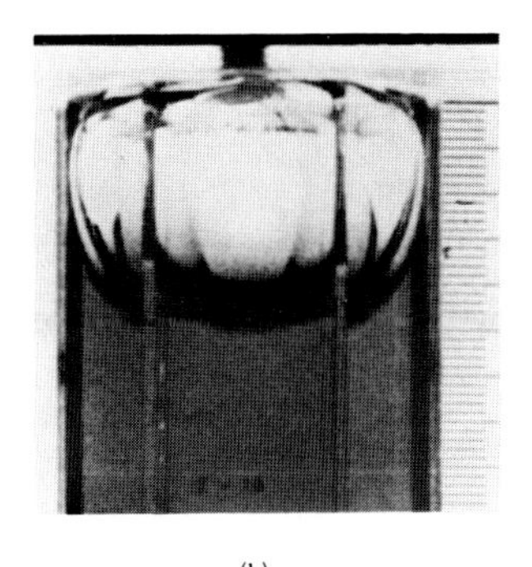

(b)

Fig. 4. *Equilibrium free surface in a hexagonal cylinder at zero gravity. (a) $\gamma \approx 48°$, (b) $\gamma \approx 25°$.*

If $\varphi(\Gamma) > 0$ holds for every Γ then a solution of (2),(3) exists.

A slight modification may be required for the case in which boundary corners are present and $\alpha + \gamma = \pi/2$. Giusti's result permits study of domains for which closed-form solutions are not available. Based on it, the following result, which does not require the testing of every Γ, was obtained in [4].

A solution exists if and only if there is a vector field $W(\mathbf{x})$ in $\overline{\Omega}$, with div $W = \Sigma/\Omega$, $\nu \cdot W = 1$ *on Σ, and $|W| < 1/\cos\gamma$ in Ω.*

The proof, which uses integration by parts over Ω^*, permits neglecting any set of Hausdorff measure zero on Σ on which the normal is not defined.

3.1 Parallelogrammatical Domains

For the parallelogram, an explicit vector field $W(\mathbf{x})$ can be constructed that has the required properties if $\alpha + \gamma \geq \pi/2$, where 2α is the smaller interior angle [5]. Denote $W = (u,v)$ and $\mathbf{x} = (x,y)$. Then for the parallelogram shown in Fig. 5 this vector field is

$$u = \frac{1}{a \sin 2\alpha}[x + \left(\frac{1}{b} - \frac{1}{a}\right) y \cot 2\alpha]$$

$$v = \frac{1}{b \sin 2\alpha} y \quad .$$

Thus one finds that the nonexistence result (6) is sharp for parallelograms, as well as for polygons circumscribing a circle.

3.2 Trapezoidal Domains

The sharpness of the nonexistence result (6) based on the corner phenomenon does not carry over to more general polygonal domains. The type of behavior that can occur is illustrated by containers with trapezoidal section. Departure from a parallelogram need not be substantial for a significant change in critical contact angle. Consider the trapezoid shown in Fig. 6 with bases

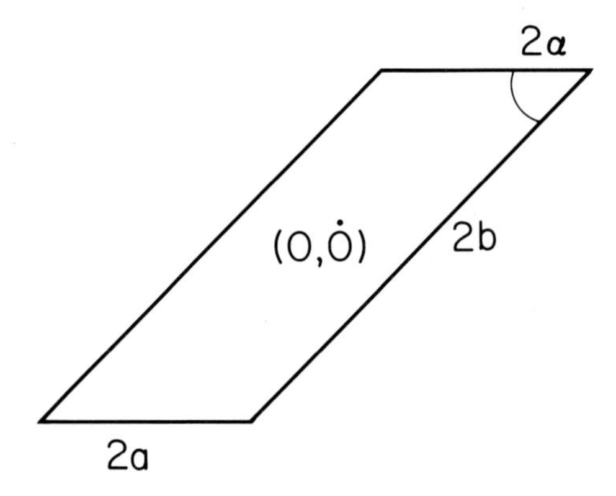

Fig. 5. *Parallelogrammatical domain.*

$b > a$ and acute angle 2α. The following result is proved in [5].

For any $\gamma < \pi/2$ and any $\varepsilon > 0$ there is a trapezoid with $|\alpha - \pi/4| < \varepsilon$, $|b/a - 1| < \varepsilon$ for which there is no solution of (2),(3).

Thus a trapezoid can be found arbitrarily close in the above sense to a rectangle, for which the critical contact angle can depart from the rectangle's $\pi/4$ to as close as desired to $\pi/2$.

The above result is obtained through a careful choice of the curve Γ that cuts off the subdomain Ω^*. The optimal choice $\Gamma = \Gamma_{cr}$, which is a circular arc of curvature $\frac{\Sigma}{\Omega}\cos\gamma$ that meets Σ with angle γ, is shown in Fig. 6. It will be discussed further in Sec. 4.

A different type of discontinuous dependence than discussed previously occurs for the trapezoid [5].

For fixed α and all sufficiently large h, there exists a unique $\gamma = \gamma_{cr}$. A solution exists if and only if $\gamma > \gamma_{cr}$. As γ approaches γ_{cr} from above, $u \to \infty$ in Ω^* and $|\nabla u| \to \infty$ on Γ_{cr}. The solution surface in $\Omega \backslash \Omega^*$ approaches the vertical circular cylinder over Γ_{cr}, as Γ_{cr} is approached from within $\Omega \backslash \Omega^*$.

To observe the manner in which the limiting behavior occurs, numerical calculations were carried out using the finite element method for a sequence of trapezoidal domains [8]. For $b = 2$, $h = 25$, $\gamma = 58°$, and the four values $a = 2$, 1.5, 1.4, 1.3, the problem (2),(3) was solved numerically using reduced biquadratic elements on a 7×75 mesh. For $a = 1.3$ the critical value of γ for the trapezoid is $\gamma_{cr} \approx 57.6°$, which is substantially larger than the one that would be yielded by the corner condition $(\pi/2) - \alpha \approx 45.4°$. In Fig. 7 are shown the numerically obtained values for $u(0,y)$ as a function y for the four trapezoids.

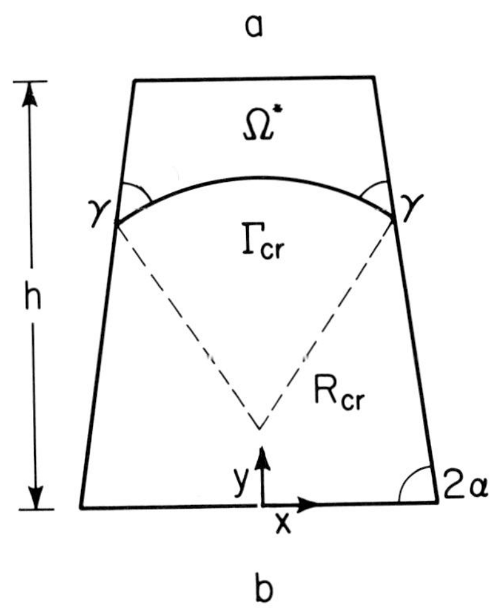

Fig. 6. *Trapezoidal domain with extremal arc Γ_{cr}*

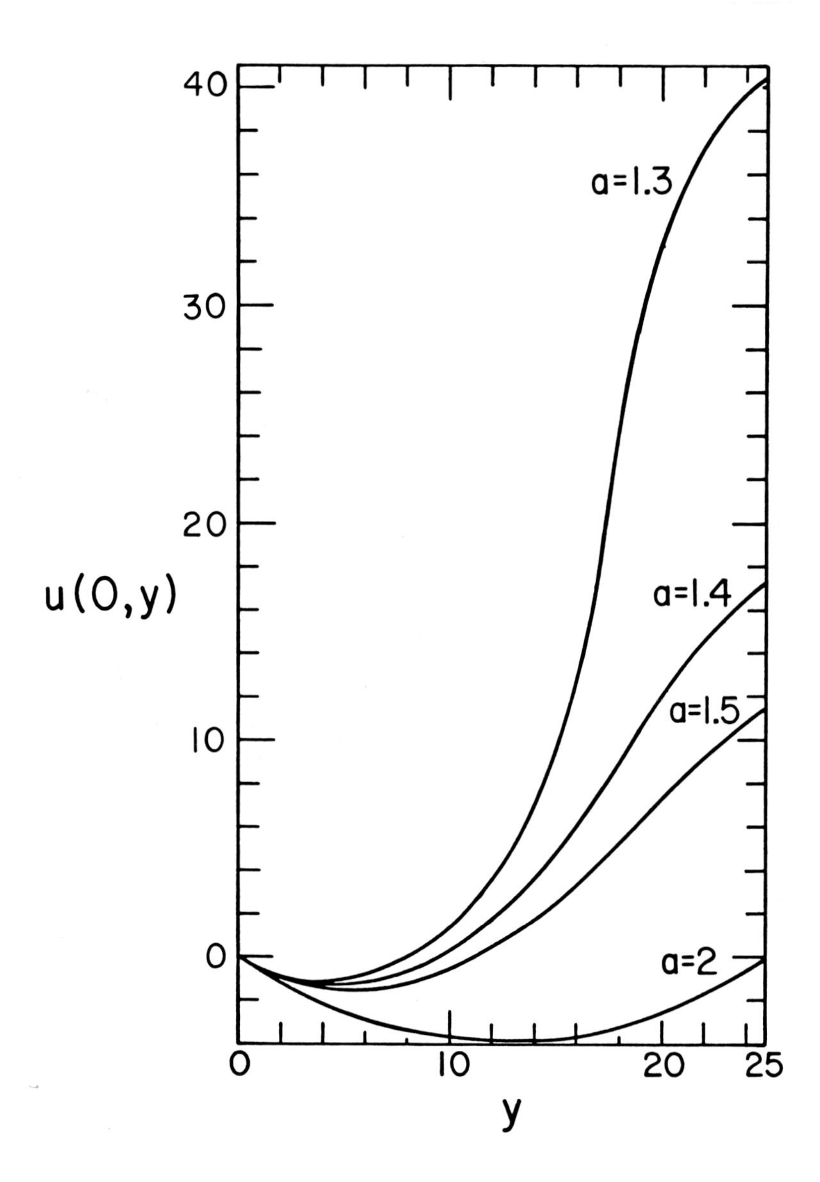

Fig. 7. *Free surface height $u(0,y)$ vs. y for four trapezoids.*

The additive constants are chosen so that $u(0,0) = 0$. The differing behavior between the near-critical configuration ($a = 1.3$) and the rectangular domain ($a = 2$) is easily observed. The near verticality of the numerically obtained curve for $a = 1.3$ is seen to be maximal at a value of y that is in good accord with the value $y \approx 17.6$ that Γ_{cr} assumes at $x = 0$ for this configuration. Experimental investigation of these properties aboard a NASA Spacelab flight is currently being planned in collaboration with D. Coles, R. Finn, and L. Hesselink, in conjunction with the NASA Lewis Research Center.

4. SUBSIDIARY VARIATIONAL PROBLEM

For general domains, determination of whether the sufficient condition for existence holds, that $\varphi(\Gamma) > 0$ for every Γ, can be approached as a subsidiary variational problem [6]. One seeks the minimum of $\varphi(\Gamma) \equiv \Gamma - \Sigma^* \cos\gamma + (\frac{\Sigma}{\Omega}\cos\gamma)\Omega^*$ over all admissible Γ. The sign of this minimum, if one exists, would then indicate existence or nonexistence of a solution to the original capillary free surface problem.

Of significance is the feature that the variational problem for minimizing $\varphi(\Gamma)$ is essentially the same as the one for the original capillary surface problem, but in one dimension lower. The capillary surface problem (2),(3) corresponds to seeking a minimum of the total surface energy subject to fixed fluid volume V, that is, to minimizing $\sigma[S - S^* \cos\gamma + (\frac{\Sigma}{\Omega}\cos\gamma)V]$, where S is the capillary free surface area and S^* is the wetted surface area of the container. Thus, as a minimizing surface S must be a surface of constant mean curvature $\frac{\Sigma}{\Omega}\cos\gamma$ intersecting the bounding cylinder walls with angle γ, so must Γ be a curve (or system of curves) of constant curvature $\frac{\Sigma}{\Omega}\cos\gamma$ intersecting Σ with angle γ. One need investigate only these specific circular arcs in seeking a minimum for $\varphi(\Gamma)$.

Detailed properties of minimizing curves Γ are given in [6]. An important feature is that:

The nonexistence of a minimizing Γ in Ω is a sufficient condition for the existence of an (energy minimizing) solution u of the capillary surface problem.

Work is currently underway to characterize geometric conditions on general domains for a minimizing Γ to exist and for a minimizing Γ to give $\varphi > 0$.

5. CONTINUOUS AND DISCONTINUOUS DISAPPEARANCE OF CAPILLARY SURFACES

As a concluding remark, emphasizing the seemingly anomalous nature of the capillary problem, we comment on the behavior at the critical contact angle γ_{cr}. In general, a solution exists for $\gamma > \gamma_{cr}$ and does not exist for $\gamma < \gamma_{cr}$. If the domain is smooth, then a solution does not exist at the critical value $\gamma = \gamma_{cr}$.

However, if the domain is not smooth and has a corner, then a solution may exist for $\gamma = \gamma_{cr}$. This feature is discussed in [3,6].

REFERENCES

1. Concus, P. and R. Finn, *On capillary free surfaces in the absence of gravity*, Acta Math., 132 (1974), pp. 177-198.
2. ________________, *On capillary free surfaces in a gravitational field*, Acta Math., 132 (1974), pp.207-223.
3. ________________, *Continuous and discontinuous disappearance of capillary surfaces*, preprint LBL-15356, Lawrence Berkeley Lab., University of California, Berkeley, CA, 1983.
4. Finn, R., *Existence and non-existence of capillary surfaces*, Manuscripta Math., 28 (1979), pp. 1-11.
5. ________, *Existence criteria for capillary free surfaces without gravity*, Indiana University Math. J., to appear.
6. ________, *A subsidiary variational problem and existence criteria for capillary surfaces*, preprint, Math. Dept., Stanford University, Stanford, CA, 1982.
7. Giusti, E., *Boundary value problems for non-parametric surfaces of prescribed mean curvature*, Ann. Scuola Norm. Sup. Pisa Cl. Sci. (4), 3 (1976), pp. 501-548.
8. Roytburd, V., *On singularities of capillary surfaces on trapezoidal domains*, Internat. J. Math. Math. Sci., to appear.

This work was supported in part by the Director, Office of Basic Energy Sciences, Engineering, Mathematical, and Geosciences Division of the U.S. Department of Energy under Contract DE-AC03-76SF00098 and by the National Aeronautics and Space Administration under grant NAG3-146.

Lawrence Berkeley Laboratory and
Department of Mathematics
University of California
Berkeley, CA 94720

CAPILLARY WAVES AND INTERFACIAL STRUCTURE

H. T. Davis

1. INTRODUCTION

We consider a planar interface separating two bulk fluid phases at equilibrium with one another. At the macroscopic level the interfacial zone is so narrow that it can be viewed as a membrane of zero thickness under tension γ. At the molecular level, densities and compositions vary continuously from their values in one bulk phase to their values in the other bulk phase.

Thermal fluctuations drive capillary waves or two-dimensional deformations of the planar interface out of the interfacial plane. The result is that the interfacial zone becomes more diffuse. The capillary wave dispersion of the interfacial zone increases as $\sqrt{\ell n A}$ in the absence of gravity and as $\sqrt{\ell n(\gamma/\Delta\rho g)}$ in the presence of gravity. A is the area transverse to the interface, g the acceleration of gravity, and $\Delta\rho$ the density difference of the bulk phases. $\sqrt{\gamma/\Delta\rho g}$ is frequently referred to as the capillary length.

Because of residual coherence even in randomly driven capillary waves, the density at one point in the interfacial zone is correlated with the density at another point in the interfacial zone. Such density-density or so-called pair correlations exist even in bulk or homogeneous fluid phases, but in that case they are short-ranged. In the interfacial zone, on the other hand, the density-density correlations are long-ranged, the characteristic range being the smaller of

ISBN 0-12-493220-7

$\sqrt{\gamma/\Delta\rho g}$ or $\sqrt{A}$. Thus in the absence of gravity (either an interface in outer space where g = 0 or a liquid-liquid interface with densities matched so that $\Delta\rho = 0$) density correlations extend across the entire interfacial zone. The magnitude of the long-ranged transverse density correlations is inversely proportional to the interfacial tension, a fact that is exploited in light-scattering determination of tension. Perpendicular to the interface, the density correlations are short-ranged, that is, density measured in the interfacial zone is independent of density measured in the bulk phase just outside the interfacial zone.

The picture given here of an interface and its capillary wave fluctuations has in recent years emerged from the statistical mechanical theory of inhomogeneous fluid. A versatile and mathematically appealing approximate theory of inhomogeneous fluid is the so-called gradient free energy theory. In what follows, we use it as the primary vehicle to expose the current status of the molecular theory of interfacial structure and capillary waves. In the last section, we show that the major implications of gradient theory can be supported by rigorous statistical mechanical principles.

2. GRADIENT THEORY OF INHOMOGENEOUS FLUID

The gradient theoretical Helmholtz free energy of one-component inhomogeneous fluid is

$$F = \int [f_o(n) + \frac{c}{2} (\nabla n)^2] d^3 r , \qquad (1)$$

where $n(\underset{\sim}{r})$ denotes the density at postion $\underset{\sim}{r}$, ∇n the gradient of $n(\underset{\sim}{r})$, $f_o(n)$ the Helmholtz free energy density of homogeneous fluid at composition n, and c the influence parameter of inhomogeneous fluid. The influence parameter is a measure of the free energy cost for maintaining density gradients in the system. Gradient theory goes back to the works of Lord Rayleigh [1] and van der Waals [2], but only lately has the influence parameter been related to fluid structure. The relation is [3,4]

$$c = \frac{k_B T}{6} \int C^o(R, n(\underset{\sim}{r})) R^2 d^3 R \qquad (2)$$

where k_B is Boltzmann's constant, T is the absolute temperature, and C^o the direct correlation function of homogeneous fluid. C^o is related to the radial distribution function by an integral transform [5].

Gradient theory has had, in recent years, numerous applications to planar interfaces as reviewed by Davis and Scriven [6]. The theory compares favorably with a computer simulation [7], with tension data on hydrocarbons and their mixtures [8], and with possibly more rigorous integral free energy theories [9]. Because of these successes and the simplicity of gradient theory we shall employ it in describing equilibrium fluid structures and fluctuations.

At equilibrium, the density distribution $n(\tilde{r})$ takes on values that minimize the grand potential $\Omega = F - \mu \int n(\tilde{r})d^3r$, where μ is the chemical potential of the fluid. From the calculus of variations, it follows that the density distribution minimizing Ω with F given by Eq. (1) obeys the equation

$$\mu = \frac{\partial f_o}{\partial n} - c\nabla^2 n - \frac{1}{2}\frac{\partial c}{\partial n}(\nabla n)^2 . \tag{3}$$

Gradient theory thus reduces the problem of determining inhomogeneous fluid equilibrium structure to solving a nonlinear second-order differential equation.

3. EQUILIBRIUM DENSITY PROFILE AND TENSION OF A PLANAR INTERFACE

The density n(x) in a planar interface between liquid and vapor phases depends only on the distance x perpendicular to the interfacial plane. Equation (3) reduces in this case to

$$\mu = \frac{\partial f_o}{\partial n} - c\frac{d^2 n}{dx^2} - \frac{1}{2}\frac{\partial c}{\partial n}\left(\frac{dn}{dx}\right)^2 . \tag{4}$$

The boundary conditions for a planar interface are that n(x) equal the bulk liquid density n_ℓ far below the interface and equal the bulk vapor density n_g far above the interface, i.e., $n(x) \to n_\ell$ or n_g as $x \to -\infty$ or ∞.

Multiplying Eq. (4) by (dn/dx)dx and integrating, we obtain

$$\frac{1}{2}c\left(\frac{dn}{dx}\right)^2 = \omega(n) - \omega_B \equiv \Delta\omega(n) , \tag{5}$$

where the quantity $\omega(n)$ is defined by

$$\omega(n) = f_o(n) - n\mu \tag{6}$$

and ω_B is its value in either bulk phase

$$\omega_B = \omega(n_\ell) = \omega(n_g) . \tag{7}$$

From the definition, it is evident that $-\omega_B = P_B$, the pressure in the bulk phases.

Equation (5) can be solved directly for the density profile n(x). The result is

$$x = \int_{n_o}^{n(x)} \sqrt{c/2\Delta\omega(n)}\, dn . \tag{8}$$

The origin of the coordinate system is chosen arbitrarily as the position of density n_o, set in value somewhere between n_g and n_ℓ.

Using Eq. (5) to eliminate $f_o(n) - n\mu$ in the grand potential Ω we obtain

$$\Omega = A \int_{-\infty}^{\infty} c\left(\frac{dn}{dx}\right)^2 dx - P_B V , \tag{9}$$

where V is the volume of the system and A the area of the interface. The interfacial tension γ, given by the thermodynamic formula $\gamma = (\partial\Omega/\partial A)_{T,V,\mu}$, is

$$\gamma = \int_{-\infty}^{\infty} c\left(\frac{dn}{dx}\right)^2 dx = \int_{n_g}^{n_\ell} \sqrt{2c\Delta\omega(n)}\, dn , \tag{10}$$

where the second, profile independent, formula is obtained by making use of Eqs. (5) and (8).

In Fig. 1, the density profile predicted by gradient theory for a 6-12 Lennard-Jones fluid is compared with the results of a computer simulation (the ripples in the computer simulation profile are statistical artifacts, removable with

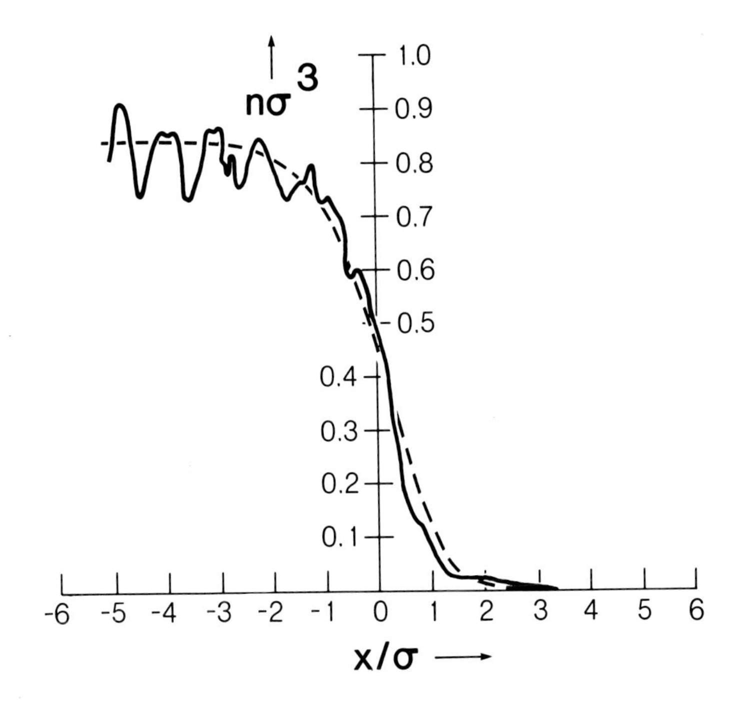

Fig. 1. Liquid-vapor density profile of 6-12 Lennard-Jones fluid. Comparison of computer simulation (solid curve) with gradient theory (dashed curve); $kT/\varepsilon = 0.703$.

long enough computer runs [10]). In the plot, density is in units of σ^{-3} and distance in units of σ, σ being the Lennard-Jones parameter representing molecular diameter. The temperature of the fluid considered in Fig. 1 is about 0.5 T_c, T_c being the critical temperature.

As is evident in Fig. 1, the interfacial width is exceedingly small (only several molecules wide) and, therefore, it is not surprising that at a macroscopic level one usually treats the interface as a mathematical surface possessing tension but no thickness. A rough estimate of interfacial width ℓ_w is obtainable from Eq. (8), namely,

$$\ell_w \simeq \frac{1}{2}\left\{\sqrt{c(n_\ell)/\partial^2 f_o(n_\ell)/\partial n^2} + \sqrt{c(n_g)/\partial^2 f_o(n_g)/\partial n^2}\right\} \quad (11)$$

The influence parameter c(n) has been shown to be a rather weak function of density. The thermodynamic relations $\mu_o = \partial f_o/\partial n$ and $d\mu_o = n^{-1}dP_o$ lead to $\partial^2 f_o/\partial n^2 = 1/n^2\kappa_o$, where $\kappa_o(\equiv n^{-1}(\partial n/\partial P_o))$ is the isothermal compressibility of homogeneous fluid. As the critical point is approached $\kappa_o(n_g)$ and $\kappa_o(n_\ell)$ diverge as [11] $|1 - T/T_c|^{-1.2}$. Thus, as T nears the critical point, the interfacial width is predicted by gradient theory to broaden according to the law

$$\ell_w \equiv \ell_o|1 - T/T_c|^{-0.6}, \quad (12)$$

where ℓ_o is a scale factor of molecular dimensions. The critical exponent predicted for ℓ_w is -0.6, to be compared with an observed value of -0.67 [12]. If ℓ_o is of the order of 5σ, then even if the temperature were within 0.05% of the critical point, the interfacial width would be only 500σ, still thin on a macroscopic scale.

As mentioned above the influence parameter is only a weak function of density [6,8]. As it greatly simplifies the analysis to follow and does not change the results appreciably, we shall simply assume that c is constant.

4. STABILITY OF FLUID AT EQUILIBRIUM

Although a candidate equilibrium fluid structure can be found by solving Eq. (3) with appropriate boundary conditions, the structure might not be thermodynamically stable. The condition of thermodynamic stability of a solution $n(\tilde{r})$ is that the free energy be a minimum, i.e., that

$$F(\{n+v\}) > F(\{n\}) \tag{13}$$

for any fluctuation $v(\tilde{r})$ that conserves particles. Since $N = \int nd^3r = \int (n+v)d^3\tilde{r}$, the constraint of particle conservation is $\int vd^3r = 0$.

The local stability condition derived from Eq. (13) is that

$$\delta^2F \equiv \frac{1}{2}\int \left[\frac{\partial^2 f_o}{\partial n^2}(n)v^2 + c(\nabla v)^2\right] d^3r > 0 \tag{14}$$

for all allowed nonzero v. We suppose that the walls of the system are impenetrable to fluid particles and so require that density fluctuations are either zero at the walls or, slightly more generally, are periodic at opposing walls. In either case, the fluctuation obeys the condition

$$\int \nabla\cdot(v\nabla v)d^3r = 0 , \tag{15}$$

a result allowing us to rearrange Eq. (14) by integration by parts to obtain

$$\delta^2F = \frac{1}{2}\int v(Lv)d^3r , \tag{16}$$

where L is an operator defined by

$$Lv \equiv -c\nabla^2 v + [\partial^2 f_o(n)/\partial n^2]v \tag{17}$$

and the boundary conditions leading to Eq. (15). L is a self-adjoint operator. Therefore, in the eigenproblem

$$Lu_\lambda = \lambda u_\lambda , \tag{18}$$

the eigenvalues λ are real and the eigenfunctions u_λ form a complete orthogonal set, which can be normalized so that

$$\int u^*_\lambda(\tilde{r})u_{\lambda'}(\tilde{r})d^3r = \delta_{\lambda\lambda'} . \tag{19}$$

The asterisk on u^*_λ denotes complex conjugate. Since the u_λ form a complete set, any density fluctuation can be expressed in the form

$$v(\tilde{r}) = \sum_\lambda a_\lambda u_\lambda(\tilde{r}) , \quad a_\lambda = \int u^*_\lambda(\tilde{r})v(\tilde{r})d^3r . \tag{20}$$

The free energy fluctuation δ^2F then reduces to

$$\delta^2F = \frac{1}{2}\sum_\lambda \lambda|a_\lambda|^2 > 0 , \tag{21}$$

for any fluctuation that conserves particles, the condition of which is

$$\sum_\lambda a_\lambda \int u_\lambda(\tilde{r})d^3r = 0 . \tag{22}$$

Stability is assured if all eigenvalues are positive and instability is definite if two or more eigenvalues are negative. Otherwise, each case must be examined separately [6].

The eigenvalue problem will be useful in what follows not only to determine stability of equilibrium structures but also to predict thermal fluctuations in equilibrium systems. Such fluctuations lead to density-density correlations and, in the case of a planar interface, to the fluctuation phenomena known as capillary waves.

5. THERMAL FLUCTUATIONS IN HOMOGENEOUS FLUID

In what follows we consider a homogeneous fluid in a cubic system of length $V^{1/3}$ on each side. We assume periodic boundary conditions, i.e., u and $\partial u/\partial x_\mu$ are the same on opposing faces of the cube for x_μ = x, y, and z. The normalized eigenfunctions of Eq. (18) are then

$$u_\lambda = \frac{e^{i\tilde{k}\cdot\tilde{r}}}{\sqrt{V}} , \tag{23}$$

with eigenvalues $\lambda = ck^2 + \partial^2 f_o(n)/\partial n^2$, where $k_\mu = 0$, $\pm 2\pi/V^{1/3}$, $\pm 4\pi/V^{1/3}, \cdots$, $\mu = x,y,z$. The fluctuation free energy becomes

$$\delta^2 F = \frac{1}{2} \sum_{\underset{\sim}{k}>0} \left[ck^2 + \frac{\partial^2 f_o}{\partial n^2}(n) \right] |a_{\lambda(k)}|^2 . \qquad (24)$$

The condition $\underset{\sim}{k} > 0$ is required for particle conservation (22). From Eq. (24) it follows that stability requires $c > 0$ and $\partial^2 f_o/\partial n^2 > 0$. The second of these is a well-known condition of stability of homogeneous fluid. The first is not commonly realized because continuous, inhomogeneous fluctuations are not usually included in stability analysis. Criteria for unstable fluid also follow from Eq. (24). Homogeneous fluids at densities such that $\partial^2 f_o/\partial n^2 < 0$ are not stable. If $c < 0$, then no homogeneous fluid is stable.

In stable fluid, thermal fluctuations cause density fluctuation $v(\underset{\sim}{r})$ about the equilibrium value n. Through second order in v, the free energy increase associated with the fluctuation is $\delta^2 F$. The probability of such a fluctuation is proportional to $\exp[-\beta\delta^2 F]$, where $\beta \equiv 1/k_B T$. Density fluctuations give rise to density-density correlations. Such density correlations are quantified by the pair density function

$$D^{(2)}(\underset{\sim}{r},\underset{\sim}{r}') \equiv \langle v(\underset{\sim}{r}) v(\underset{\sim}{r}')^* \rangle , \qquad (25)$$

where $\langle\alpha\rangle$ denotes the thermal average of α over all fluctuations. If $v(\underset{\sim}{r})$ is expressed in the eigenfunction expansion

$$v(\underset{\sim}{r}) = \sum_{\underset{\sim}{k}>0} a_{\underset{\sim}{k}} e^{i\underset{\sim}{k}\cdot\underset{\sim}{r}}/\sqrt{V} , \qquad (26)$$

then

$$\delta^2 F = \sum_{\underset{\sim}{k}>0} \left[ck^2 + \frac{\partial^2 f_o}{\partial n^2}(n) \right] |a_{\underset{\sim}{k}}|^2 . \qquad (27)$$

The coefficients a_k can be treated as random variables and the average in Eq. (25) can be calculated as

$$D^{(2)}(\underset{\sim}{r},\underset{\sim}{r}') = \sum_{\underset{\sim}{k},\underset{\sim}{k}'>0} \frac{e^{i\underset{\sim}{k}\cdot\underset{\sim}{r}-i\underset{\sim}{k}'\cdot\underset{\sim}{r}'}}{V} \frac{\int\cdots\int e^{-\beta\delta^2 F} a_{\underset{\sim}{k}} a_{\underset{\sim}{k}}^* \prod_{\underset{\sim}{k}>0} da_{\underset{\sim}{k}}}{\int\cdots\int e^{-\beta\delta^2 F} \prod_{\underset{\sim}{k}>0} da_{\underset{\sim}{k}}}$$

$$= \sum_{\underset{\sim}{k}>0} \frac{e^{i\underset{\sim}{k}\cdot(\underset{\sim}{r}-\underset{\sim}{r}')}}{\beta V[ck^2 + \partial^2 f_o/\partial n^2]} . \qquad (28)$$

As expected for an isotropic fluid $D^{(2)}$ depends on $\underset{\sim}{r}$ and $\underset{\sim}{r}'$ only through the difference $\underset{\sim}{r} - \underset{\sim}{r}'$. The Fourier transform in Eq. (28) can be carried out to yield the well-known result [13]

$$D^{(2)}(\underset{\sim}{r}-\underset{\sim}{r}') = \frac{1}{4\pi\beta c\,|\underset{\sim}{r}-\underset{\sim}{r}'|} \exp[-|\underset{\sim}{r}-\underset{\sim}{r}'|/\ell_{cor}] \qquad (29)$$

where

$$\ell_{cor} \equiv \sqrt{c/(\partial^2 f_o/\partial n^2)} . \qquad (30)$$

ℓ_{cor} is a length characterizing the range of the correlation of the thermally induced density fluctuations. It is interesting that the ratio of the influence parameter to the curvature $\partial^2 f_o/\partial n^2$ of the free energy density determines the range of density-density correlations as well as the width of a planar liquid-vapor interface. As pointed out in Section 4, the rhs of Eq. (30) will be of the order of molecular dimensions sufficiently far from a critical point. Thus, the density-density correlations are short-ranged sufficiently far from a critical point. On the other hand, sufficiently near a critical point, it follows from the arguments presented in Section 4 that $\ell_{cor} \simeq \ell_o |1 - T/T_c|^{-0.6}$, and so the correlations become longer and longer ranged as the critical point is approached. These long-ranged correlations are responsible for critical opalescence of a homogeneous fluid as it approaches the critical point [14].

As seen in Eq. (29), gradient theory as detailed here predicts a density-density correlation function that is singular at the origin, i.e., $D^{(2)}(\underset{\sim}{r}-\underset{\sim}{r}') \rightarrow \infty$ as $|\underset{\sim}{r}-\underset{\sim}{r}'| \rightarrow 0$. The reason for this is that in carrying out the thermal average in Eq. (28) we assumed that the coefficients a_k of the eigenfunction expansion of fluctuation v can be treated as independent variables, i.e., as degrees of freedom in the free energy. If the range of $\underset{\sim}{k}$ is not restricted the free energy would have more degrees of freedom than the number of degrees of freedom (3N) of the N molecules in the system. Since a density fluctuation involves positioning of molecules, it is inconsistent with the molecular nature of matter to let the number of degrees of freedom exceed 3N. Thus, a cutoff should be introduced for k. We should restrict k to values smaller than k_{max}, where $3N = \sum_{k<k_{max}} \Delta n_x \Delta n_y \Delta n_z = \frac{V}{(2\pi)^3} \sum_{k<k_{max}}$

$\Delta k_x \Delta k_y \Delta k_z \simeq \frac{4\pi}{(2\pi)^3} \int_0^{k_{max}} k^2 dk$, or $k_{max} \simeq (6\pi^2 N)^{1/3}$. With this restriction

$$D^{(2)}(0) = \sum_{\underset{\sim}{k}>0}^{k_{max}} \frac{1}{\beta V[k^2 c + \partial^2 f_o/\partial n^2]} \simeq \frac{1}{2\pi^2 \beta} \int_0^{k_{max}} \frac{k^2 dk}{[k^2 c + \partial^2 f_o/\partial n^2]}, \tag{31}$$

a finite quantity.

Even though the cutoff eliminates the undesirable singularity, the short-range behavior predicted by gradient theory is qualitatively wrong. The reason is that gradient theory incorrectly estimates $\delta^2 F$ for short wavelength (large k) fluctuations. As is discussed in Section 7, the rigorous behavior of $\lambda(k)$ is $\lambda(k) \rightarrow (\beta n)^{-1}$ as $k \rightarrow \infty$, not $\lambda(k) \rightarrow ck^2$ as predicted by gradient theory. (Actually, an alternative way to choose k_{max} would be to equate $\lambda(k_{max})$ of gradient theory to $(\beta n)^{-1}$.) However, the long-ranged behavior of the density-density correlation function predicted by gradient theory agrees with that predicted by rigorous free energy theory. Thus, as long as it is restricted to long-ranged fluctuations (i.e., small-k terms in Eq. (28)) gradient theory is accurate. This is an important point for the capillary wave theory developed in the next section.

6. FLUCTUATIONS OF A PLANAR INTERFACE: CAPILLARY WAVES

At the macroscopic level a planar interface has been viewed traditionally as a stretched, flat membrane. Such a membrane can be distorted from the interfacial plane by thermal fluctuations, and these fluctuations, thermally driven capillary waves, can induce density-density correlations in the interfacial plane and can cause a diffuseness of the interfacial zone beyond the intrinsic width of a planar interface. The most striking features are that in the absence of gravity, the range of density-density correlations in the interfacial plane and the capillary dispersion of the width of an interface both increase without bound as the interfacial area approaches infinity. These features have been verified with continuum models [15,16], a gradient free energy model [17,6] and, most rigorously, with density-functional statistical mechanics [16,18,19]. To its credit, gradient free energy theory agrees with the most rigorous approaches to capillary wave theory. In what follows, we first develop a somewhat heuristic gradient free energy theory of capillary waves. Then we demonstrate that the same results are produced by the mathematically correct version of the gradient theory.

The position $\xi(y,z)$ of the surface of a membrane distorted from its flat, lowest energy configuration can be represented as a superposition of harmonic surface waves, i.e.,

$$\xi(\underset{\sim}{s}) = \sum_{h>0} \xi_{\underset{\sim}{h}} e^{i\underset{\sim}{h}\cdot\underset{\sim}{s}} , \tag{32}$$

where $\underset{\sim}{s} = y\hat{j} + z\hat{k}$ is position in the interfacial plane and $\underset{\sim}{h} = h_y\hat{j} + h_z\hat{k}$ and ξ_h are the wavevector and amplitude of the surface wave $\exp(i\underset{\sim}{h}\cdot\underset{\sim}{s})$. We suppose the membrane has periodic boundaries (fixed boundaries lead to the same conclusions as obtained with periodic boundaries) so that

$$h_\mu = 2\pi n_\mu / A^{1/2}, \quad n_\mu = 0, \pm 1, \pm 2, \cdots, \quad \mu = y \text{ and } z , \qquad (33)$$

where for convenience we assume a square interface of area A. The free energy consistent with the macroscopic view of an interface vibrating as an intact membrane is the assumption that the interface deforms without distorting its intrinsic density profile, i.e., that the local density is $n(x - \xi(\underset{\sim}{s}))$, where $n(x)$ is the density profile. Since the free energy of a surface wave is proportional to h^2, capillary waves of large wavelength, small h, are more likely thermal fluctuations. The gentle curvatures associated with large wavelength will not distort the intrinsic interface, and so the approximation $n(x - \xi)$ should be accurate for the most populous capillary waves.

The free energy corresponding to $n(x - \xi)$, when the gravitational field is included, is

$$F = \int \{f_o(n(x - \xi)) + \frac{c}{2} [1 + (\nabla\xi)^2][n'(x - \xi)]^2 + mgxn(x - \xi)\} d^3r , \qquad (34)$$

where m is the mass of a fluid particle, g the acceleration of gravity, and $n' \equiv dn/dx$. Far below the interface $n(x) = n_\ell$ and $n'(x) = 0$ and far above it $n(x) = n_g$ and $n'(x) = 0$. These properties plus the condition $\int \xi d^2s = 0$ imply

$$\int G(n(x - \xi))dx = \xi[G(n_\ell) - G(n_g)] + \int G(n(x))dx$$

$$\int (x - \xi)G(n(x - \xi))dxd^2s = \int \left\{\frac{1}{2} \xi^2 [G(n_g) - G(n_\ell)] + \int xG(n(x))dx\right\}ds^2 \qquad (35)$$

for any function $G(n)$ and that Eq. (34) reduces to the form

$$F = \int\{f_o(n(x)) + \frac{c}{2} [n'(x)]^2 + mgxn(x)\}d^3r + \frac{1}{2} \int [g\Delta\rho\xi^2 + \gamma(\nabla\xi)^2]d^2s , \qquad (36)$$

where $\Delta\rho \equiv m(n_\ell - n_g)$ and $\gamma(= \int c(n'(x))^2 dx)$ is the tension of the interface. The first term on the rhs of Eq. (36) is the free energy F^P of the undisturbed planar interface. Using Eq. (32) for ξ and carrying out the integration over d^2s in the second term on the rhs of Eq. (36) we obtain for the free energy $\Delta F \equiv F - F^P$ of the fluctuation ξ the following equation

$$\Delta F = \frac{1}{2} \sum_{\underset{\sim}{h}>0} A[g\Delta\rho + \gamma h^2]|\xi_{\underset{\sim}{h}}|^2 . \quad (37)$$

The density-density correlations associated with thermally induced capillary waves can be computed by treating the amplitudes $\xi_{\underset{\sim}{h}}$ as random variables and averaging $[n(x - \xi(\underset{\sim}{s})) - n(x)][n(x' - \xi(\underset{\sim}{s}')) - n(x)]$ with respect to the distribution $\exp[-\beta\Delta F]$. Because of the form of ΔF the small amplitude waves are most likely and so $n(x - \xi) - n(x) \simeq -\xi n'(x)$. With this aproximation, we obtain

$$D^{(2)}(x,x',\underset{\sim}{s}-\underset{\sim}{s}') = \sum_{h>0} \frac{e^{i\underset{\sim}{h}\cdot(\underset{\sim}{s}-\underset{\sim}{s}')} n'(x)n'(x')}{\beta A[\gamma h^2 + g\Delta\rho]} , \quad (38)$$

a result derived previously with a continuum free energy model as well as from density-functional arguments [16]. As discussed in Section 5, the fact that a finite number of particles is involved in a fluctuation imposes a physical upper limit h_{max} on the wavevector h. Also, as pointed out in Section 5, gradient theory fails in the large h limit.

From Eq. (38) it follows that the density-density correlations caused by capillary waves are short-ranged in the longitudinal direction (i.e., in the x-direction) since $n'(x)$ vanishes when x lies outside the interface. Wertheim [18] has established this property from rigorous statistical mechanical principles.

The two-dimensional Fourier transform of $D^{(2)}(x,x',\underset{\sim}{s}-\underset{\sim}{s}')$ is

$$\tilde{D}^{(2)}(x,x',\underset{\sim}{h}) = \frac{n'(x)n'(x')}{\beta\gamma[h^2 + \alpha^2]} \quad (39)$$

where

$$\alpha \equiv \sqrt{\Delta\rho g/\gamma} . \quad (40)$$

In a gravity-free ($\alpha = 0$) computer simulation, Kalos et al. [16] verified that $\tilde{D}^{(2)}$ obeys Eq. (39) at low h. The implication of Eq. (39) is that in the absence of gravity the range of transverse correlations increases without limit as the interfacial area A approaches infinity. And in the presence of gravity, the characteristic range of transverse correlations is the capillary length $\alpha^{-1} \equiv \sqrt{\gamma/\Delta\rho g}$ in large systems ($A >> \alpha^{-2}$). For water at 25°C, $\alpha^{-1} \simeq 2.7$ mm. As the critical point is approached α^{-1} approaches zero as $|1 - T/T_c|^{0.48}$. Thus, in contrast to correlations in bulk phase, transverse correlations in the interface become shorter ranged as the critical point is approached. It is worth noting that the linearity of $1/\tilde{D}^{(2)}$ with $\beta\gamma h^2$ provides the basis of the determination of surface tension by light scattering.

The assumption that $n = n(x - \xi)$ for a fluctuation ξ implies that density has the value $n(x_o)$ everywhere that $x - \xi(y,z) = x_o$. Thus, the probability that the density has the value $n(x_o)$ at x is the same as the probability that $x - x_o = \xi(y,z)$. The quantity

$$\sigma_\xi^2 \equiv \langle (x-x_o)^2 \rangle = \langle \xi^2(y,z) \rangle , \quad (41)$$

where $\langle \cdots \rangle$ denotes a thermal average with respect to $\exp[-\beta\Delta F]$, represents the mean square deviation of the position of the density value $n(x_o)$ about its mean position x_o. Carrying out the average shown in Eq. (41), we find

$$\sigma_\xi^2 = \sum_{\underset{\sim}{h}>0} \frac{1}{\beta A\gamma[h^2 + \alpha^2]} \simeq \frac{1}{4\pi\beta\gamma} \ln \left[\frac{h_{max}^2 + \alpha^2}{(2\pi)^2/A + \alpha^2} \right] . \quad (42)$$

In effect, the capillary waves widen the interfacial zone by amount σ_ξ. In the absence of gravity, σ_ξ increases as $\ln A$ as the interfacial area increases. This means that in the small, gravity free systems that are generally considered in a computer simulation the width of the interfacial zone will depend on system size. The widening of the interface with system size which was observed by Chapela et al. [20] in computer simulations was shown by Davis [17] to obey the trend predicted by Eq. (42).

To estimate the value of σ_ξ the cutoff h_{max} must be specified. Buff et al. [15] suggested $h_{max} = 2\pi/\sigma_\xi$, arguing that anything described as a capillary wave should not have a wavelength shorter than the thickness of the interfacial zone. It is probably more logical to take $h_{max} = 2\pi/\ell_w$, ℓ_w being the intrinsic width of the planar interface, although a completely rigorous theory of thermal fluctuations would avoid the need for an ad hoc assignment of a cutoff. For a 1 cm^2 interface of water at 25°C, we estimate σ_ξ = 3.92 and 3.83Å, for the choices $h_{max} = 2\pi/\sigma_\xi$ and $2\pi/\ell_w$, respectively (assuming the reasonable value ℓ_w = 8Å for the intrinsic interfacial width). If instead of 1 cm^2 the area of the interface is 10^{-6} cm^2, then we estimate σ_ξ = 3.07 and 2.93Å for the two assignments of h_{max}.

As the critical point is approached, γ goes to zero as $|1 - T/T_c|^{1.3}$ (see Ref. 11), so σ_ξ diverges as $|1 - T/T_c|^{-0.65}$. This compares well with the observed exponent of -0.67 for interfacial width near the critical point and is similar to the exponent of -0.6 predicted above for the intrinsic width of a planar interface.

Actually, even though we have restricted our attention to one-component fluids, Eq. (42) is equally valid for multi-component fluids, with $\alpha^2 \equiv \Delta\rho g/\gamma$, where $\Delta\rho$ is the difference between the mass densities of the coexisting phases. The interesting consequence of this is that chemical components can be chosen to form a liquid-liquid interface with matched densities and so $\alpha = 0$. Then $\Delta\sigma_\xi^2 \equiv \sigma_\xi^2(A_2) - \sigma_\xi^2(A_1) = (4\pi\beta\gamma)^{-1} \ln(A_2/A_1)$. If γ = 25 dyne/cm and T = 25°C, $\sqrt{\Delta\sigma_\xi^2}$ = 3.46Å if A_2 = 100 cm^2 and A_1 = 1 mm^2. Such an effect ought to be measurable. Incidentally, for this situation $\sigma_\xi(A_2)$ = 7.07Å.

We now leave the heuristic version of gradient theory of capillary waves and develop the mathematically rigorous version of gradient theory. Since c is assumed constant and $\partial^2 f_0(n(x))/\partial n^2$ depends only on x, the eigenfunctions of Eq. (18) are of the form

$$u_\lambda(x,\underset{\sim}{s}) = \tilde{u}_\lambda(x,\underset{\sim}{h}) \frac{e^{i\underset{\sim}{h}\cdot\underset{\sim}{s}}}{\sqrt{A}}, \tag{43}$$

where $\tilde{u}_\lambda(x,\underset{\sim}{h})$ satisfies

$$\tilde{L}\tilde{u}_{\lambda} \equiv -c \frac{d^2 u_{\lambda}}{dx^2}(x,\underset{\sim}{h}) + \frac{\partial^2 f_o}{\partial n^2}(n(x))\tilde{u}(x,\underset{\sim}{h}) = \zeta\, \tilde{u}(x,\underset{\sim}{h}) , \qquad (44)$$

where $\zeta \equiv \lambda - ch^2$ and $\tilde{u}_{\lambda}(x,\underset{\sim}{h})$ satisfies periodic boundary conditions.

The quantity $n'(x)$ has been shown to be an eigenfunction of Eq. (44) corresponding to $\zeta = 0$. Since $\tilde{L}$ is a Sturm-Liouville operator and $n'(x)$ is bounded from below by zero, it follows that $\zeta = 0$ is the lowest eigenvalue of $\tilde{L}$. Thus, the solutions to Eq. (44), $\tilde{u}_1 = n'(x)$, $\tilde{u}_2(x)$, $\tilde{u}_3(x)$,$\cdots$ are independent of h, form a complete orthonormal set

$$\int \tilde{u}_m(x)\tilde{u}_{m'}(x)dx = \delta_{mm'} ,$$

and correspond to the eigenvalues $\zeta_1 = 0 < \zeta_2 < \zeta_3, \cdots$. The eigenfunctions of L are then

$$u_{\lambda_m(h)}(x,\underset{\sim}{s}) = u_m(x)\,\frac{e^{i\underset{\sim}{h}\cdot\underset{\sim}{s}}}{\sqrt{A}} ,$$

with eigenvalues $\lambda_m(h) = \zeta_m + ch^2$, $m = 1,2,\cdots$ and $\underset{\sim}{h}$ ranging as indicated at Eq. (33).

An arbitrary density fluctuation $v(\underset{\sim}{r})$ can be expressed in the form

$$v(\underset{\sim}{r}) = \sum_{\underset{\sim}{h}}\sum_{m} a_{\lambda_m}(\underset{\sim}{h})\, u_m(x)\,\frac{e^{i\underset{\sim}{h}\cdot\underset{\sim}{s}}}{\sqrt{A}} , \qquad (45)$$

which yields for $\delta^2 F$ the equation

$$\delta^2 F = \frac{1}{2}\sum_{\underset{\sim}{h}}\sum_{m} \lambda_m(h)\,|a_{\lambda_m}(\underset{\sim}{h})|^2 . \qquad (46)$$

Although $\lambda_1(0) = 0$, the eigenfunction $u_{\lambda_1}(0) = n'(x)/\sqrt{A\int(n')^2dx}$ does not satisfy particle conservation, since $\int n'(x)d^3r \neq 0$, and so we conclude from the properties of $\lambda_m(h)$ ($\lambda_m(h) > \lambda_m(0)$ and $\lambda_m(0) > \lambda_1(0)$, $m = 2,3,\cdots$) that $\delta^2 F$ is positive for any allowed density fluctuation. Thus, a planar interface is stable.

The density-density correlation function can be computed by averaging $v(\underset{\sim}{r})v(\underset{\sim}{r}')$ with respect to the distribution $\exp[-\beta\delta^2 F]$ treating the coefficients $a_{\lambda_m}(\underset{\sim}{h})$ as random variables. The result is

$$D^{(2)}(\underset{\sim}{r},\underset{\sim}{r}') = \sum_{\underset{\sim}{h}} \sum_{m} e^{i\underset{\sim}{h}\cdot(\underset{\sim}{s}-\underset{\sim}{s}')} \frac{\tilde{u}_m(x)\tilde{u}_m(x')}{\beta A \lambda_m(h)}$$

or

$$D^{(2)}(x,x',\underset{\sim}{s}-\underset{\sim}{s}') = \sum_{\underset{\sim}{h}>0} e^{i\underset{\sim}{h}\cdot(\underset{\sim}{s}-\underset{\sim}{s}')} \frac{n'(x)n'(x')}{\beta A \gamma h^2} + \sum_{h} \sum_{m>1} e^{i\underset{\sim}{h}\cdot(\underset{\sim}{s}-\underset{\sim}{s}')} \frac{\tilde{u}_m(x)\tilde{u}_m(x')}{\beta A[\zeta_m + ch^2]} . \tag{47}$$

For simplicity we have considered the gravity free case. Its effect here would be the same as it was in the heuristic treatment. The first term on the rhs of Eq. (47) is identical to the gravity-free correlation function (38) obtained in the heuristic treatment. This is not surprising since we used there $v(\underset{\sim}{r}) = \sum_{\underset{\sim}{h}} \xi_h n'(x) e^{i\underset{\sim}{h}\cdot\underset{\sim}{s}}$, the same form as $\sum_{\underset{\sim}{h}>0} a_{\lambda_1}(\underset{\sim}{h}) u_{\lambda_1(\underset{\sim}{h})}(x,\underset{\sim}{s})$. The remaining terms on the rhs of Eq. (47) give only short-ranged transverse correlations since $\zeta_m > 0$ for $m > 1$. Actually $\partial^2 f_o(n(x))/\partial n^2$ behaves as a relatively shallow potential well, as illustrated in Fig. 2, and so only the first few (maybe only the first) eigenfunctions, $\tilde{u}_1, \tilde{u}_2, \cdots$, will be strongly localized in the interfacial region. After all, far from the interface Eq.(47) must transform to the bulk correlation function, Eq. (28). Thus, for sufficiently large m, the eigenfunctions $\tilde{u}_m(x)$ must approach $e^{ik_x x}/V^{1/3}$ and the eigenvalues ζ_m must approach $k_x^2 + \partial^2 f_o/\partial n^2$.

Next let us consider the capillary wave broadening of the interfacial zone. Equation (45) can be rewritten in the form

$$v(\underset{\sim}{r}) = \sum_{m} \xi^{(m)}(\underset{\sim}{s})\tilde{u}_m(x) , \tag{48}$$

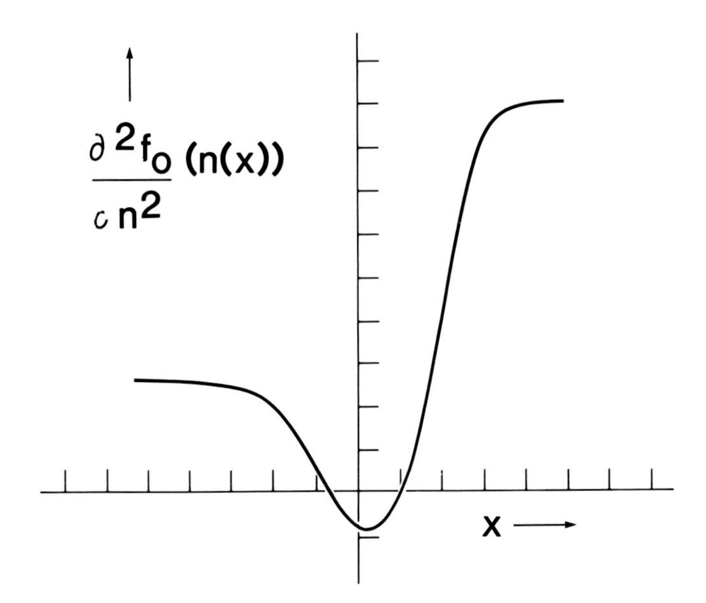

Fig. 2. Typical behavior of $\partial^2 f_o(n(x))/\partial n^2$ as a function of position x in a planar interface. Arbitrary units.

where

$$\xi^{(m)}(\underset{\sim}{s}) \equiv \sum_{\underset{\sim}{h}} \frac{a_{\lambda_m}(\underset{\sim}{h})}{\sqrt{A}} e^{i\underset{\sim}{h}\cdot\underset{\sim}{s}} . \tag{49}$$

For small amplitude, long wavelength capillary waves, we argued earlier that $v(r) = -\xi(y,z)n'(x)$, so that $\int v^2(r)dx/\int(n')^2dx = \tilde{\xi^2}$. It is difficult to find from Eq. (48) the surface $\xi(y,z)$ along which $n(x) + v(x,y,z) = n(x_o)$ in the interface. And since we have not actually found the eigenvalues $\tilde{u}_m$ for $m > 1$, we do not know which of these eigenfunctions correspond to capillary waves, i.e. we do not know which $\tilde{u}_m$ are localized in the interfacial zone. However, the quantity $\int v^2(r)dx/\int(n')^2dx$ is equal to $\xi^2(y,z)$ for the special case that $v = n(x - \xi) - n(x) \simeq -\xi(y,z)n'(x)$. Thus, we shall <u>define</u> a mean square dispersion length as

$$\sigma_\xi^2 \equiv \langle v^2(\underset{\sim}{r})dx/\int(n')^2dx\rangle = \frac{1}{\int(n')^2dx}\sum_m \langle[\xi^{(m)}]^2\rangle \tag{50}$$

Performing the average with respect to $\exp[-\beta\delta^2F]$, we find

$$\sigma_\xi^2 = \sum_{\underset{\sim}{h}>0} \frac{1}{\beta A\gamma h^2} + \sum_{\underset{\sim}{h}} \sum_{m>1} \frac{1}{\beta A\gamma[h^2 + \zeta_m/c]} . \tag{51}$$

The first term on the rhs of Eq. (51) is the field-free capillary wave mean-square dispersion obtained heuristically (Eq. (42)). This term increases as ℓnA as the interfacial area increases. The other terms on the rhs of Eq. (51) go to a finite limit as $A \to \infty$. Most of these terms represent bulk fluctuations, since for m sufficiently large the eigenfunctions $\tilde{u}_m(x)$ are not localized in the interface.

7. <u>RIGOROUS DENSITY FUNCTIONAL FREE ENERGY THEORY</u>

According to the general statistical mechanical theory of inhomogeneous fluid, the Helmholtz free energy $F(\{n\})$ is a functional of the density distribution $n(\underset{\sim}{r})$. The first functional derivative of F is the chemical potential,

$$\mu = \frac{\delta F}{\delta n(\underset{\sim}{r})} , \tag{52}$$

and the second functional derivative is [5]

$$\frac{\delta^2 F}{\delta n(\underset{\sim}{r})\delta n(\underset{\sim}{r}')} = [\beta n(\underset{\sim}{r})]^{-1}\,\delta(\underset{\sim}{r}-\underset{\sim}{r}') - \beta^{-1}\,C(\underset{\sim}{r},\underset{\sim}{r}') \equiv K(\underset{\sim}{r},\underset{\sim}{r}')\;, \quad (53)$$

where δ is the Dirac delta function and $C(\underset{\sim}{r},\underset{\sim}{r}')$ the direct correlation function.

In a homogeneous fluid

$$K(\underset{\sim}{r},\underset{\sim}{r}') = K(\underset{\sim}{r}-\underset{\sim}{r}') \equiv (\beta n)^{-1}\,\delta(\underset{\sim}{r}-\underset{\sim}{r}') - \beta^{-1}\,C^{o}(|\underset{\sim}{r}-\underset{\sim}{r}'|,n)\;, \quad (54)$$

and stability is determined by the eigenvalues of the equation

$$-\beta^{-1}\int C^{o}(|\underset{\sim}{r}-\underset{\sim}{r}'|,n)\,u_{\lambda}(\underset{\sim}{r}')d^3r' + (\beta n)^{-1}u_{\lambda}(\underset{\sim}{r}) = \lambda u_{\lambda}(\underset{\sim}{r})\;. \quad (55)$$

The eigenfunctions of the equation are

$$u_{\lambda}(\underset{\sim}{k}) = \frac{e^{i\underset{\sim}{k}\cdot\underset{\sim}{r}}}{\sqrt{V}}\;, \quad (56)$$

corresponding to the eigenvalues

$$\lambda(k) = -\beta^{-1}\,\tilde{C}^{o}(k,n) + (\beta n)^{-1}\;, \quad (57)$$

where $\tilde{C}^{o}(k)$ is the finite Fourier transform of $C^{o}(|\underset{\sim}{r}-\underset{\sim}{r}'|)$, i.e.,

$$\tilde{C}^{o}(k) = \int C^{o}(R)e^{i\underset{\sim}{k}\cdot\underset{\sim}{R}}d^3R\;. \quad (58)$$

In the small-k range, a Taylor series of $\tilde{C}^{o}(k)$ in k^2 yields

$$\lambda(k) = \frac{\partial^2 f_o}{\partial n^2}(n) + ck^2 + O(k^4)\;, \quad (59)$$

where the thermodynamic relation [13] $-\beta^{-1}\int C^{o}(R,n)d^3R + (\beta n)^{-1} = \partial^2 f_o/\partial n^2$ has been used. Thus, we see that for long wavelength (small k) fluctuations, gradient theory, Eq. (24), is in exact agreement with the rigorous theory. However, as $k \to \infty$, $\tilde{C}^{o}(k) \to 0$ and $\lambda(k) \to (\beta n)^{-1}$. Thus, short wavelength fluctuations (and, therefore, short-ranged fluctuations) are incorrectly treated by gradient theory.

Consider next a planar interface. For a fluctuation $v(\underset{\sim}{r})$ of the interface, we compute the free energy to second order in v by expanding F in a functional Taylor series about $n(\underset{\sim}{r}) = n(x)$. The result is

$$F = F^P + \frac{1}{2} \int K(\underset{\sim}{r},\underset{\sim}{r}')v(\underset{\sim}{r})v(\underset{\sim}{r}')d^3rd^3r' , \tag{60}$$

where F^P is the Helmholtz free energy of the unperturbed planar interface. The quantity $K(\underset{\sim}{r},\underset{\sim}{r}')$ is evaluated at the planar density n(x). K defines a symmetric operator which is isotropic in the yz-plane, i.e., it depends on $\underset{\sim}{s}$ and $\underset{\sim}{s}'$ only through $|\underset{\sim}{s}-\underset{\sim}{s}'|$.

Stability of the planar interface is determined by the eigenvalues of the equation

$$\int K(\underset{\sim}{r},\underset{\sim}{r}')\ u_\lambda(\underset{\sim}{r}')d^3r' = \lambda u_\lambda(\underset{\sim}{r}) . \tag{61}$$

One solution to this equation is [6]

$$u_1 = n'(x)/\int(n'(x))^2dx , \tag{62}$$

corresponding to $\lambda_1 = 0$. Because of the isotropy of K in the yz-plane, the eigenfunctions of Eq. (61) are of the form

$$u_{m,\underset{\sim}{h}}(\underset{\sim}{r}) = \tilde{u}_m(x,h) \frac{e^{i\underset{\sim}{h}\cdot\underset{\sim}{s}}}{\sqrt{A}} , \tag{63}$$

where the $\tilde{u}_m$ satisfy

$$-\beta^{-1}\int\tilde{C}(x,x',h)\tilde{u}_m(x',h)dx' + [\beta n(x)]^{-1}\tilde{u}_m(x',h) = \lambda_m(h)\tilde{u}_m(x,h) , \tag{64}$$

and $\tilde{C}(x,x',h)$ is the two-dimensional finite Fourier transform,

$$\tilde{C}(x,x',h) = \int C(x,x',S)e^{i\underset{\sim}{h}\cdot\underset{\sim}{S}}d^2S , \tag{65}$$

m being an integer indexing a particular eigenvalue.

Since capillary waves are the long wavelength or small $\underset{\sim}{h}$ fluctuations, we need only the eigenfunctions of Eq. (64) corresponding to small h. Thus, we expand $\tilde{K}$, $\tilde{u}_\lambda$, and λ in a Taylor series in h^2 and truncate at first order. This reduces Eq. (64) to the form $a + bh^2 = 0$. Equating coefficients a and b independently to zero we find

$$\int \tilde{K}(x,x',0)\ \tilde{u}_\lambda^{(0)}(x',0)dx' = \lambda^{(0)}\tilde{u}_\lambda^{(0)}(x,0) \tag{66}$$

and

$$\int \tilde{K}(x,x',0)\tilde{u}_\lambda^{(1)}(x',0)dx' - \lambda^{(0)}\tilde{u}_\lambda^{(1)}(x,0)$$

$$= -\int \tilde{K}^{(1)}(x,x',0)\tilde{u}_\lambda^{(0)}(x',0)dx' + \lambda^{(1)}\tilde{u}_\lambda^{(0)}(x,0) \ . \tag{67}$$

Equation (62) is the solution to Eq. (66) for $\lambda_1^{(0)} = 0$. The Fredholm alternative theorem for solvability of Eq. (67) then fixes the value of $\lambda_1^{(1)}$, namely

$$\lambda_1^{(1)} = \int \tilde{K}^{(1)}(x,x',0)n'(x)n'(x')dxdx'/\int(n')^2dx \ . \tag{68}$$

Since $\lambda_1(h) = h^2\lambda_1^{(1)}$ and since $\int u_{m,\underset{\sim}{h}}(\underset{\sim}{r})d^3r = 0$, $h \neq 0$, it follows that for a stable interface $\lambda_1^{(1)}$ is positive and $\lambda_1(h{=}0) = 0$ is the smallest eigenvalue of K. The eigenvalues of Eq. (64) can then be ordered as $\lambda_1(h) < \lambda_2(h), \cdots$, and for small values of h the above perturbation analysis yields $\lambda_1(h) = h^2\lambda_1^{(1)}$ and $\lambda_m(h) = \lambda_m^{(0)} + h^2\lambda_m^{(1)}$, with $\lambda_m^{(0)} > 0$ and $\lambda_m^{(1)} > 0$, $m > 1$. Expressing a fluctuation in the form

$$v(\underset{\sim}{r}) = \sum_{\underset{\sim}{h}>0} \sum_m a_{m,\underset{\sim}{h}} u_{m,\underset{\sim}{h}}(\underset{\sim}{r}) \ , \tag{69}$$

we find for the fluctuation free energy

$$\delta^2F = \frac{1}{2}\sum_{\underset{\sim}{h}} \sum_m \lambda_m(h)|a_{m,\underset{\sim}{h}}|^2 \ . \tag{70}$$

The thermal average of $v(\underset{\sim}{r})v(\underset{\sim}{r}')$ with respect to the distribution $\exp[-\beta\delta^2F]$ leads to the density-density correlation function

$$D^{(2)}(x,x',\underset{\sim}{s}-\underset{\sim}{s}') = \frac{1}{A}\sum_{\underset{\sim}{h}>0}\sum_{m} e^{i\underset{\sim}{h}\cdot(\underset{\sim}{s}-\underset{\sim}{s}')}\frac{\tilde{u}_m(x,h)\tilde{u}_m(x',h)}{\lambda_m(h)}$$

$$\simeq \sum_{\underset{\sim}{h}>0} e^{i\underset{\sim}{h}\cdot(\underset{\sim}{s}-\underset{\sim}{s}')}\frac{[n'(x)n'(x') + 0(h^2)]}{\beta A h^2\lambda_1^{(1)} \int (n')^2 dx}$$

$$+ \frac{1}{A}\sum_{\underset{\sim}{h}>0}\sum_{m>1} e^{i\underset{\sim}{h}\cdot(\underset{\sim}{s}-\underset{\sim}{s}')}\frac{\tilde{u}_m(x,h)\tilde{u}_m(x,h)}{\lambda_m^{(0)} + h^2\lambda_m^{(1)} + 0(h^4)} , \tag{71}$$

a form quite similar to the gradient theoretical result, Eq. (47). In fact, it has been established that the tension of a planar interface is given rigorously by [21,22]

$$\gamma = \frac{1}{4\beta}\int (\underset{\sim}{s}-\underset{\sim}{s}')^2 C(\underset{\sim}{r},\underset{\sim}{r}')n'(x)n'(x')dxdx'd^2s . \tag{72}$$

From this, it follows that $\lambda_1^{(1)} \int (n')^2 dx = \gamma$, and so the Fourier transform of $D^{(2)}$ for small h is

$$\tilde{D}^{(2)}(x,x',h) = \frac{n'(x)n'(x')}{\beta\gamma h^2} + 0(1) , \tag{73}$$

a rigorous result with which gradient theory, Eq. (47), agrees exactly. The only difference is that in gradient theory the tension is computed from Eq. (72) by approximating $C(\underset{\sim}{r},\underset{\sim}{r}')$ by the homogeneous fluid direct correlation function $C^o(|\underset{\sim}{r}-\underset{\sim}{r}'|,n)$ and approximating $n'(x')$ by $n'(x)$.

We can also determine the dispersion length defined in Eq. (50). The result is

$$\sigma^2 = \sum_{\underset{\sim}{h}>0}\frac{1}{\beta A\gamma h^2} + \sum_{\underset{\sim}{h}}\sum_{m>1}\frac{1}{\beta A\gamma[\lambda_m^{(0)} + h^2\lambda_m^{(1)} + 0(h^4)]} . \tag{74}$$

Again the rigorous theory confirms the small h prediction of gradient theory (Eq. (51)).

The quantity $\tilde{C}(x,x',h) \to 0$ as $h \to \infty$, and so $\lambda_m(h)$ is bounded (by $(\beta n_g)^{-1}$) from above as $h \to \infty$. Thus we see again that gradient theory fails in the small wavelength (large h) limit.

Although its derivations [21,22] were not based on density functional free energy theory, the rigorous formula, Eq. (72), for tension can be established quite easily with the free energy theory. Consider a small amplitude, extremely slowly varying deformation $\xi(y,z)$ of a planar interface. Macroscopically, the free energy of this interface, viewed as a membrane, is

$$F = F^B + \gamma A \; \gamma \int_A \sqrt{1 + (\nabla\xi)^2}\, dydz = F^P + \frac{\gamma}{2} \int (\nabla\xi)^2 dydz + 0(\xi^4), \tag{75}$$

where F^B is the free energy of bulk fluid and F^P that of the planar interface. For the deformation envisioned $v = n(x - \xi) - n(x)$ and the free energy functional $F(\{n(x) + v(r)\})$ can be expanded to yield

$$F = F^P + \delta^2 F + 0(\xi^4) \;, \tag{76}$$

where

$$\delta^2 F = \frac{1}{2} \int \{[\beta n(x)]^{-1} \delta(\underset{\sim}{r}-\underset{\sim}{r}') - \beta^{-1} C(\underset{\sim}{r},\underset{\sim}{r}')\} n'(x) n'(x') \xi(\underset{\sim}{s}) \xi(\underset{\sim}{s}') d^3r d^3r' + 0(\xi^4) \;. \tag{77}$$

The quantity $C(\underset{\sim}{r}-\underset{\sim}{r}') = C(x,x',|\underset{\sim}{s}-\underset{\sim}{s}'|)$ since it is the direct correlation function of a planar system. We have arbitrarily chosen $\xi(\underset{\sim}{s})$ to be an extremely slowly varying deformation, and so $\xi(\underset{\sim}{s}')$ can be well approximated by $\xi(\underset{\sim}{s}) + (\underset{\sim}{s}'-\underset{\sim}{s})\cdot\nabla\xi(\underset{\sim}{s}) + (1/2)(\underset{\sim}{s}'-\underset{\sim}{s})(\underset{\sim}{s}'-\underset{\sim}{s}) : \underset{\sim}{\nabla}\nabla\xi(s)$. Inserting this into Eq. (77), we find

$$\delta^2 F = [\frac{1}{8\beta} \int C(x,x',\underset{\sim}{S}) S^2 n'(x) n'(x') dx dx' d^2S] \int (\nabla\xi)^2 d^2s + 0(\xi^4) \;, \tag{78}$$

which when compared with the corresponding term in Eq. (75), yields the exact formula, Eq. (72), for tension. It should be noted that this formula is valid independently of the nature of intermolecular forces among fluid particles. In particular, the result is not restricted to pair additive forces. To arrive at Eq. (78), we have noted that $n'(x')$ is the zero eigenvalue of the integrand in curly brackets of

Eq. (77), have noted that by the planar isotropy of $C(\tilde{r},\tilde{r}')$ the term linear in $\nabla\xi$ vanishes and $(\tilde{s}'-\tilde{s})(\tilde{s}'-\tilde{s}): \nabla\nabla\xi$ can be replaced by (1/2) $(\tilde{s}'-\tilde{s})^2\nabla^2\xi$, and have performed an integration by parts to get $\int \xi\nabla^2\xi d^2 s = -\int (\nabla\xi)^2 d^2 s$.

In closing, it is appropriate to address the question of why an approximation such as gradient theory is retained when rigorous theory yields the same results. There are several compelling reasons to continue development of gradient theory: (1) it is mathematically relatively simple, (2) it agrees quite well with rigorous theory where comparisons can be made, (3) it provides a means of calculating interfacial profiles and tensions, an accomplishment thus far escaping rigorous theory, and (4) it is easily applicable to a variety of other interesting inhomogeneous fluid systems, examples being thin films and layered structures [23,24] and spherical drops and bubbles [24,25].

REFERENCES

1. Rayleigh, Lord, Phil. Mag. 33, 208 (1892).
2. van der Waals, J. D. and P. Kohnstamm, Lehrbuch der Thermodynamik, Vol. 1, Maas and van Suchtelen, Leipzig, 1908.
3. Bongiorno, V., L. E. Scriven, and H. T. Davis, J. Colloid and Interface Sci. 57, 472 (1976).
4. Yang, A. J. M., P. D. Fleming III, and J. H. Gibbs, J. Chem. Phys. 64, 3732 (1976).
5. Percus, J. K., The Equilibrium Theory of Classical Fluids, H. L. Frisch and J. L. Lebowitz, Eds., Benjamin, N.Y., 1963.
6. Davis, H. T. and L. E. Scriven, Adv. in Chemical Physics 49, 357 (1982).
7. Bongiorno, V. and H. T. Davis, Phys. Rev. A 12, 2213 (1975).

8. Carey, B. S., L. E. Scriven, and H. T. Davis, Am. Inst. Chem. Eng. J. 24, 1076 (1978); 26; 705 (1980).

9. McCoy, B. F. and H. T. Davis, Phys. Rev. A 20, 1201 (1979).

10. Lee, J. K., J. A. Barker, and G. M. Pound, J. Chem. Phys. 60, 1976 (1974); F. F. Abraham, D. E. Schreiber, and J. A. Barker, J. Chem. Phys. 62, 1958 (1975).

11. Ma, S.-K., Modern Theory of Critical Phenomena, Benjamin, New York, 1976.

12. Huang, J. S. and W. W. Webb, J. Chem. Phys. 50, 3677 (1969).

13. See for example, A. Münster, Statistical Thermodynamics, Vol. I, Academic Press, N.Y., 1969.

14. Brumberger, H. in Critical Phenomena, Proceedings of a Conference held in Washington, D.C., April 1965, Edited by M. S. Green and J. V. Sengers. NBS Miscellaneous Publication 273.

15. Buff, F. P., R. A. Lovett, and F. H. Stillinger, Jr., Phys. Rev. Lett. 15, 621 (1965).

16. Kalos, M. H., J. K. Percus, and M. Rao, J. Stat. Phys. 17, 111 (1977).

17. Davis, H. T., J. Chem. Phys. 67, 3636 (1977); erratum in J. Chem. Phys. 70, 600 (1979).

18. Wertheim, M. S., J. Chem. Phys. 65, 2377 (1976).

19. Evans, R., Advances in Physics 28, 143 (1979).

20. Chapela, G. A., G. Saville, S. M. Thompson, and J. S. Rowlinson, J. Chem. Soc. Trans. Faraday II 73, 1133 (1977).

21. Triezenberg, D. G. and R. Zwanzig, Phys. Rev. Lett. 28, 1183 (1972).

22. Lovett, R., P. W. DeHaven, J. J. Vieceli, Jr., and F. P. Buff, J. Chem. Phys. 58, 1880 (1973).

23. Davis, H. T. and L. E. Scriven, J. Stat. Phys. 24, 243 (1981).

24. Falls, A. H., L. E. Scriven, and H. T. Davis, J. Chem. Phys. 75, 3986 (1981).

25. Falls, A. H., L. E. Scriven, and H. T. Davis, J. Chem. Phys. (to appear 1983).

This work was supported by a grant from the Department of Energy.

Department of Chemical Engineering & Materials Science
University of Minnesota - Mpls.
Minneapolis, MN 55455

INTERFACIAL AND CRITICAL PHENOMENA IN MICROEMULSIONS

M. W. Kim, J. Bock, and J. S. Huang

ABSTRACT

The interfacial and bulk properties of microemulsions exhibit critical behavior in certain regions of the phase diagram. To study these compositions, two techniques based on dynamic laser light scattering have been used. The first technique involves interfacial scattering to determine the interfacial tension between a saturated microemulsion and its excess phase. The second provides a measure of the correlation length in the bulk microemulsion phase. The dependence of the interfacial tension (σ) and the correlation length (a) on salinity can be described in terms of a simple power law of the distance from a critical salinity, $|S_c - S|$. Based on these scaling laws the interfacial tension between a saturated microemulsion and its excess water phase is found to be inversely proportional to the square of the correlation length in the microemulsion: $\sigma \sim a^{-2}$. This is in accordance with scaling predictions. Furthermore, the interfacial and critical phenomena of microemulsions as described by these scaling relationships is analogous to a binary system near its critical consolute point.

ISBN 0-12-493220-7

1. Introduction

A microemulsion[1] is a homogeneous liquid that contains hydrocarbon, water, and one or more surface active agents called surfactants. When a small amount of surfactant is mixed with oil, the surfactant molecules may form aggregates to expose only their lipophilic moiety in order to mininize the free energy. This is known as a micellar solution. If we should further add a small amount of water to this oil external micellar solution, the water molecules are solubilized into the hydrophilic core of the micelles. This solution containing water solubilized in micelles is called a microemulsion. It is possible to formulate saturated microemulsions to coexist with either or both water and oil phases. It has been noted that under certain conditions, the interfacial tension between a microemulsion and its excess phase diminishes, resembling the property of the interface in a binary liquid mixture near its critical consolute point. We have studied the salinity dependence of interfacial tension (IFT), and the correlation length of the microemulsion phase in a two-phase microemulsion system (microemulsion/excess water) by optical homodyne spectroscopy.

The measurement of surface tension by optical methods was first demonstrated by Katyl and Ingard[2] for the studies of a free liquid surface. Then Huang and Webb[3] took advantage of the vanishing interfacial tension and density difference in a binary liquid mixture near the critical consolute point to develop a correlation spectroscopy technique. Recently, the technique has been utilized for studies of surfactant monolayers[4] and microemulsion systems.[5]

The advantages of using an optical probe to determine IFT over the conventional means involving direct or indirect measurements of the mechanical restoring forces are: (1) higher sensitivity, and (2) complete equilibrium of the surface under study. The dynamic laser light scattering technique originally was developed to study bulk fluid scattering[6], by measuring the characteristic decay time of the intensity-intensity correlation function. This decay time provided a link between the bulk properties of liquids and their interfacial behavior, a relation that van der Waals[7] recognized about one hundred years ago. More recently, the relationship between the bulk and the interface near the critical consolute point has been predicted by Widom[8] using scaling concepts and has been shown by experiments in a binary system.[9]

2. Theory

A) Light Scattering From Interfacial Capillary Waves

The interface between two liquids in a gravitational field without any excitations is a flat plane in an equilibrium position. However, the random thermal motion of molecules produces fluctuations on the interface. These fluctuations are thought of as interfacial waves (capillary waves)[10] with gravity and interfacial tension providing the restoring force and viscosity providing the damping. Mathematically, the capillary waves are described by the solutions of the Navier-Stokes equation in a uniform gravitational field, subject to boundary conditions at the fluid interface and infinity. A detailed derivation of the dispersion ratio can be found in Ref. 11.

A major conclusion is that for a low tension interface, the interfacial wave is over-damped wave with a time constant τ given by the following expression:

$$\tau^{-1} = \frac{q}{2(\eta+\eta')} \left(\sigma + \frac{\rho-\rho'}{q^2}\right) \tag{1}$$

where σ is the interfacial tension, g is the gravitational acceleration constant, $q = \frac{2\pi}{\Lambda}$ is the wave number of the capillary wave with wave length Λ, and ρ, η and ρ', η' are the density and the shear viscosity of the lower and upper phases, respectively. It is assumed that there is no interfacial shear viscosity, the fluids are incompressible, the depth(h) of the fluids is larger than the capillary wavelength (i.e. $qh \gg 1$), and no flow. Interfacial (surface) fluctuations at any given instance t, can be analyzed by its Fourier components (normal modes). Each of the normal modes characterized by the wave number q, can be considered as a diffraction grating that scatters light into a particular direction Θ_q as prescribed by the Bragg law. The scattered intensity as measured by a photomultiplier is proportional to the amplitude of that q- component. For the over-damped waves, it can be shown [3] that the time-averaged intensity-intensity correlation function:

$$C(\Delta t) = \frac{1}{t_o} \int_0^{t_o} \langle I(t).I(t + \Delta t) \rangle dt \qquad (2)$$

is an exponential function with a characteristic damping time, τ. In principle, by determining τ at various angles, the interfacial tension and the sum of the bulk viscosities can be deduced through Equation (1). Measurements at a fixed scattering angle can be used to determine σ, when the viscosities (η, η') and the densities (ρ, ρ') have been determined by independent measurement.

B) Dynamic Light Scattering From Microemulsion Phase

As we described previously, the microemulsion contains small droplets with a surfactant layer. These small droplets maintain a given spatial configuration during the characteristic time (τ) before it is completely altered by the random Brownian motion.[6] The essence of the dynamic light scattering technique is the determination of the characteristic time from the scattering intensity-intensity correlation function as defined by Equation (2). Furthermore, if the scatterers are mono-dispersed and have random motions, then $C(\Delta t)$ can be expressed by a simple exponential function, $\exp(-\Gamma \Delta t)$. Since the spatial configurations, measured on the scale of the inverse scattering wave number k, where $k = 4\pi/\lambda \, \mathrm{Sin}(\Theta/2)$ with λ representing the wavelength of light in the medium, Θ representing the scattering angle, we could anticipate the relation

$$\Gamma = Dk^2 \qquad (3)$$

where D is the diffusion constant. From the diffusion constant D, the Stoke-Einstein relation is used to evaluate a mean hydrodynamic radius $\underline{a}$:

$$\underline{a} = \frac{k_b T}{6\pi\eta D} \cdot \qquad (4)$$

where k_b is the Boltzmann constant, T is the absolute temperature, η is the solvent shear viscosity, and $\underline{a}$ is the equivalent of the dynamic correlation length in an ordinary critical fluid mixture.

C) Scaling Relation

From the point of view of critical phenomena in microemulsion,[12] the interfacial tension (σ) and the correlation length ($\underline{a}$) can be expressed by a simple power law with respect to the proper scaling variables. The exponent (μ) that describes the diminishing of the interfacial tension has a simple relation with other critical exponents such as γ, β and ν describing the singularities of the isothermal compressibility, order parameter and correlation length, respectively. From the thermodynamic scaling relations, Widom[8] developed a theory of the interface which predicts

$$\gamma = \mu + \nu - 2\beta \, . \qquad (5)$$

If the theory of scaling laws is carried into full correlation scaling, the number of independent exponents is reduced by one, yielding the following relationships

$$\mu = \gamma = 2\nu \, , \qquad (6)$$

and

$$\frac{\mu}{\nu} = 2 \, . \qquad (7)$$

Theoretical calculation[13] for ν, and μ gives values of 0.64 and 1.28, respectively.

3. Experiment

A). Sample Preparation and Physical Properties

The surfactant used for this study was the monoethanol amine salt of dodecyl orthoxylene sulfonate ($C_{12}OXS$). This surfactant contained 86% actives. 98% pure tertiary amyl alcohol (TAA) and 99 + % pure n-decane, purchased from MC/B and Aldrich Chemical Company respectively, were used without further purification. The micro-emulsions were composed of 2% surfactant, 1% TAA, mixed with a 1 to 1 volume ratio of brine (X% NaCl in doubly distilled water) and decane. The salinity of the brine was varied from 1.7% to 4% NaCl.

The samples were kept in a constant temperature bath controlled at 25°C to within ± 0.1°C. The viscosities of the micro-emulsion phase and the equilibrated excess phases were measured at a shear rate of 1.285 sec^{-1} with a Contraves Rheometer (LS 30) at 25°C. The results are shown in Figure 1.

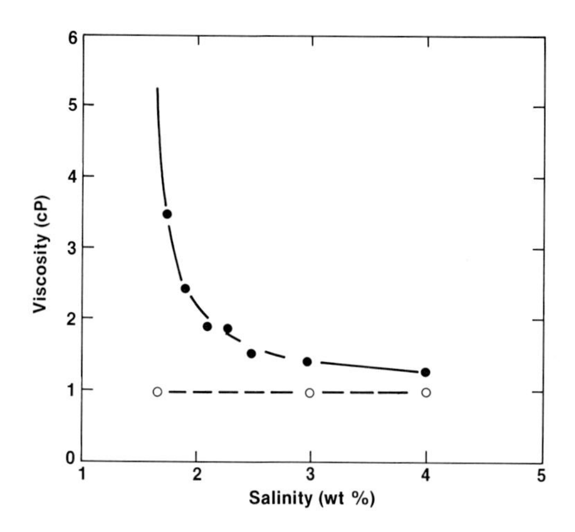

Figure 1. Viscosities of Microemulsion Phase (.) and its Excess Phase (o) as a Function of NaCl Concentration

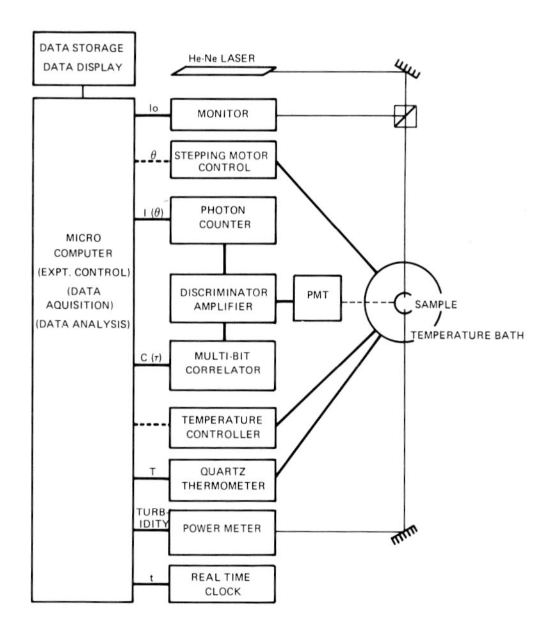

Figure 2. Schematic Setup for the Bulk Scattering Experiments

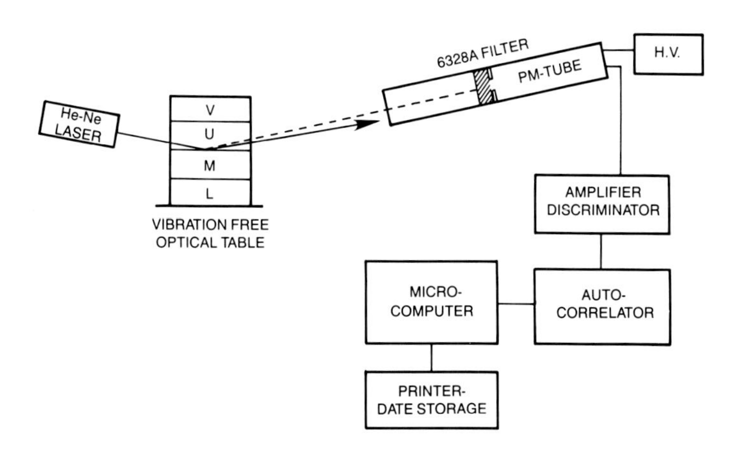

Figure 3. Schematic Setup for the Interface Scattering Experiments

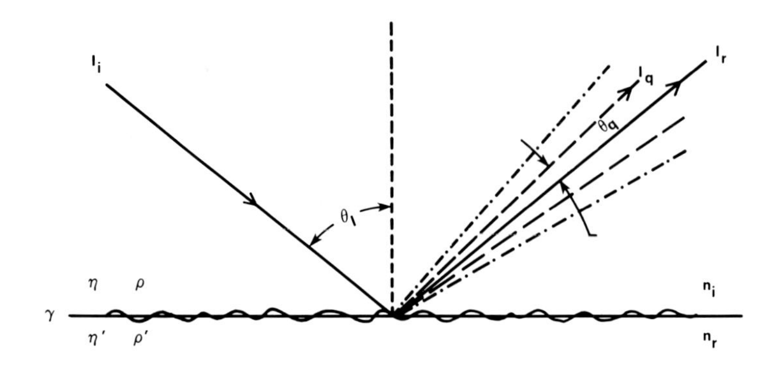

Figure 4. Scattering Geometry for Interfacial Scattering

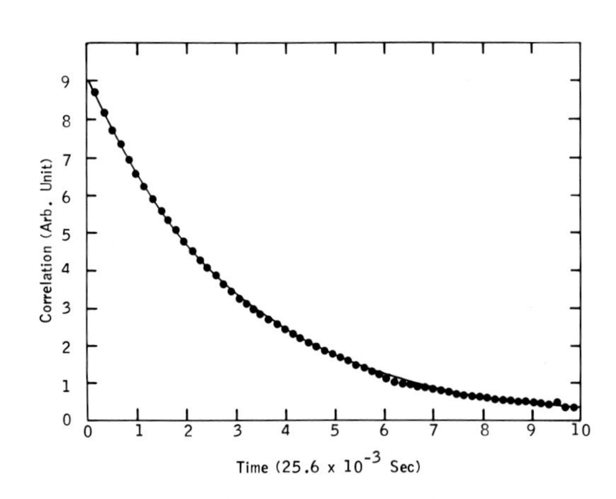

Figure 5. A Typical Correlation Function

B) Photon-Correlation Spectroscopy

An HP-9845B microcomputer was employed to control the photon correlation spectrometer to measure the linewidth of the bulk scattering from a microemulsion phase as well as and the linewidth of the interfacial scattering from the interface between a microemulsion phase and the excess water phase. Other major components used in our experiments consisted of a He-Ne laser, Malvern 128 channel multi-unit correlator, and a PAR 1112 photon processor. The schematic set up of our light scattering experiments for bulk scattering and interface scattering is shown in Fig 2, and Fig. 3 respectively.

1) Interface Scattering

The scattering geometry is sketched in Figure 4. The incident light I_i impinges on the interface at an angle Θ_i with respect to the specular direction normal to the surface, the scattered angle Θ_q is measured with respect to the reflected light I_r. The wave number q is related to the scattering angle by:

$$q = \frac{2\pi}{\lambda} n_i \left[\sin\left(\theta_i - \theta_q\right) - \sin\theta_i\right] \tag{8}$$

where λ is the wavelength of light in vacuo and n_i is the index of refraction of the incident phase. A typical correlation function is shown in Figure 5.

The interfacial tension, σ, can be expressed in terms of the best fit single exponential decay time, τ, to the measured intensity correlation function by inverting Equation (1):

$$\sigma = \frac{2(\eta+\eta')}{\tau q} - g\,\frac{\Delta\rho}{q^2}\,. \tag{9}$$

2) Bulk Scattering

The linewidth of the Rayleigh component in the spectrum of scattered light was determined using the experimental set up shown in Fig. 2. The photon correlator calculates the auto correlation function of the scattering intensity, $C(\Delta t) \equiv \langle I(t)I(t+\Delta t)\rangle$, where

$I(t)$ is the scattering intensity at time t, and Δt is the delay time. Since scattering intensity flucturations decay exponentially with time, the auto correlation function $C(\Delta t)$ has a simple exponential form:

$$C(\Delta t)=\langle I(t)I(t+\Delta t)\rangle = e\langle I\rangle\delta(\Delta t) + \langle I\rangle^2 + \beta\langle I\rangle^2 e^{-2\Gamma\Delta t} \quad (10)$$

The correlation function in Eq. (11) contains three components. The first, $e\langle I\rangle$ is the shot noise term. The second, $\langle I\rangle^2$, is the dc component. The third term is an exponential with decay rate 2Γ, which is the quantity of interest. The prefactor β, of order unity, accounts for the spatial aperature of the detectors. The linewidth Γ was obtained by the cummulants method(13) for each of the scattering angles ranging from 45° to 135°. The average center-of-mass diffusion constant, D_0, is obtained by extrapolating Γ/k^2 to $k = 0$. The mean hydrodynamic droplet radius, a, can be evaluated from the diffusion constant D_0 by the use of the Stokes-Einstein relationship expressed by Eq. 4.

4. Results

A. Interfacial Tension

The salinity dependence of interfacial tension between an upper phase microemulsion and excess water phase was studied at 25°C for the C_{12}OXS system in a two phase salinity range (1.7% NaCl - 4.0% NaCl) by the optical probe technique. The results are shown in Fig. 6. The critical salinity, S_c, is defined to be the salinity at which the interfacial tension between the microemulsion (oil-external) and excess water-phase vanishes. A power law dependence of σ on the relative salinity, $|S_c-S|$, was found as follows:

$$\sigma \sim |S-S_c|^{\mu}$$

with $\mu = 1.38$. This relationship described the microemulsion/excess brine interfacial tension as shown in Figure 7.

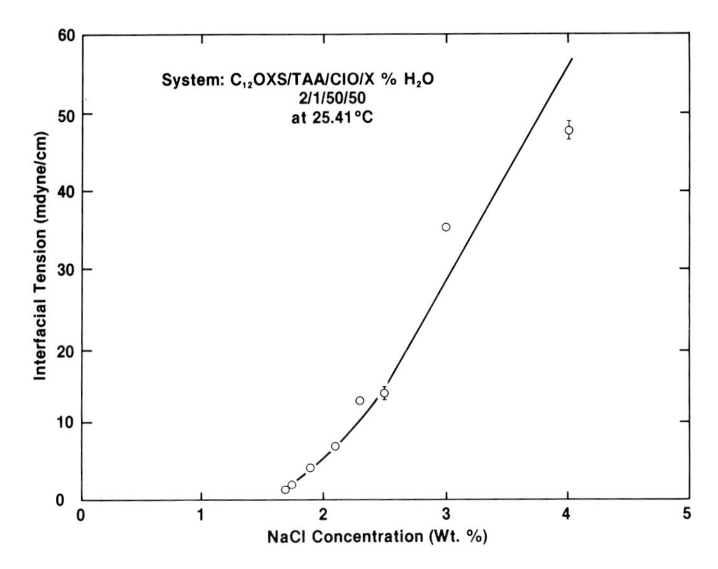

Figure 6. The Salinity Dependence of IFT at 25°C

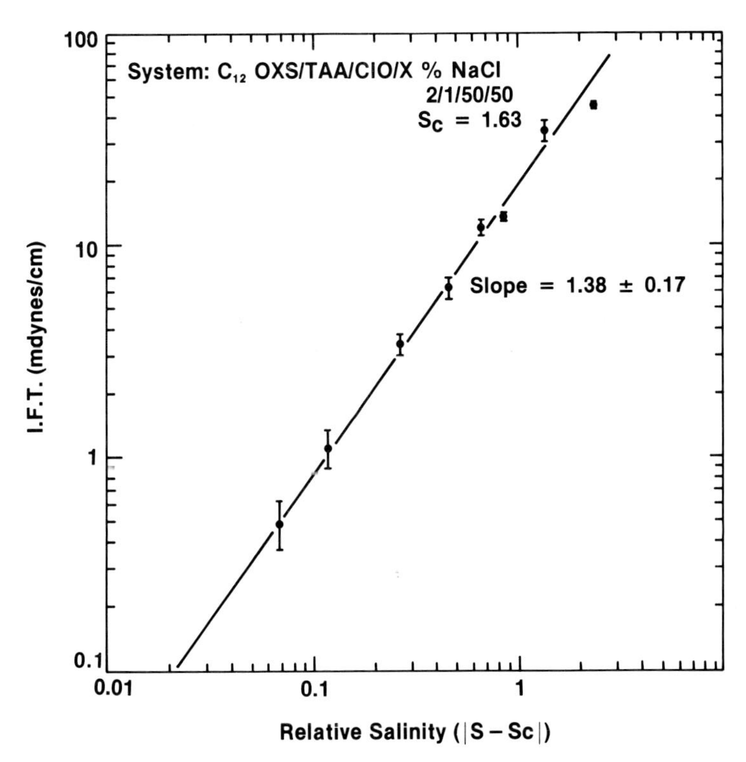

Figure 7. The Power Law Dependence of IFT on the Relative Salinity $|S_C-S|$

B. Correlation Length

The salinity dependence of the correlation length was studied for $C_{12}OXS$ system in the same two-phase region. The hydrodynamic droplet radius (correlation length) was deduced from the diffusion constant by equation (14). The result is shown in Figure 8. The correlation length was found to scale with salinity using the same extrapolated critical salinity S_c determined in the interfacial tension study. A power law dependence of the correlation length on the relative salinity, $|S_c - S|$, was found as follows:

$$\underline{a} \sim |S-S_c|^{-\nu}$$

with $\nu = 0.73$. This relation described the correlation length as shown in Fig. 9.

C. Correlation Between Bulk and Interface Properties

When the critical point in a binary fluid mixture is approached from the 2-phase region, the properties of the two co-existing phases approaches each other, causing the interfacial tension to vanish. The distance to the critical point can be measured by the correlation length in the bulk fluid. Thus, it is entirely reasonable to correlate the surface critical behavior i.e. the vanishing of the interfacial tension, with the bulk correlation length. In fact, the scaling laws[8] predicts that $\sigma \simeq k_B T/\xi^2$, a relation expected on the grounds of dimension arguments alone. This prediction was confirmed by our measurements as shown in Fig. 10 since the slope of the best fit line of $\underline{a}$ vs σ on the log-log scale is 0.52.

5. Conclusion

Dynamic light scattering was employed to study the interface and the bulk properties of microemulsion systems in an oil external microemulsion two-phase region. We have analyzed the intensity

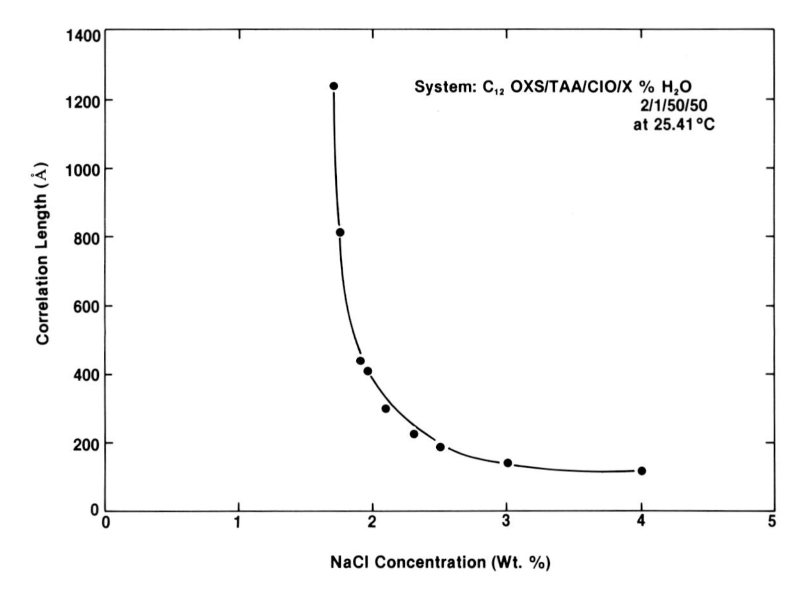

Figure 8. The Salinity Dependence of Correlation Length

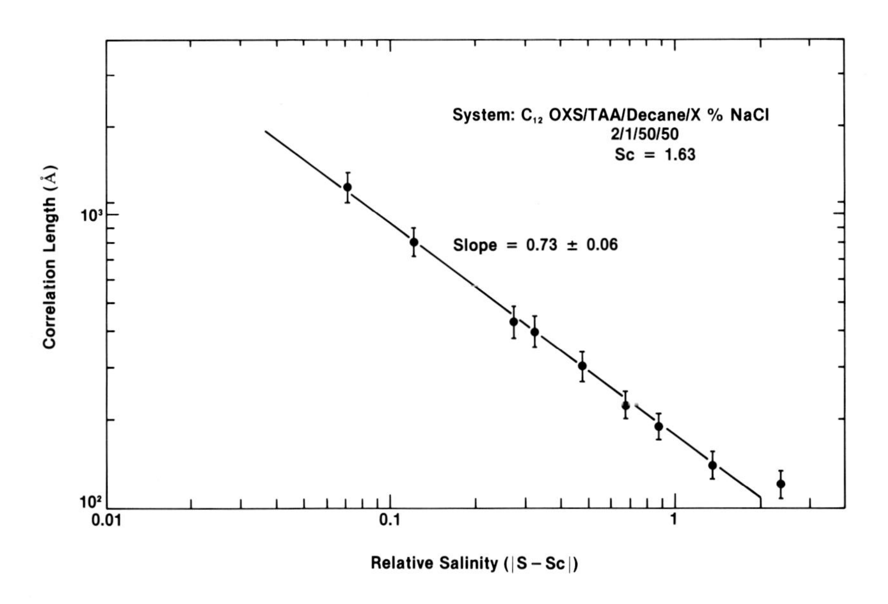

Figure 9. The Power Law Dependence of Correlation Length on the Relative Salinity $|S_C - S|$

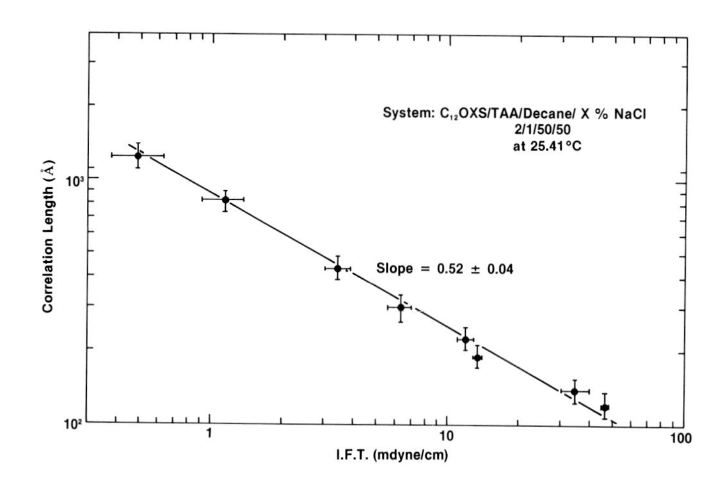

Figure 10. The Correlation Length vs. Interfacial Tension in a Log-Log Scale

auto correlation functions to determine the interfacial tension and a hydrodynamic radius. We have found that the interfacial tension can be described by a power law relationship of the form $|S-S_c|^{\mu}$ with $\mu = 1.38$, where S_c is the extrapolated critical salinity. The hydrodynamic radius was also found to vary $|S-S_c|^{-\nu}$ with $\nu = 0.73$, with the same S_c. The behavior was found to be consistent with the scaling behavior of other microemulsion systems. The value of critical exponent ν of this system was consistent with that of the oil external three component microemulsion system.[15] Furthermore, the interfacial tension between the microemulsion and its excess brine phase was found to be proportional to the inverse of the square of the hydrodynamic radius. This is in agreement with the theoretical calculation and scaling prediction.

References

1. Robbins, M. L.; "Micellization, Solubilization, and Microemulsion," edited by K. L. Mittal (Plenum, New York, 1977) Vol. II.
2. Katyl, R. H. and Ingard, U.; phys. Rev. Lett., 19, 64 (1967).
3. Huang, J. S. and Webb, W. W.; Phys. Rev. Lett., 23, 160 (1969).
4. Hard, S. and Neuman, R.; Jourl. of Coll. and Int. Sci., 83, 315 (1981).
5. Kim, M. W., Huang, J. S. and Bock, J. SPE/DOE 10788 (1982), Pouchelon, A., Meunier, J., Langevin, D. and Cazabat, A. H. J. Phys. Lett., 41, 239, (1980).
6. Berne, B. and Pecora, R. "Dynamic Light Scattering," John Wiley & Sons, Inc. N. Y., 1976.
7. Van der Waals, J. D., Z. Physik Chen., (Leipzig) 13, 667, (1894).
8. Widom, B.; J. Chem. Phys., 43, 3892, (1965).
9. Huang, J. S. and Webb, W. W.; J. Chem. Phys., 50, 3677, (1969), Warren, C. and Webb, W. W.; J. Chem. phys., 50, 3694, (1969).
10. Landau, L. D. and Lifshitz, E. M.; "Fluid Mechanics," Pergamon Press, Inc., New York, (1959).
11. Huang, J. S.; PhD Thesis, Cornell Univ. (1969).
12. Huang, J. S. and Kim, M. W., Phys. Rev. Lett., 47, 1462, (1981).
13. Stanley, H. E., "Introduction to Phase Transition and Critical Phenomena," Oxford Univ. Press, N. Y., 1971.
14. Miller, A. M., Hwan, R-N., Benton, W. J. and Fort, T. Jr., Jourl. of Coll. and Int. Sci., 61, 554, (1977).
15. Huang, J. S. and Kim, M. W. "Scattering Technique Applied to Supramolecular and Non-Equilibrium Systems," Ed. S. H. Chen, B. Chu, and R. Nassal (Plenum Press, 1981).

Exxon Research and Engineering Co.
Corporate Research - Science Labs.
P. O. Box 45
Linden, New Jersey 07036

ELECTROHYDRODYNAMIC SURFACE WAVES

J. Melcher

1. BACKGROUND

For those concerned with interfacial phenomena, it is not necessary to review the variety of configurations that can be assumed by even a pair of fluids. Examples are drops, bubbles, jets and dispersions. Because of the widely ranging electrical properties of fluids, this variety becomes far greater when an electric field is applied. Fortunately, practical applications are also widely ranging.

Fundamental to interactions, regardless of the configuration, is the dynamics "in the small", on a scale where the interface at least starts out essentially planar. With attention focused on the dynamics of planar interfaces subjected to piece-wise uniform perpendicular fields, the following sections begin as a review of interfacial interactions in the absence of electrical shear stresses. Research aimed at understanding waves and instabilities with shear stresses is then described. Highlighted first is the identification of surface-dilatation modes that account for the tendency of highly insulating interfaces supporting trapped charge to behave as equipotentials. Exemplified is the "imposed ω-k" technique of studying the sinusoidal steady state electrical response with the wave-number imposed.

Instabilities are the most dramatic result of applying fields to planar interfaces and recent developments have shown that these can be used to print. For either relatively

ISBN 0-12-493220-7

conducting or insulating inks, a discussion is given on how the required resolution and growth rate is achieved.

A related "imposed ω-k" technique is then described in which steady streaming results from the imposition of a traveling wave. Shear-stress interactions are reviewed and new experimental results for traveling-wave interactions with trapped-charge modes on insulating interfaces and gravity-capillary field coupled modes on equipotential interfaces compared to predictions.

Work concerned with the internal electrohydrodynamic structure of interfaces is then the concluding theme. Here, the "imposed ω-k" technique is applied to both sensing the electrical response to interactions with a homogeneous semi-insulating liquid and to inducing steady streaming of the liquid. In these cases, the surface is pinned down by an insulating conduit wall so that the dominant effects result because of the internal structure of the interface. Practical impetus for an interest in the electrohydrodynamics of weak double layers comes from the need to understand electrical breakdown caused by the convection of highly insulating liquids that are used to cool and insulate high voltage equipment.

1.1 Volume and Surface Force Densities of Electrical Origin

The electrical force that causes the motion of a fluid is the result of a transfer of momentum from the charged particles acted on by the electric field $\bar{E}$ to the largely neutral fluid molecules. In Fig. 1.1, two types of forces are distinguished by the way in which this is accomplished. The free-charge force density, the first term in the following equation, results as positive and negative carriers collide as individuals with the surrounding neutral particles.

$$\bar{F} = \rho_f \bar{E} - \frac{1}{2} E^2 \nabla \varepsilon \quad (1.1)$$

As depicted in Fig. 1.1a, the negative particles tend to undo the force from the positive particles [1 Sec. 3.2].

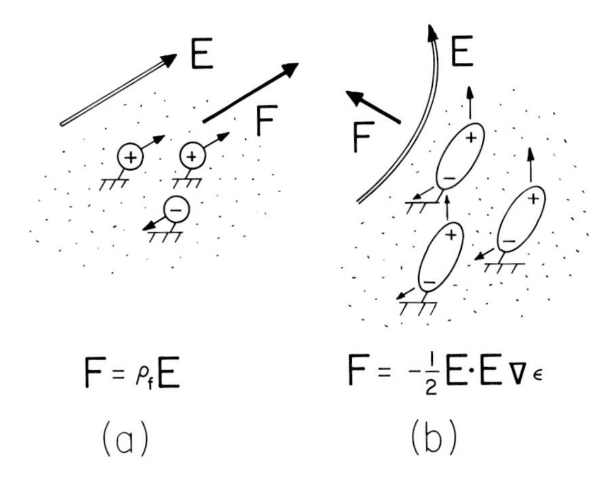

Fig. 1.1. Schematic Representation of Free-charge and Polarization Force-densities.

Thus, it is the net charge that is represented by the density ρ_f.

Typically, the majority of particles in a liquid are polarizable, to a degree represented by the polarization density $\bar{P}$. As illustrated by Fig. 1.1b, in a nonuniform electric field, the dipoles are subject to an electric force. Because these transfer this electric force to the fluid as individuals, there is a net force density, $\bar{P}\cdot\nabla\bar{E}$, acting on the fluid [1 Sec. 3.6]. Strictly, this is the force density on tenuous dipoles. The second term in Eq. 1 represents the polarization force density in a fluid having a polarization density that is linear in the electric field, $\bar{P}=(\varepsilon-\varepsilon_o)\bar{E}$ where the permittivity ε is a property of the fluid. It is derived using thermodynamic arguments and the assumption that deformations of interest are incompressible [1 Sec. 3.7]. Although it has a distribution that is very different from $\bar{P}\cdot\nabla\bar{E}$, the two force densities differ only by the gradient of a pressure. Provided care is taken to use a self-consistent force density and stress tensor, incompressible motions of a fluid are not altered by the addition of a force density which is the gradient of a scalar "pressure" [1 Secs. 3.7 and 8.3]. The stress tensor consistent with the Korteweg-Helmholtz force density of Eq. 1 is

$$T_{ij} = \varepsilon E_i E_j - \frac{1}{2}\varepsilon\delta_{ij}E_k E_k \qquad (1.2)$$

Fluids are often piece-wise uniform in their properties, so the polarization force density written as the second term in Eq. 1 is very convenient. In such systems, the effects of polarization are then clearly confined to interfaces.

1.2 Polarization and Conduction Constitutive Relations

In writing Eq. 1, we have already assumed the fluid to be electrically linear. Especially in considering interfacial interactions on a scale large compared to that typical of the interfacial thickness, an Ohmic conduction model gives a useful picture of the free charge dynamics. Thus, the current density $\bar{J}'=\sigma\bar{E}'$, where σ is the electrical conductivity, a property of the fluid, and the primes denote variables evaluated in a frame of reference fixed to the fluid.

In systems where the electrical conductivity is piecewise uniform, the free-charge accumulates at the interfaces. At a given point in a region of uniform conductivity, the charge density is finite only if the fluid element at that point can be traced back along a particle line to a point where it was given an initial charge [1 Sec. 5.10]. In piecewise uniform systems, not only is the polarization force density of Eq. 1 confined to interfaces, the free-charge force density usually is as well.

1.3 Electric Flux-Potential Relations

The electrohydrodynamics of the conventional fluid is electroquasistatic. Thus, the electric field $\bar{E}$ is irrotational and can be represented in terms of the potential Φ, $\bar{E}=-\nabla\Phi$. In the volume of an initially static fluid region having uniform permittivity and conductivity, the charge density is zero. It follows from Gauss' law that $\bar{E}$ is also solenoidal so that Φ obeys Laplace's equation.

Of use here is the solution of Laplace's for the planar region of thickness Δ shown in Fig. 1.2. Solutions are taken as having the canonical form $\Phi=\mathrm{Re}\hat{\Phi}(x)\exp j(\omega t-ky)$. Then with $(\varepsilon\hat{E}_x^\alpha,\hat{\Phi}^\alpha)$ and $(\varepsilon\hat{E}_x^\beta,\hat{\Phi}^\beta)$ respectively taken as the amplitudes of the electric flux and potential at the upper and lower surfaces, it follows from Laplace's equation that [1, Sec. 2.16]

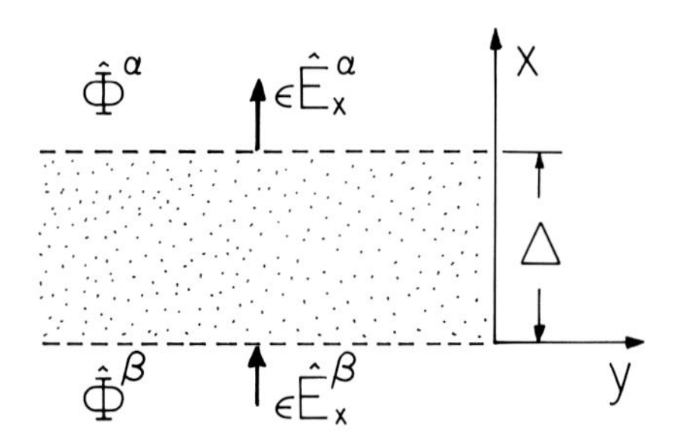

Fig. 1.2. Planar Dielectric Layer Showing Flux-potential Surface Variables.

$$\begin{bmatrix} \varepsilon \hat{E}_x^{\alpha} \\ \varepsilon \hat{E}_x^{\beta} \end{bmatrix} = \varepsilon k \begin{bmatrix} -\coth k\Delta & \dfrac{1}{\sinh k\Delta} \\ -\dfrac{1}{\sinh k\Delta} & \coth k\Delta \end{bmatrix} \begin{bmatrix} \hat{\Phi}^{\alpha} \\ \hat{\Phi}^{\beta} \end{bmatrix} \qquad (1.3)$$

These flux-potential relations hold not only for complex amplitudes, but when the variables are Fourier amplitudes or Fourier transforms as well.

1.4 Stress-velocity Relations for Newtonian Fluid

Again in one region of a piece-wise uniform fluid system, the mechanical mass density ρ and viscosity η are constants. Motions are described by the Navier-Stokes equations and the continuity condition that the fluid velocity v be solenoidal. This latter condition is met by introducing a vector potential such that $\bar{v}=\nabla \times \bar{A}$.

For the planar region shown in Fig. 1.3, where small amplitude motions are initiated from a static equilibrium, $\bar{A} = A_v \bar{i}_z$, where A_v is the stream function and it follows from the Navier-Stokes equations that

$$\nabla^2 \left(\rho \frac{\partial A_v}{\partial t} - \eta \nabla^2 A_v\right) = 0 \qquad (1.4)$$

The mechanical stress is related to $\bar{v}$, and hence to A_v, by

$$S_{ij} = -p\ \delta_{ij} + \eta\left(\frac{\partial v_i}{\partial x_j} + \frac{\partial v_j}{\partial x_i}\right) \qquad (1.5)$$

where p is the pressure.

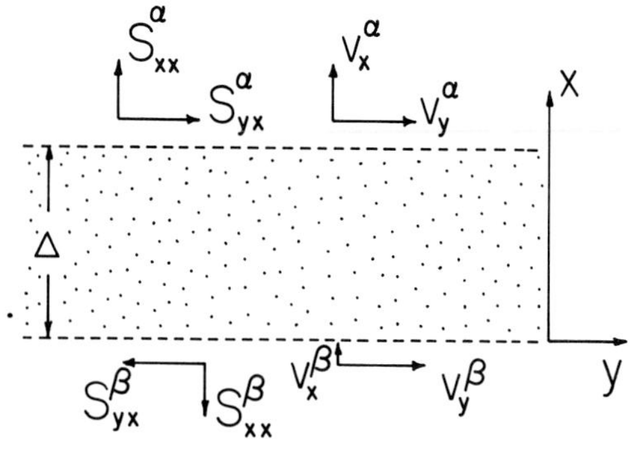

Fig. 1.3 Planar Layar of Fluid Showing Stress-velocity Surface Variables.

For complex amplitudes having the same form as for Eq. 1.3, it follows that if the normal and shear velocities in the top and bottom planes are $(\hat{S}^{\alpha}_{xx}, \hat{S}^{\alpha}_{yx})$ and $(\hat{S}^{\beta}_{xx}, \hat{S}^{\beta}_{yx})$ respectively, then the normal and shear stresses in these planes are given by the velocity-stress relations [1 Sec. 7.19]

$$\begin{bmatrix} \hat{S}^{\alpha}_{xx} \\ \hat{S}^{\beta}_{xx} \\ \hat{S}^{\alpha}_{yx} \\ \hat{S}^{\beta}_{yx} \end{bmatrix} = \eta [P_{ij}] \begin{bmatrix} \hat{v}^{\alpha}_{x} \\ \hat{v}^{\beta}_{x} \\ \hat{v}^{\alpha}_{y} \\ \hat{v}^{\beta}_{y} \end{bmatrix} \tag{1.6}$$

Only two of the 16 coefficients P_{ij} are needed here. These are

$$P_{11} = k_n [1-(\phi/\theta)^2][(\phi/\theta)\cosh\phi \sinh\theta - (\phi/\theta)^2 \cosh\theta \sinh\phi]/F$$

$$P_{13} = -P_{31} = jk_n \{(\phi/\theta)[3+(\phi/\theta)^2][1-\cosh\theta \cosh\phi] + [1+3(\phi/\theta)^2] \sinh\theta \sinh\} \phi/F \tag{1.7}$$

$$P_{33} = k_n [(\phi/\theta)^2 - 1][(\phi/\theta)\sinh\theta \cosh\phi - \sinh\phi \cosh\theta]/F$$

where

$$F = (2\phi/\theta)(1-\cosh\phi\cosh\theta) + \sinh\phi\sinh\theta\,[(\phi/\theta)^2 + 1]$$

$$\phi = \Delta\sqrt{k_n^2 + j\omega\rho/\eta} \quad ; \quad \theta = k_n\Delta$$

Summarized in the relations of Eq. 1.6 are the "pressure" modes and viscous diffusion modes that are apparent in Eq. 1.4. It is the latter that bring in the viscous skin depth $\delta = \sqrt{2\eta/\omega\rho}$ that appears in Eq. 1.7.

2. Qualitative View of Field-Coupled Surface Waves With No Electric Shear Stress

In terms of the Korteweg-Helmholtz stress tensor, Eq. 1.2, the i'th component of the surface force density acting on interfaces between uniform fluid regions is

$$T_i = [\![T_{ij}]\!]\, n_j \tag{2.1}$$

where $[\![B]\!] \equiv B^a - B^b$ and the unit normal $\bar{n}$ is directed from region (b) into region (a).

For two types of fluid interfaces, $\bar{T}$ acts normal to the interface. It is only in these limiting cases that a self-consistent physical picture does not have to include mechanical mechanisms, such as viscosity, to equilibrate an electrical shearing surface force density [1 Sec. 8.2].

For the first of these the interface is an equipotential. Charges relax on such an interface sufficiently rapidly that the tangential electric field remains essentially zero. Thus, at least in one of the fluids or within the interface itself, the effective charge relaxation time must be short compared to times of interest. Also, one of the fluids must be much less conducting than the other.

Consider what happens when an initially flat section of an equipotential interface stressed by an initially uniform electric field is deformed as shown in Fig. 2.1. A perturbation surface charge density is induced that adds to the original density where the conducting fluid projects into the insulating one and subtracts where the insulating fluid extends into the conducting one. As a result, the upward surface force density that was originally uniformly distributed over the

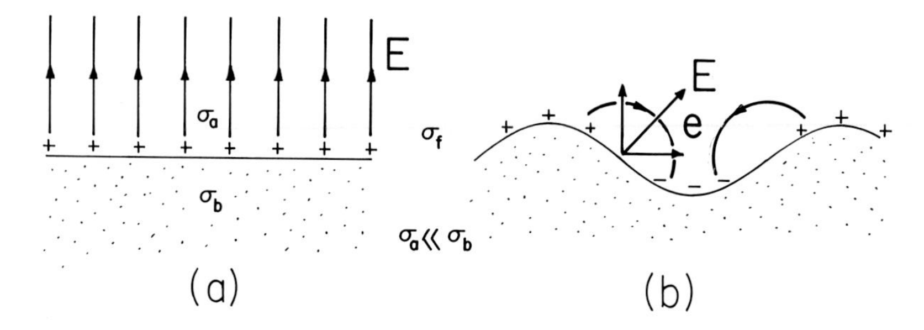

Fig. 2.1. Equipotential Interface Between Uniformly Conducting Fluids.

interface is enhanced on the upward peaks and reduced in the valleys. The electrical surface force density tends to counteract the effects of surface tension. Thus, it decreases the phase velocity of gravity-capillary surface waves [2, 1 Sec. 8.10].

The second type of interface that is not subject to a shearing surface force density is perfectly insulating and supports no free surface charge density. In this case, there is no implication for the orientation of a field applied to the interface, for it is the polarization force density that acts on the interface and it is clear from either Eq. 1.1 with $\rho_f=0$ or Eq. 2.1 with $\bar{n}\cdot\varepsilon_a\bar{E}_n^a=\bar{n}\cdot\varepsilon_b\bar{E}_n^b$ that there is no shearing component of $\bar{T}$.

Consider what happens to the surface force density as the interface deforms in the presence of applied fields. In Fig. 2.2, the applied field is perpendicular to the interface and, for purposes of illustration, the fluid below has the greater permittivity. Thus, the field above tends to have a reduced tangential field at the interface and the alterations in the original field are much as for the equipotential interface. The surface charges are, of course, polarization rather than free charges. The imposed field is augmented at the peaks and reduced in the valleys. It follows from the surface polarization force density, Eq. 1.1, that where the field is augmented the originally upward surface force density is as well, so the force encourages a further deformation. Regardless of which permittivity is the greater, the

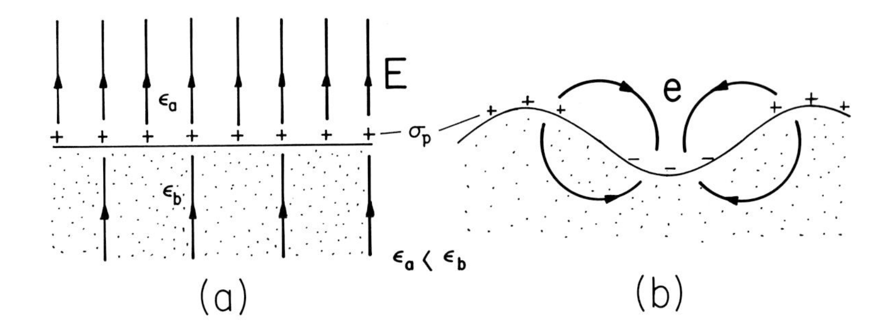

Fig. 2.2. Interface Between Dielectric Fluids that Supports No Free Surface Charge.

field coupling again reduces the phase velocity of interfacial waves and can result in instability [4].

In a tangential field, the velocity of surface waves propagating along the field lines is increased while that across the field lines is not influenced. Thus, if it has any effect, the tangential field tends to stabilize the interface [2,1 Sec. 8.11]. We confine interest here to coupling dominated by the perpendicular field.

3. DYNAMICS OF FREE-SURFACE MONOLAYERS

3.1 Self-consistent Linear Electrohydrodynamics

Many standard techniques for measuring liquid motions are difficult to apply to the study of electrohydrodynamic effects in highly insulating fluids. Particles used as tracers are likely to become charged, and hence to migrate relative to the fluid. Optical techniques for measuring the transverse deflection of an interface are relatively insensitive to the interfacial dilatations of interest here. However, the same electric field that makes it difficult to apply conventional approaches can be used to "see" the motions of a pure liquid with a clean interface. This is accomplished here using the apparatus shown in Fig. 3.1. (Only an abbreviated account of this work has been given previously [5].)

Shown in cross-section at the bottom of Fig. 3.1 is a metal reservoir of circular cross-section that is filled with the insulating liquid. In the experiments reported here, hexane is used. The equilibrium interface is flush with the upper surface of the reservoir, with the level maintained

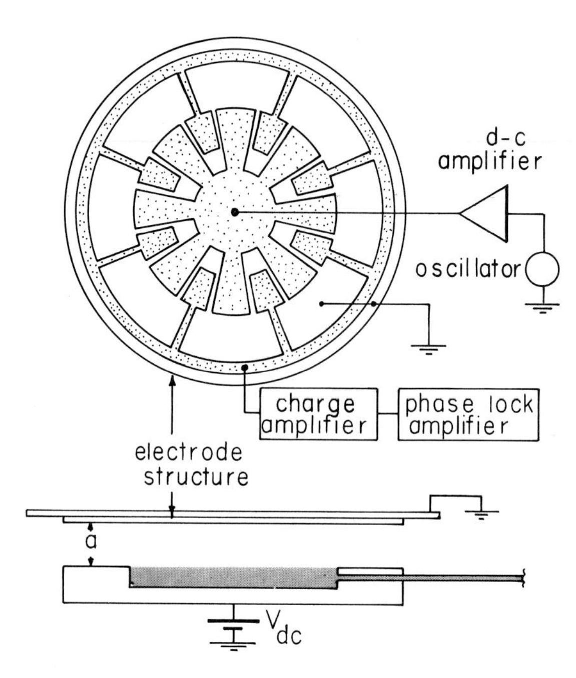

Fig. 3.1. Interdigital Electrode System for Driving and Sensing Motions of Charge Monolayer on Liquid Interface.

against evaporation and electrical mechanisms of fluid loss by adding liquid through the conduit shown. At a distance a above the interface is an electrode. In the absence of electrical excitations that will be described, this electrode is essentially grounded. A dc electric field is then applied to the interface by raising the reservoir to the potential V_{dc}.

One of two methods is used to create the ion-pairs in the liquid required to form a uniform monolayer at the interface. Either a source of alpha radiation or sharply pointed electrodes are positioned in the bottom of the reservoir. Charges of the appropriate sign are then pulled to the interface, and the equilibrium picture of the field is then no different then that for a conducting fluid. The electric field $V_{dc}/a=E_o$ is terminated by the interfacial surface charge and hence confined to the air-gap.

The electrode structure is used to both drive the interface and detect its response. Shown in bottom view at the top of Fig. 3.1, its surface is divided into three types of electrodes that together form an interdigital electrode system. The periodicity is azimuthal with the "active" region of the annulus formed by the eight interdigital pairs of electrodes, alternately connected to the oscillator-amplifier driving system and to the charge amplifier detection system. The grounded electrodes represented by the clear regions minimize direct coupling between the edges of the driving and detection electrodes.

The driving electrodes have a modest voltage (100 volts or less) at the angular frequency ω while the detection electrodes are at essentially ground potential. Thus, with a small uniform field component subtracted out, the azimuthal perturbation potential distribution takes the form of a standing wave.

The charge induced on the detection electrodes is in a large part the image of charge on the driving electrodes. To focus on the small part imaging charge on the interface, a phase-lock amplifier is used. This makes it possible to detect that part of the signal that is out of phase with the driving signal. This quadrature component of the detected charge is entirely due to the interfacial dynamics.

3.2 Response to Standing Waves Induced by Interdigital Electrode

The cross-section of the fluid layer, air-gap and electrode structure is shown in Fig. 3.2. The y direction is azimuthal in Fig. 3.1. Thus, the radial direction in Fig. 3.1 is perpendicular to the paper in the developed model represented by Fig. 3.2. At the top of this figure is an instantaneous distribution of the electrode potential, with the potential between the grounded guard electrodes and the driving electrodes approximated as being linear.

The potential is now used to represent the perturbation part of the electric field intensity.

$$\bar{E} = E_o \bar{i}_x + \bar{e} \quad ; \quad \bar{e} = -\nabla\Phi \tag{3.1}$$

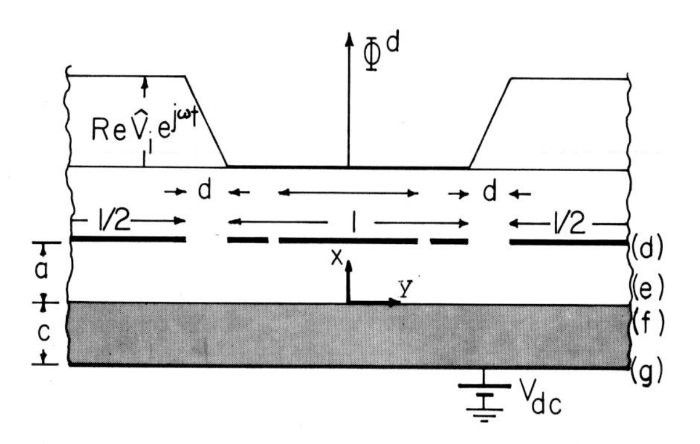

Fig. 3.2. Schematic Cross-section of Model for System of Fig. 3.1 showing locations of Surface Positions.

Note the designation of positions at the right in Fig. 3.2. It follows from Fig. 3.2 that the potential in the plane of the electrode is related to the driving potential V_i and electrode dimensions by

$$\Phi^d = Re \sum_{n=-\infty}^{+\infty} \hat{\Phi}_n^d e^{j(\omega t - k_n y)} \qquad (3.2)$$

where $k_n = n\pi/(\ell+d)$ and

$$\hat{\Phi}_o^d = \frac{1}{2}\hat{V}_i \; ; \quad \hat{\Phi}_n^d = \frac{\hat{V}_i}{n\pi d k_n}[\cos k_n(\frac{\ell}{2} + f) - \cos \frac{k_n \ell}{2}]$$

The complex amplitude of the detected charge is then

$$\frac{\hat{q}}{\hat{V}_i} = Nw \int_{-\frac{h}{2}}^{\frac{h}{2}} \hat{e}_x dz = - Nwh\varepsilon_o \hat{e}_{xo}^d - 4Nw \sum_{n=1}^{\infty} \varepsilon_o \hat{e}_{xn}^d \sin(\frac{k_n h}{2}) \qquad (3.3)$$

The field distributions in the air-gap and in the liquid are conveniently represented by Eq. 1.3, applied to the respective regions.

$$\begin{bmatrix} \varepsilon_o \hat{e}_x^d \\ \varepsilon_o \hat{e}_x^e \end{bmatrix} = \varepsilon_o k \begin{bmatrix} -\coth ka & \sinh ka \\ -\frac{1}{\sinh ka} & \coth ka \end{bmatrix} \begin{bmatrix} \hat{\Phi}^d \\ \hat{\Phi}^e \end{bmatrix} \qquad (3.4)$$

$$\begin{bmatrix} \varepsilon_o \hat{e}_x^f \\ \\ \varepsilon_o \hat{e}_x^g \end{bmatrix} = \varepsilon k \begin{bmatrix} -\coth kc & \sinh kc \\ \\ -\dfrac{1}{\sinh kc} & \coth kc \end{bmatrix} \begin{bmatrix} \hat{\Phi}^f \\ \\ \hat{\Phi}^g \end{bmatrix} \tag{3.5}$$

Here, we look forward to linearizing the interfacial boundary conditions in such a way that all variables are referenced to the plane of the unperturbed interface.

The liquid layer is similarly described by Eq. 1.6 with appropriate identification of variables.

$$\begin{bmatrix} \hat{S}_{xx}^f \\ \hat{S}_{xx}^g \\ \hat{S}_{yx}^f \\ \hat{S}_{yx}^g \end{bmatrix} = \eta[P_{ij}] \begin{bmatrix} \hat{v}_x^f \\ \hat{v}_x^g \\ \hat{v}_y^f \\ \hat{v}_y^g \end{bmatrix} \tag{3.6}$$

If effects of the fluid above were appreciable, these same relations could be applied to that region as well.

Boundary conditions relate the amplitudes appearing in these relations. In Eq. 3.4a, $\hat{\Phi}^d$ is the driving potential given by Eq. 3.2. At the bottom, the perturbation potential is zero, so $\hat{\Phi}^g=0$ in Eq. 3.5b. Also, on the bottom, the tangential and shear velocities must be zero, so $\hat{v}_x^g=0$ and $\hat{v}_y^g=0$ in Eqs. 3.6b and 3.6d respectively.

The physics of interest is represented by the interfacial boundary conditions. Two of these represent the effect of the interfacial deformation and convection on the field and two account for the effect of the field on the mechanics. First, there is the kinematic relation between the interface position, Fig. 3.3, and the velocity of the fluid.

$$\xi = \xi(y,t) \quad ; \quad \hat{\xi} \simeq \frac{\hat{v}_x}{j\omega} \tag{3.7}$$

Fig. 3.3. Definition of Interface Deflection.

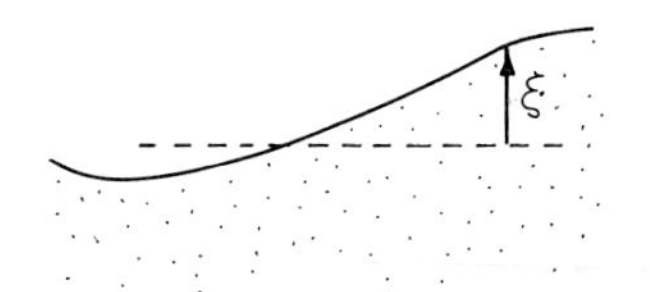

Then, consistent with the electrical laws are the continuity conditions on the tangential electric field intensity

$$\bar{n} \times [\![\bar{E}]\!] = 0 \qquad (3.8)$$

and on the conservation of charge. In the model proposed here, surface charge is lost from an incremental volume enclosing the interface through three mechanisms. Two of these involve surface currents, convective or migration, while the third is due to volume conduction. It will be found that for fluids such as hexane, the first of these can dominate the other two. But, in general the charge conservation continuity condition requires that at the interface

$$\frac{\partial \sigma_f}{\partial t} = - \nabla_\Sigma \cdot (\sigma_f \bar{v} + \sigma_o b_s \bar{E}) - \bar{n} \cdot [\![\sigma \bar{E}]\!]$$

$$\sigma_f = \bar{n} \cdot \varepsilon \bar{E} \qquad (3.9)$$

In this expression, Gauss' continuity condition relates the surface charge density to the electric field intensity.

Force equilibrium for the interface requires that the sum of the fluid-mechanical and electric stresses acting on the interface be balanced by the surface force density due to surface tension, γ.

$$[\![S_{ij}]\!] n_j + [\![T_{ij}]\!] n_j + n_i \gamma \frac{\partial^2 \xi}{\partial y^2} = 0 \quad i = 1,2 \qquad (3.10)$$

To linear terms, the normal and shear stress equilibria are represented by this expression with i=x and i=y respectively.

These four interfacial conditions are linearized and the bulk relations, Eqs. 3.4-3.6, used to eliminate the $\hat{e}_x$'s and $\hat{S}_{ix}$'s. Thus, tangential field continuity, charge conservation, normal stress equilibrium and shear stress equilibrium are respectively represented by the following four equations.

$$\begin{bmatrix} 1 & -1 & \frac{jE_o}{\omega} & 0 \\ j\omega\varepsilon_o k_n \coth k_n a & [(j\omega\varepsilon+\sigma)k_n \coth k_n c + k_n^2 \sigma_o b_s] & 0 & -jk_n\sigma_o \\ k_n\sigma_o \coth k_n a & 0 & [\frac{j}{\omega}(\rho g + k_n^2\gamma) - \eta P_{11}] & -\eta P_{13} \\ jk_n\sigma_o & 0 & [-\frac{k_n\sigma_o E_o}{\omega} - \eta P_{31}] & -\eta P_{33} \end{bmatrix} \begin{bmatrix} \hat{\Phi}_n^e \\ \hat{\Phi}_n^f \\ \hat{v}_{xn}^f \\ \hat{v}_{yn}^f \end{bmatrix} = \begin{bmatrix} 0 \\ \frac{j\omega\varepsilon_o k_n \hat{\Phi}_n^d}{\sinh k_n a} \\ \frac{k_n \sigma_o \hat{\Phi}_n^d}{\sinh k_n a} \\ 0 \end{bmatrix} \tag{3.11}$$

The sensing electrode charge, Eq. 3.3, is now found by using these expressions to determine $\hat{\Phi}^e$ and in turn Eq. 3.4a to evaluate $\varepsilon_o \hat{e}_x^d$. The response takes the form

$$\hat{q} = \sum_{n=1}^{\infty} \frac{g(\omega, k_n)\hat{V}_i}{D(\omega, k_n)} \tag{3.12}$$

With the amplifier of Fig. 3.1 phase-locked to the driving signal, the detected signal is proportional to the imaginary part of Eq. 3.12. This is exemplified as a function of the frequency $f=\omega/2\pi$ by the solid curves in Fig. 3.4. The response is typified by two resonances. If the modes that these represent are essentially uncoupled, it is possible to approximate the resonance frequencies by simple expressions. The sharp high-frequency resonance of Fig. 3.4a comes in the neighborhood of

$$f_{gc} = \frac{1}{2\pi}\left[\frac{g\rho k + \gamma k^3 - k^2\sigma_o E_o \coth ka}{\rho \coth ka}\right] \tag{3.13}$$

and involves a field-coupled form of gravity-capillary waves. In a way familiar from the dynamics of such surface waves on equipotential interfaces stressed by fields, [2] the effect of the field is to reduce the phase velocity of gravity-capillary waves, and so the high frequency resonance in Fig. 3.4 shifts downward as the surface charge density is raised.

The relatively broad low frequency resonance of Fig. 3.4a comes near the frequency

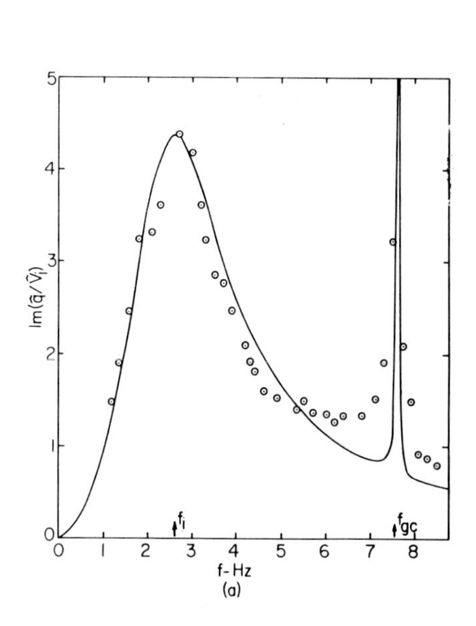

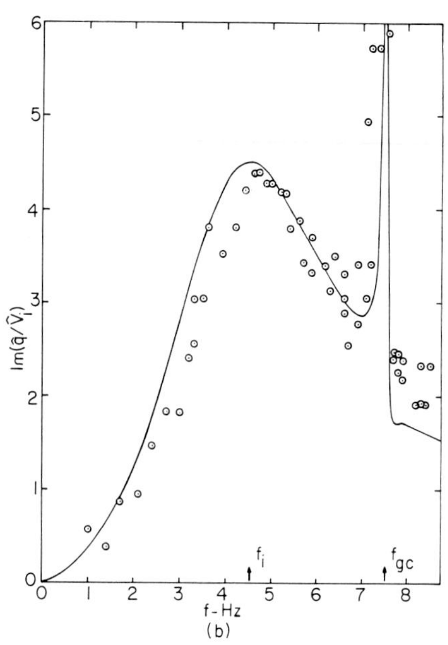

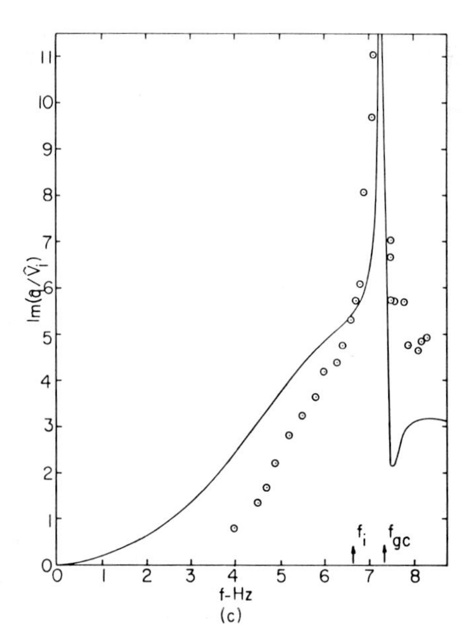

Fig. 3.4. Frequency Response of Quadrature Component of Charge for Three Monolayer Charge Densities.
(a) $\sigma_o = 2.15 \times 10^{-6}$ C/m^2,
(b) $\sigma_o = 3.22 \times 10^{-6}$ C/m^2,
(c) $\sigma_o = 4.30 \times 10^{-6}$ C/m^2.
Other parameters are ε = permittivity of liquid = 1.89 ε_o, $\rho = 0.69 \times 10^3$ kg/m^3, $\eta = 0.374 \times 10^{-3}$ newt - sec/m^2, $\gamma = 1.89 \times 10^{-2}$ newt/m, ℓ = 1.375 cm, h = 0.95 cm. d = 0.1 cm, w = 1.6 cm, a = 0.453 cm and c = 1.2 cm.

$$f_i = \frac{1}{2\pi}\left\{ \frac{k\sigma_o^2}{\varepsilon_o\sqrt{\rho\eta}} [\coth ka + \frac{\varepsilon}{\varepsilon_o} \coth kc] \right\}^{2/3} \quad (3.14)$$

In writing this expression, it has been assumed that the liquid is perfectly insulating. Thus, the surface mobility and bulk electrical conductivity have both been assumed to be zero. It is this mode that the experiment was designed to probe, for it would exist even in the limit where gravity was so strong as to maintain a flat interface.

The physical picture that goes with the resonance at frequency f_i is one in which charge trapped on the interface undergoes surface dilatations. Compression of surface charge results in an electrostatic repulsion in the interfacial plane. Thus, the effect of the charge monolayer is similar to that of a monomolecular film [6]. It is the potential electric energy stored in the surface dilatations of the interface that interplays with the kinetic energy of the liquid below to give rise to the resonance. Viscous stresses are the only mechanism for the interface to couple to the inertia of the liquid, so the resonance is intrinsically one involving dissipation and hence not the sharp resonance associated with the transverse motions of the interface. A traveling-wave approach to identifying these modes has been described elsewhere [7]. The standing-wave technique reported here is an improvement on that approach.

3.3 Experimental Demonstration of Interfacial Dilatational Trapped-Charge Waves

Measured responses are the data points also shown in Fig. 3.4. Physical parameters for these experiments are given in the figure caption. The predicted downward frequency shift of the gravity-capillary field-coupled mode and the upward frequency shift of the trapped-charge mode are clearly observed. Of course, as the frequencies of these modes approach each other, it is not possible to speak of one mode or the other.

If the surface mobility is taken as being typical of bulk ionic mobilities in hydrocarbons having the same viscosity as hexane [8], the predicted response curves do not differ appreciably from those shown. Thus, it is not surprising that the surface charge behaves as though it is trapped in the interface. The electrohydrodynamic mechanism of charge transport characterized by f_i tends to render the interface an equipotential. Thus, at least under incipient conditions, interfacial instability appears to be very similar to what would be expected of a "perfect" conductor rather than a highly insulating fluid [9].

What has been described is an approach to sorting out physical processes that occur at the interfaces of highly insulating liquids. It is clear that there remains the need for a systematic study of interfaces using this technique. In the experiments described here, surface charge could have been immobilized by surface contaminants and failure to obtain a more refined agreement between theory and experiments can be traced to difficulties in maintaining a steady equilibrium. Relatively little is known about the macroscopic characterization of charged particles in free insulating interfaces, and experiments should be carried to the point where failures to predict observed responses can be used to refine the basic interfacial model.

3.4 Instabilities Used to Print

The salient aspect of the dynamics resulting when an electric field is applied to a layer of liquid is instability. Thus, the dispersion equation, $D(\omega,k)=0$, (defined as the denominator of Eq. 3.12) has been used to define conditions for incipient instability and growth-rates of instability, at least for layers of "infinite" depth [9]. Typically, in such configurations the wavelength of instability is on the order of the Taylor wavelength [2,10].

By contrast with the Rayleigh-Taylor instabilities driven by gravity, electrohydrodynamic instabilities usually culminate in sharply pointed "processes" that transport fluid into the region of high field intensity. These can terminate in continuous jets, in drops, in local breakdown and ion generation or in combinations of these [11,12]. Make the liquid layer ink and replace the upper electrode with an image in the form of surface charge and it is natural to ask if the instabilities could be the basis for printing. Practical image resolution requires instabilities typified by wavelengths that are at most a fraction of a millimeter. With the Taylor wavelength typical of that for instability as the field strength E_o is raised, it seems unlikely that the required resolution could be achieved.

Nevertheless, commercial developments have demonstrated practical printing resolution using the electrohydrodynamic instability of both water-based (highly conducting) and hydrocarbon-based (relatively insulating) inks [13,14]. Key to this success has been the use of films that are of a thickness on the order of the desired spatial resolution. The thickness of the film has little to do with the conditions for incipient instability, but with the field raised beyond this condition, it has a strong influence on the wavelength for maximum rate of growth.

To understand the physical processes contributing to the printing, the dispersion equation can be normalized in terms of the characteristic times defined in Table 3.1. For hydrocarbon-based inks, the ordering of these times is typically as shown in Fig. 3.5. In water-based inks, τ_e is much shorter than all of the other times. Times of interest are longer than the electro-viscous time $\tau_{E\eta}$ but much shorter than the gravity-viscous time $\tau_{g\eta}$. If the viscosity were somewhat higher, the viscous diffusion time τ_v would be much shorter than all of the others, and a creep-flow limit would be appropriate. This is the approximation used in generating the normalized growth-rate curves of Figs. 3.6 and 3.7 [15]. Here, the growth rate is normalized to the electroviscous time $\tau_{E\eta}$ while the wavenumber is normalized to the thickness of the liquid layer. Dimensionless parameters are summarized in Table 3.2. For field strengths less than breakdown in air, these curves demonstrate a growth-rate and resolution commensurate with the observed printing. Note that because of the charge-trapping dynamics, there is little difference between the curves for highly conducting and highly insulating interfaces.

In the printing process it could be argued that the charge representing the image simply pulls portions of the liquid upward. Is an instability actually involved? A high speed photograph of the gap is not particularly helpful in

TABLE 3.1. Characteristic Times

$\tau_{\gamma\eta}$	=	$c\eta/\gamma$	Surface tension acting against viscous drag
$\tau_{E\eta}$	=	$\eta/\varepsilon E^2$	Electric stress against viscous drag
$\tau_{g\eta}$	=	$\eta/\rho g c$	Gravity against viscous drag
τ_e	=	ε/σ	Surface charge redistribution through conduction
τ_v	=	$\rho c^2/\eta$	Viscous diffusion time

TABLE 3.2. Characteristic Numbers

$\underline{\rho}$	=	$\frac{\rho g c^2}{\gamma} = \tau\gamma\eta/\tau_{g\eta}$	"normalized mass density"
r	=	$\tau_e/\tau_{\gamma\eta}$	"normalized charge relaxation time"
$j\underline{\omega}$	=	$j\omega\tau_{\gamma\eta} = \underline{s}$	"normalized growth constant"
$\underline{u}$	=	$\tau_{\gamma\eta}/\tau_{E\eta}$	"normalized square of bias field"
$\underline{\lambda}$	=	$\lambda/c=2\pi/kc=2\pi/\underline{k}$	"normalized wavelength"
R_V	=	$\tau_V/\tau_{\gamma\eta}=\rho c\gamma/\eta^2$	"normalized viscous diffusion time"
$\underline{a}$	=	a/c	"normalized air-gap spacing"

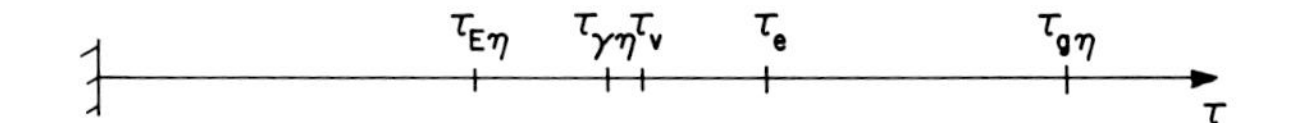

Fig. 3.5. Hierarchy of Characteristic Times for Instability Used in Printing Process.

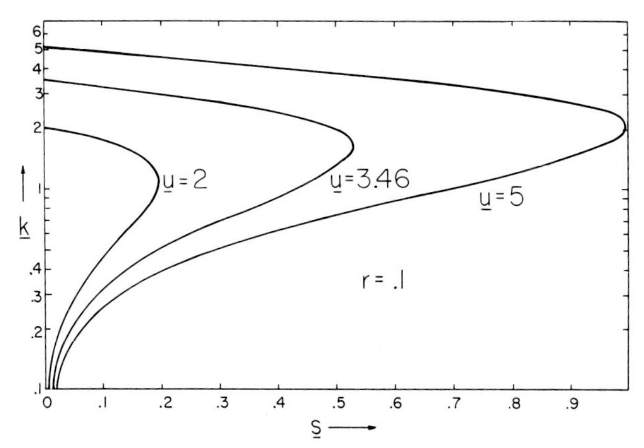

Fig. 3.6. Normalized wavenumber as a Function of Normalized Growth-Rate with Normalized Square of Applied Field as a Parameter. The Fluid is Relatively Conducting.

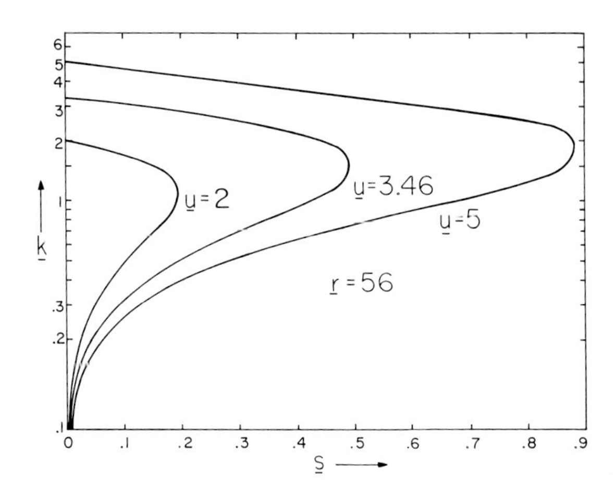

Fig. 3.7. Normalized Wavenumber as a Function of Normalized Growth-rate with Normalized Square of Applied Field as a Parameter. The Fluid is Relatively Insulating.

this regard. Figure 3.8 shows the development of a "pseudopod" as it leads to the transport of ink across a 1 mil air-gap. A more satisfying answer is given by printing dots that are larger in diameter than one of the pseudopods. A magnified view of the printed image, Fig. 3.9, shows that the interface has responded to a row of continuous dots with several processes within each dot. Thus, there seems to be a natural wavelength which determines the spatial resolution and it is this length that should be on the order of the wavelength for maximum rate of growth. Typical printing is as shown in Fig. 3.10.

It would be expected that effects of fluid inertia not included in the evaluation of growth-rates summarized by Figs. 3.6 and 3.7 would have some influence on both the growth rates and the wavenumber for maximum growth-rate. In the printing system, the ink only comes into the vicinity of the charge-image as the two approach each other on the surfaces of almost touching rollers. Ignored here is the process by which the equilibrium charges relax to the interface through the hydrocarbon-based inks as they approach the region where the ink faces the image across the air-gap.

4. TRAVELING-WAVE INDUCED INTERFACIAL PUMPING

In Sec. 3, the "imposed ω-k" technique that is exemplified involves the electrical measurement of the sinusoidal steady-state electromechanical response to a drive that takes the form of a standing wave. In this section the "imposed ω-k" excitation takes the form of a traveling wave

$$\Phi^d = \text{Re}\,\hat{V}\,e^{j(\omega t-ky)} \tag{4.1}$$

Otherwise, the configuration is again the liquid layer shown in cross-section by Fig. 3.2. However, now the response is the steady streaming of the liquid. This results from space-average electrical stresses acting at the interface.

Streaming of the layer in the y direction is pictured as resulting from an electrical surface force density acting in the y direction on a control volume that encloses the interface and has a wavelength in the y direction, as shown in Fig. 4.1. Because the fields and hence the interfacial

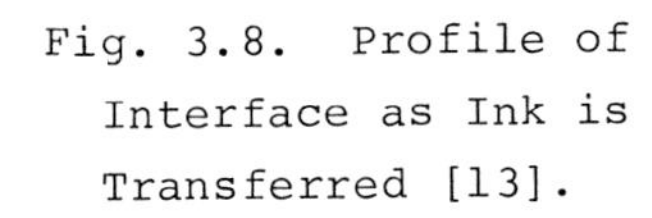

Fig. 3.8. Profile of Interface as Ink is Transferred [13].

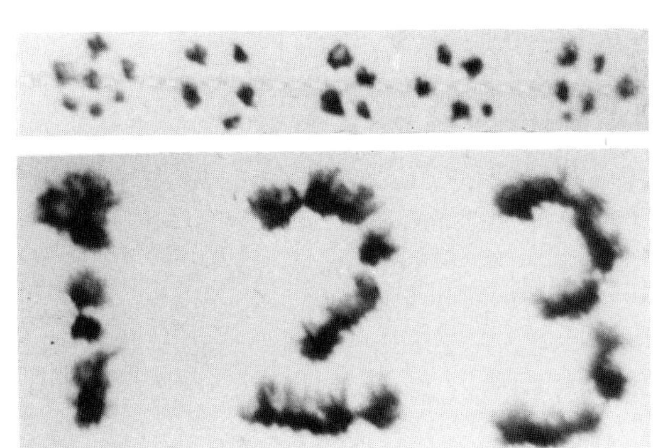

Fig. 3.9. Large Dots are Printed as Four or Five Dots, Giving Evidence that Printing Process Involves Instability [13].

Fig. 3.10. Typical Letters Printed by Means of Instabilities [13].

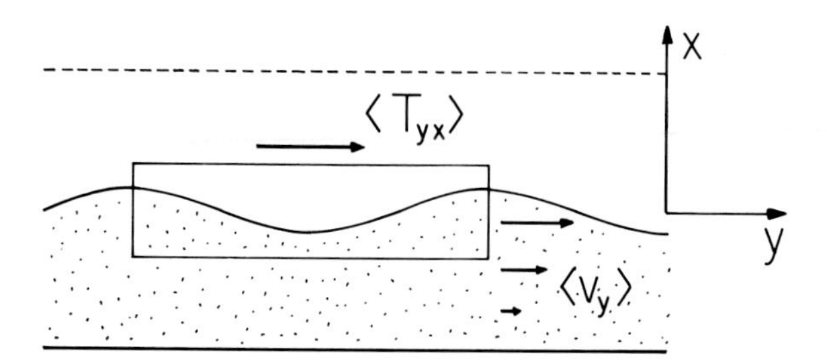

Fig. 4.1. Control Volume Used to Determine Space-Average Shearing Surface Force Density Acting on Inter-Face Subjected to Traveling-Wave Excitation.

response are spatially periodic, contributions to the integration of the stress over the surface of this control volume from the surfaces having y directed normals cancel.

$$\langle T_y\rangle_y = \langle \varepsilon_o E_x E_y \rangle_y = \tfrac{1}{2}\varepsilon_o \mathrm{Re} \; (\hat{e}_x)^* \hat{e}_y = \tfrac{1}{2}\varepsilon_o \mathrm{Re} \; (\hat{e}_x)^* jk\hat{\Phi} \quad (4.2)$$

If Eqs. 3.4a (with $\hat{\Phi}^d=\hat{V}$) and 3.5b (with $\hat{\Phi}^g=0$) are substituted into this expression, it follows that to first order in perturbation amplitudes the space (and time) average surface force density is simply

$$\langle T_y\rangle_y = -\frac{\varepsilon_o k^2}{\sinh ka} \mathrm{Re}\, j\, \hat{V}\hat{\Phi}^e \quad (4.3)$$

Previous observations of streaming induced by this type of interaction have depended on following the convection of particles entrained in the fluid. The pendulum apparatus of Fig. 4.2 makes it possible to measure the momentum transferred to the liquid by the interaction with the interface without observing the streaming.

The liquid is again contained by a circular-cylindrical reservoir milled from a metal disk. The disk is suspended from a wire. Thus, the liquid (with its interface flush with the upper surface of the disk) and the reservoir are the inertial element of a torsion pendulum. Damping is provided, as shown in the figure, so that torsional excursions of the disk can be conveniently observed by means of a telescope.

The structure used to impose the traveling wave of electric field is suspended just above the interface. Radial

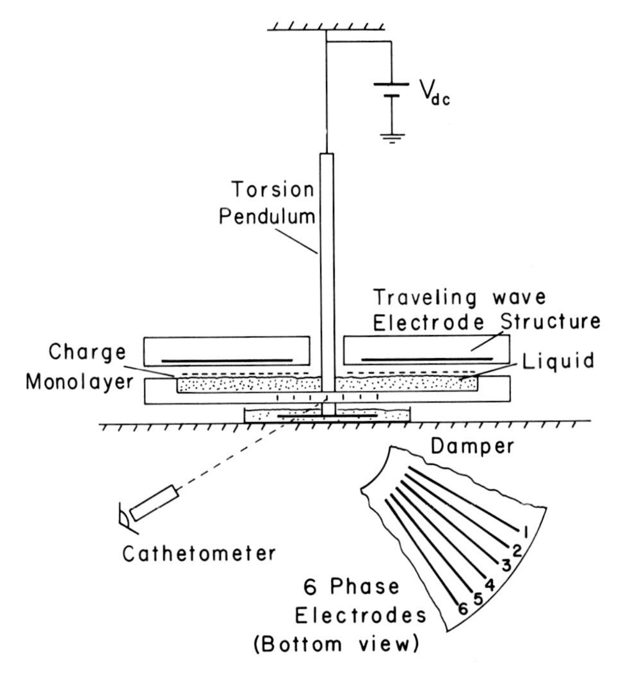

Fig. 4.2. Cross-section of Traveling-wave Excited Pendulum System for measuring Stresses Induced by Traveling Wave of Electric Field.

electrodes are driven by a six-phase variable-frequency source. The space-average electric surface force density results in a steady azimuthal convection of the liquid. The viscous shear stresses on the bottom of the reservoir pass the resulting torque to the pendulum. Thus, given the torsion spring constant of the pendulum and geometric factors, the steady rotational displacement of the pendulum can be used to infer the shearing surface force density acting at the liquid interface.

The streaming induced by this surface force density and even represented by Eq. 4.3 can have diverse origins. For example, even in the absence of an equilibrium interfacial charge density (with no dc voltage applied to the reservoir) in fluids that can be modeled by a bulk conductivity σ, charges relaxing to the interface with the characteristic time ε/σ find themselves spatially lagging their images on the electrodes used to produce the traveling wave, as shown in Fig. 4.3a. The resulting streaming of the fluid can be predicted by a theory that ignores all mechanical responses except the steady streaming [16, 17, 18]. Formally, this is done by setting the normal and shear perturbation velocities in Eq. 3.11 to zero and using only the first two of the four equations, those representing tangential field continuity and interfacial charge conservation. With $b_s=0$ in the theory, measurements made using this apparatus corroborate those many measurements previously made using tracer particles. Typically, particle migration velocities in semi-insulating liquids are much less than the fluid velocities of interest.

The trapped charge surface waves shown schematically by Fig. 4.3b can also be the basis for inducing streaming. In this case, bulk conduction is negligible. The steady convection resulting when both the traveling-wave frequency and the frequency f_i of the trapped mode resonance are well-removed

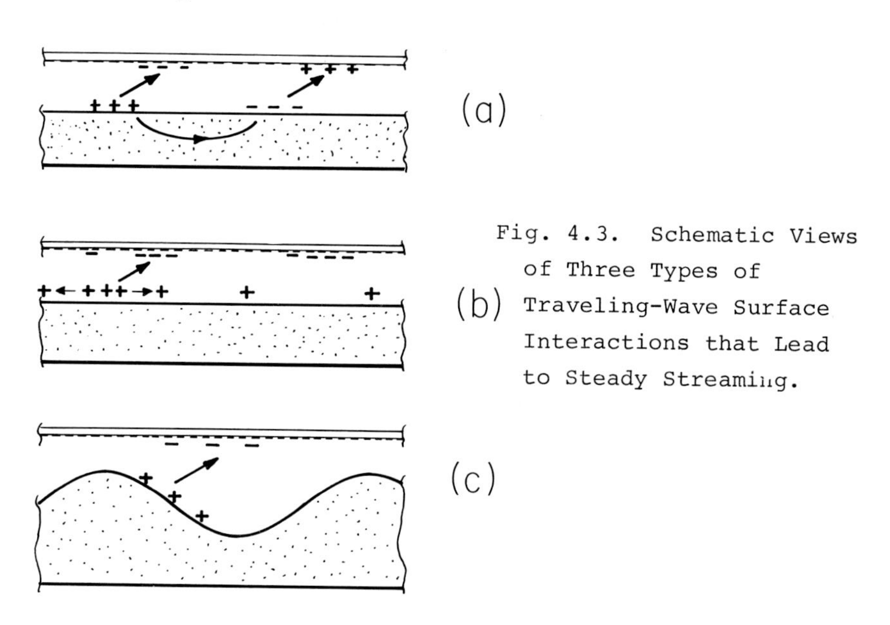

Fig. 4.3. Schematic Views of Three Types of Traveling-Wave Surface Interactions that Lead to Steady Streaming.

from that for the field coupled gravity-capillary wave has been discussed elsewhere [5]. The apparatus of Fig. 4.2 has been used to corroborate certain aspects of what is predicted. An example is shown by Fig. 4.4, where the dependence of the pendulum angular deflection proves to have the square-law dependence on the surface charge density predicted by Eq. 4.3.

Fig. 4.4 Pendulum Deflection as a function of Variable Proportional to Equilibrium Surface Charge. Interaction is with Monolayer of Charge on an Insulating Liquid.

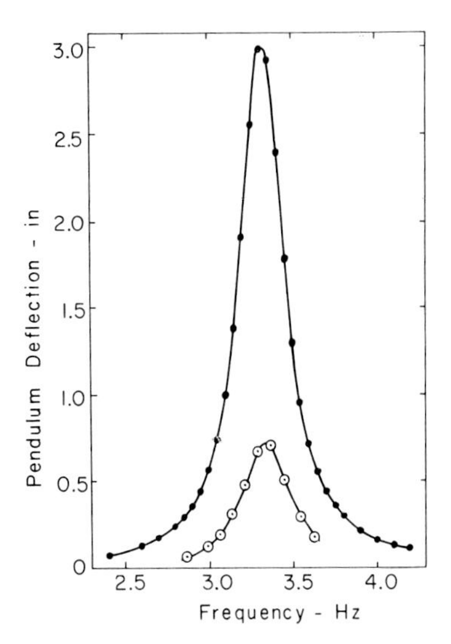

For these experiments, the fluid was transformer oil and a 60Hz traveling-wave was used. For the obtainable surface charge densities, the frequency f_i is much less than the applied frequency. Thus, the space-average electric stress should be theoretically a quadratic function of V_{dc} if the dynamics is dominated by electrohydrodynamics, but should be a linear function of V_{dc} if dominated by interfacial charge migration. The observed dependence of the angular deflection on V_{dc} is found to be clearly quadratic, so again the transport of charge at the interface is dominated by the electrohydrodynamics.

The configuration shown schematically by Fig. 4.3c takes yet another extreme. Here, the fluid is highly conducting water. Thus, each section of the interface is locally free of shear stress. The space-average shearing surface force density now results because of the saliency of the interface. The traveling wave interacts in a synchronous fashion with the field-coupled gravity-capillary wave. Deflections measured in the pendulum apparatus in this case are compared to the theoretical predictions in Fig. 4.5 [18]. The qualitative agreement suggests that the model identifies the correct physics, but also makes it clear that there is a need to refine the model. The failure to predict the magnitude of the stress might be attributed to using too large an

Fig. 4.5. Pendulum Deflection as a Function of Excitation Frequency. Fluid is Highly Conducting So Interaction is with Gravity-capillary Wave.

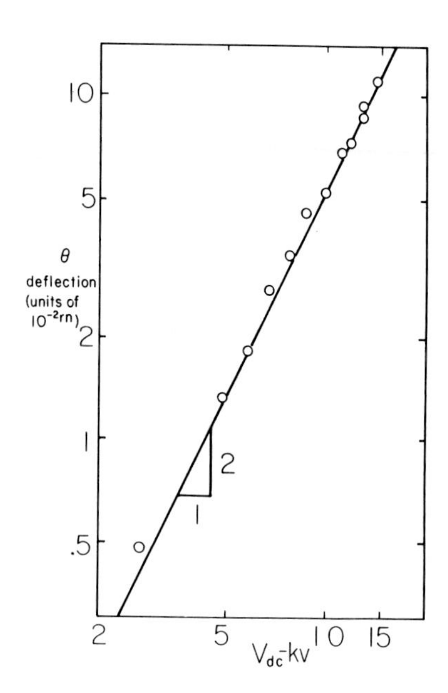

excitation. However, measurements of the deflection as a function of driving voltage show the expected quadratic dependence over a wide range of voltages.

5. INTERNAL ELECTROHYDRODYNAMICS OF INTERFACES

5.1 Non-equilibrium Dynamics of Weak Double-Layers

Interfaces are often viewed as being only a few molecules thick. If charged particles are involved, the Debye length is one characteristic of the interfacial thickness that can be considerably larger. This is the length over which diffusion can counteract the tendency of self-fields to make oppositely charged particles neutralize each other.

In highly insulating liquids, the Debye length can be on the order of micrometers. In this case, the electric field within the double layer, typically the thermal voltage of 25 mv divided by the thickness of the layer, becomes weak enough that it can be dominated by fields applied externally. It seems clear that the usual thermodynamic picture of the double layer must be replaced by one that recognizes not only electrohydrodynamic phenomena, but the possibility that the system is not in chemical equilibrium as well.

5.2 Standing-Wave Sensor as a Probe of Weak Double-Layers

The cross-section of what might be termed an ac electrokinetic flow-meter for highly insulating liquids is shown in Fig. 5.1. An insulating conduit of circular cross-section is shown with eight conductors wound in helical fashion on its periphery. Four adjacent conductors are driven by a sinusoidal voltage. The other four are essentially at zero potential. Thus, with the voltage distribution shifted by a uniform potential equal to half of the applied potential, the winding subjects the fluid to a standing-wave of potential.

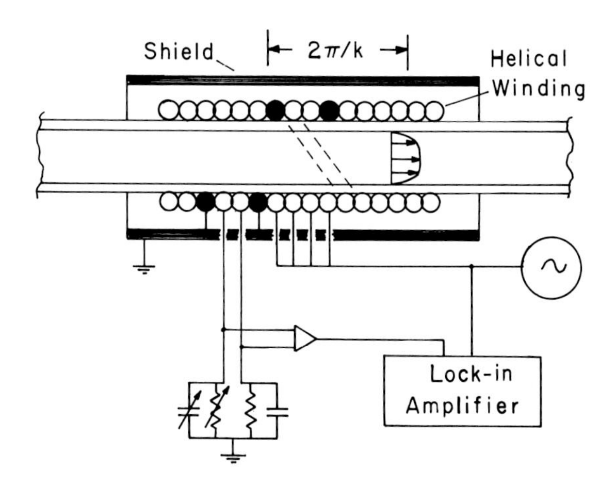

Fig. 5.1. Schematic Cross-sectional View of Standing-wave Sensor for Study of Charges Induced at Interface Between Insulating Solid and Slightly Conducting Liquid.

Of the four essentially zero potential conductors, the outside ones are actually grounded while the center pair are connected to ground through balancing resistors and capacitors. In the absence of flow, the resulting voltage is nulled. Thus, the signal measured reflects the difference in charge induced on the adjacent electrodes because of the interaction of the field with charge in the neighborhood of the liquid-solid boundary that is influenced by the convection.

The scheme is similar to one developed for the non-invasive sensing of the motion and electrical properties of free semi-insulating interfaces [19]. Experiments have demonstrated a measured response to liquid convection that has a frequency dependence similar to that predicted for and found in these earlier interfacial experiments. These experiments will be described in the doctoral thesis of Mr. Steven Gasworth, currently being carried out under the supervision of Professor Markus Zahn and the author.

5.3 Traveling-Wave Pumping as a Probe of Weak Double-Layers

Also, part of the Gasworth research project is an electrical-to-mechanical experiment (first described in a thesis by Ehrlich [20]) that is intended to probe the double-layer, again using the "imposed ω-k" technique. Here, the interface is also pinned down by a solid insulating boundary. The fluid is within an insulating duct, as shown in Fig. 5.2. Six conductors, wrapped around this conduit in a helical fashion, are driven by a variable frequency six-phase source. Thus, the homogeneous liquid is driven by a wave of potential that travels in a somewhat helical but predominantly axial direction.

Measurements have been made of the pressure rise induced by the presumably fully developed flow caused by the pumping. It is found that depending on the frequency and magnitude of the applied traveling wave, the pumping can be forward or backward. Attempts to model the pumping in terms of a binary electrolyte with finite rates of ion generation and recombination explain some features of what is observed [21]. It seems clear that at high field strengths, ion migration dominates diffusion. Complicating the picture is the possibility that ions are injected into the fluid from the walls. Typically, in these experiments an antistatic agent is used to control the species dominant in the conduction phenomena. The experiment will hopefully contribute to an understanding of how these agents behave at liquid-insulating solid interfaces under electric stress.

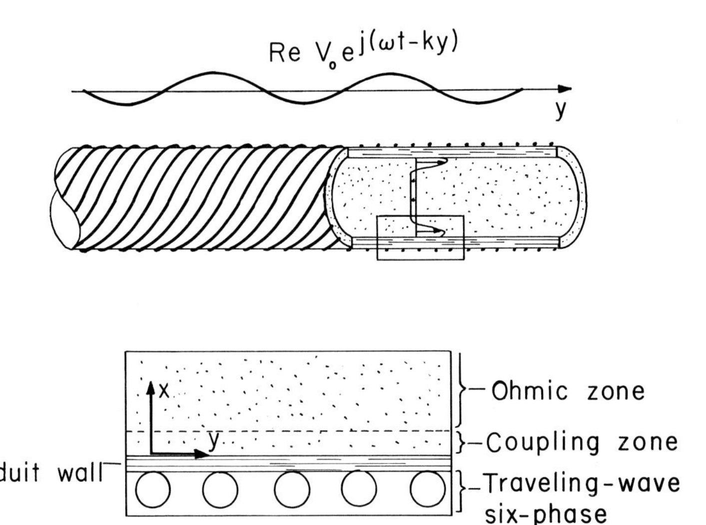

Fig. 5.2. Traveling-wave of Potential Produced by System of Wires Wrapped Around Insulating Conduit Induces Forces and Hence Streaming in the Neighborhood of the Interface Between the Liquid and the Insulating Wall of the Conduit.

With the need for replacing transformer oils with other liquids that can simultaneously serve insulation and heat transfer functions comes an interest in systems using freons and other highly insulating relatively inviscid liquids. In these applications, the transport of charge in insulating conduits has been found to sometimes generate unexpected electrical stresses. The objective of these last two projects is a basic understanding of these transport processes, especially as they involve interfaces between homogeneous liquids and insulating solids.

6. SUMMARY REMARKS

The previous sections are variations on a theme of the electrohydrodynamics of a planar layer of liquid subjected to a dominantly perpendicular electric field. Both as a technique for studying the sinusoidal steady-state response and the steady streaming induced by electrical means, the "imposed ω-k" approach has been shown to be useful in sorting out physical phenomena contributing to the dynamics of interfaces. Trapped-charge surface waves on interfaces of highly insulating liquids have been especially emphasized. The methods exemplified in describing these waves are typical of those applied to a wide range of problems in a recently published book, intended for use both as a text and as a reference [1]. Hopefully, the applications discussed give some hint as to the diverse practical interests that give an engineering dimension to the subject.

REFERENCES

1. Melcher, J. R., Continuum Electromechanics, M.I.T. Press, Cambridge, MA, 1982.

2. Melcher, J. R., Field-Coupled Surface Waves, M.I.T. Press, Cambridge, MA, 1963.

3. Taylor, G. I. and McEwan, A. D., The Stability of a Horizontal Fluid Interface in a Vertical Electric Field, Jrl. Fluid Mechanics, 22 (1965) 1-15.

4. Devitt, E. B. and Melcher, J. R., Surface Electrohydrodynamics with High Frequency Fields, Phys. Fluids, 8 (1965) 1193-1194.

5. Melcher, J. R., Oscillations and Traveling-Wave-Induced Streaming of Charge Monolayers on Insulating Liquid Interfaces, 1976 Annual Report of Conference on Electric Insulators and Dielectric Phenomena, National Academy of Sciences.

6. Zelazo, R. E. and Melcher, J. R., Dynamic Interactions of Monomolecular Films with Imposed Electric Fields, Phys. Fluids, 17 (1974) 61-72.

7. Melcher, J. R., Dynamics of Charge Monolayers on Insulating Liquid Interfaces, Jrl. Fluid Mechanics, 60 (1973), 417-431.

8. Adamczewski, I., Ionization, Conductivity and Breakdown in Dielectric Liquids, Taylor and Francis, London, (1969), 224-225.

9. Melcher, J. R. and Smith, C. V., Electrohydrodynamic Charge Relaxation and Interfacial Perpendicular-Field Instability, Physics of Fluids, 12 (1969), 778-790.

10. Chandrasekhar, S., Hydrodynamic and Hydromagnetic Stability, Dover Publications, Inc. (1981), 441-453.

11. Melcher, J. R. and Warren, E. P., Electrohydrodynamics of a Current-Carrying Semi-Insulating Jet, Jrl. Fluid Mechanics, 47 (1971) 127-143.

12. Hoburg, J. F. and Melcher, J. R., Current-driven, Corona-terminated Water Jets as Sources of Charged Droplets and Audible Noise, IEEE Transactions on Power Apparatus and Systems, PAS-94 (1975) 128-136.

13. Case, C. E., Greenberg, A. and Klavan, I. L., Pseudopod Formation and Gap Jumping Development of Latent Electrostatic Images, Fourth SPSE International Conference on Electrophotography, Washington, D.C., Nov. 16-18, 1981.

14. Klavan, I. L. et al, United States Patents 4,202,620 (1980), 4,202,913 (1980), 4,268,597 (1981).

15. These curves were computed by Mr. K. G. Rhoads [1, Prob. 8.16.3].

16. Melcher, J. R., Traveling-Wave-Induced Electroconvection, Phys. Fluids, 9 (1966), 1548-1555.

17. Melcher, J. R. and Taylor, G. I., Electrohydrodynamics: A Review of the Role of Interfacial Shear Stresses, Chap. in the First Review of Fluid Mechanics, Annual Reviews, Inc., Palo Alto, CA, 111-146 (1969).

18. Hinson, D. P., Synchronous Pumping of a Highly Conducting Fluid by an Electric Field, B. S. Thesis, Dept. of Elec. Eng. and Comp. Science, M.I.T., Cambridge, MA (1976).

19. Melcher, J. R., Charge Relaxation on a Moving Liquid Interface, Phys. Fluids, 10 (1967), 325-332.

20. Ehrlich, R. M., Non-equilibrium Electrokinetic Pumping of a Semi-Insulating Liquid, M. S. Thesis, Dept. of Elec. Eng. and Comp. Sci., M.I.T., Cambridge, MA (1979).

21. Ehrlich, R. M. and Melcher, J. R., Bipolar Model for Traveling-Wave Induced Non-Equilibrium Double-Layer Streaming in Insulating Liquids, Phys. Fluids, 25 (1982) in publication.

Department of Electrical Engineering
Massachusetts Institute of Technology
Cambridge, MA 02139

SOLITARY WAVES ON DENSITY INTERFACES

T. Maxworthy

1. INTRODUCTION.

In many photographs of the sea surface taken from aircraft or satellites using passive or active sensors the surface signatures of internal waves can be clearly seen. In figure 1 we show one such photograph of the Gulf of California taken from the Seasat SAR archives at J.P.L. These waves are visible because the particle motions they create as they propagate produce a straining field at the surface

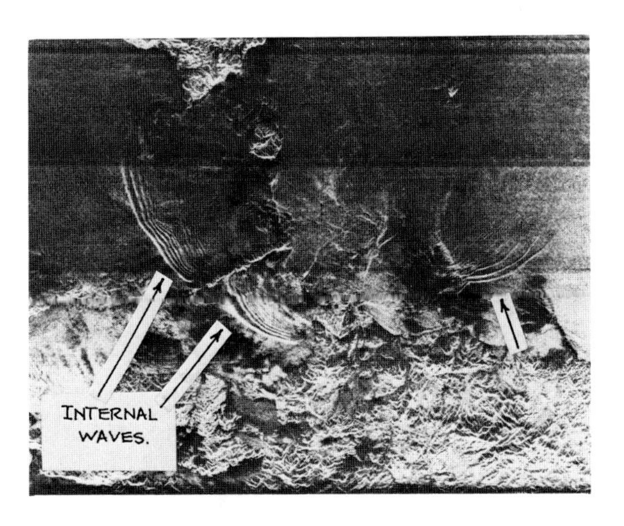

Fig. 1. Internal wave signatures on the sea surface in the Gulf of California. Radar image provided by J.P.L.

ISBN 0-12-493220-7

which modifies the reflective properties of the surface-wave field. It is now well established that such internal waves are most often formed by the interaction of the flow created by the barotropic tide with bottom topography, Lee and Beardsley (1974), Maxworthy (1979), Apel and Holbrook (1980), Farmer and Smith (1980), etc. In what follows we discuss this generation mechanism in some detail and extend our previous understanding using the more recent experiment and theoretical results of Lansing (1980) and Lansing and Maxworthy (1983). Once these waves have been formed they propagate and evolve into a sequence of finite amplitude, solitary waves, ordered by amplitude, which then slowly separates in space. Such a property is a general one for systems which can support solitary waves and has been discussed for the present case by Maxworthy (1980), Liu (1981) among many others. However, the effect of the earth's rotation on these waves has not been considered up to now and in our final section we discuss the extension of these ideas to the generation and propagation of solitary, Kelvin waves in a laboratory tank and their application to observations in nature (Maxworthy 1983).

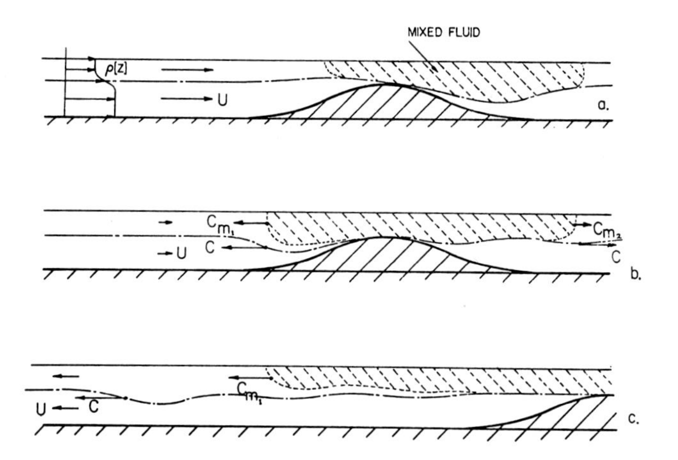

Fig. 2. Generation of internal waves by tidal flow over bottom topography. A quasi-steady lee wave forms at high enough Froude number and mixing can occur. As the tidal flow is reduced wave evolution and mixed region collapse generate internal waves.

2. TIDAL GENERATION OF INTERNAL SOLITARY WAVES:

a) The Physical Model:

In figure 2 we show a diagrammatic view of the process of internal wave generation proposed in Maxworthy (1979), representing a modification of the original work of Lee and Beardsley (1974). It shows the two major effects operating, lee wave formation and collapse of the fluid region mixed by wave breaking and shear flow instability. The latter may or may not be important depending on the values of the relevant independent variables. In summary, in figure 2a as the stratified fluid is transported over the bottom topography by a tidal flow, which is increasing in velocity, it reaches a critical state at which a lee-wave depression begins to form, as the flow continues this depression grows in amplitude and under some circumstances one or more waves may form behind it. If the wave amplitudes become large enough or if there is sufficient shear across the density interface mixing can occur which produces a region of intermediate density. As the tidal flow slows the lee waves propagate upstream, to the left in our diagram, figure 2b, and evolve into a sequence of solitary waves, figure 2c. This process may be enhanced by the collapse of the region of mixed fluid which can also produce weak waves propagating to the right, as in figure 2b.

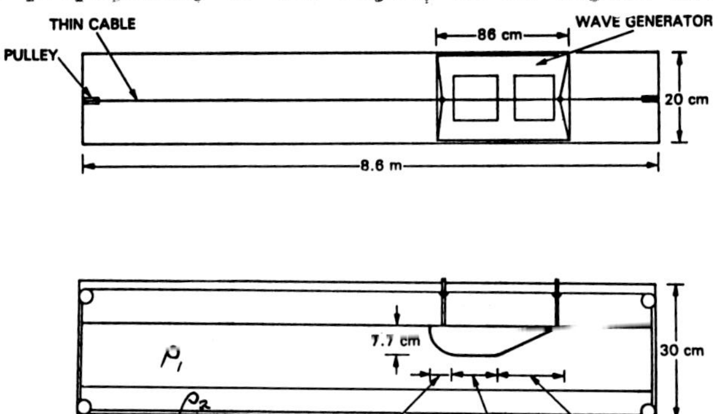

Fig. 3. Apparatus. For experiment convenience the geometry of the oceanic prototype has been inverted. The topographic feature floats in the upper surface while the equivalent of the mixed layer rests on the bottom of the tank. Tidal flow is simulated by towing the obstacle back and forth using a simple crank-connecting rod arrangement.

The general validity of these ideas was confirmed by a series of experiments reported in Maxworthy (1979) for flow over a three-dimensional ridge. More recently Lansing (1981) and Lansing and Maxworthy (1983) have extended these results and, in addition, have produced a useful numerical model which can be used to reproduce the major characteristics of these flows under a variety of different circumstances.

b) The Experimental Model:

The experimental arrangement is shown in figure 3. For experimental convenience the geometry of the oceanic prototype has been inverted so that the topography floats at the surface of the upper fluid and is towed back and forth, using a crank-connecting rod arrangement to simulate the tidal motion. The fluid is stratified by running a layer of heavier fluid beneath the upper water layer. The geometrical factors were varied mainly by changing only the "tidal"

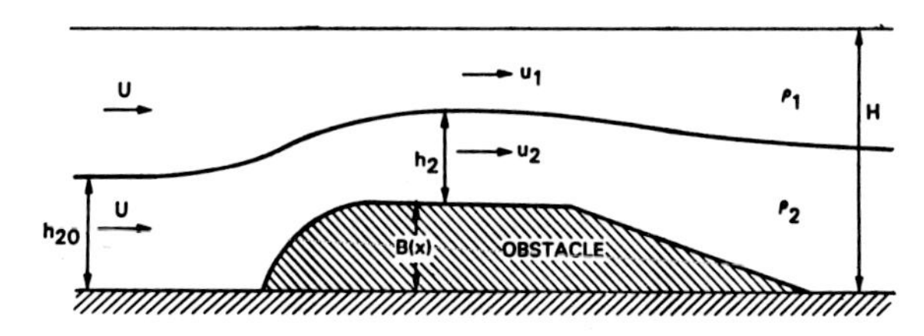

Fig. 4. Hydrostatic, steady model of two layer flow over an obstacle.

period and not its excursion. Distortions of the interface were recorded photographically at known time intervals after the start of a tidal cycle and these results will be shown after presentation of the theoretical procedures, at which stage they can be compared quantitatively.

c) The Quasi-Steady Theoretical Model:

To model the generation mechanism of the interfacial waves theoretically, we started by using a modification of the steady model developed by Long (1954). In his study of the motion of the steady flow of a two fluid system over an obstacle, as shown in Figure 4, the assumptions of a very small density difference and a hydrostatic pressure distribution were used. We assumed that this steady model was valid at each instant of time in our unsteady flow.

Introducing the following parameters, $H' = \frac{(H-h_{20})h_{20}}{H}$ the equivalent depth of the flow; $Fr^2(t) = \frac{\rho U^2(t)}{(\rho_2-\rho_1)gH'}$, the

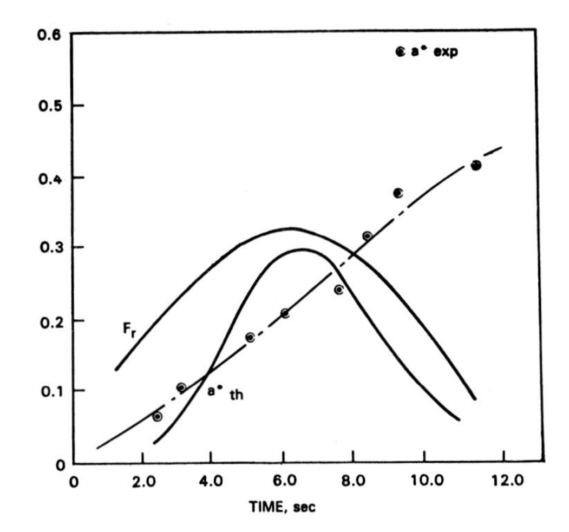

Fig. 5. Comparison of the experimentally determined maximum amplitude, $a^*_{exp.}$, and the quasi-steady value as a function of time for R = 0.9 and maximum $F_r = 0.33$. Curve of Froude Numer as a function of time is also plotted.

Froude number; $h^*(x) = \frac{H-B(x)}{H'}$ dimensionless depth of the obstacle at any location; $a^* = \frac{h_{20}-h_2(x)-B(x)}{H-h_{20}}$ the dimensionless amplitude of the disturbance at any location (x) and finally, $R = \frac{h_{20}}{H}$ the ratio of the two fluid layers, into the continuity and energy equation, yield upon some manipulation.

$$h^* = \frac{(1+a^*)}{R}\left[1 \pm \frac{R}{\left\{1 + \frac{2a^*(1+a^*)^2}{RF_r^2}\right\}^{\frac{1}{2}}}\right]$$

To compare the results obtained from this steady model with the experimental results we assumed that this solution could be applied at each instant of time, i.e. at each instantaneous value of Froude number, and at this instant the maximum amplitude was found. In figure 5 we present one case just to demonstrate the differences between the quasi-steady solution and the experimental results, as a function of time. During the formation of the depression it seems that the experimental amplitude increases almost linearly with time, while the theoretical amplitude must take a shape that reflects the variations of Froude number. From this and other cases not presented in detail here it is clear that while the theory might agree with the experiments for small times or for small Froude numbers it can never reproduce the continuously increasing amplitude that was found even when the Froude number was decreasing.

In order to improve the theoretical model we adopted the vortex-point method to solve the problem of both the generation and the evolution of interfacial waves numerically.

d) The Theoretical Model based on the Vortex-point Representation of the Fluid Interface:

Rosenhead (1931) developed the vortex point method in order to study Kelvin-Helmholtz instability. The motion was two-dimensional and the common interface was a vortex sheet across which the tangential velocity jumped from one value to another. In this method the vortex sheet was then replaced by a distribution of elemental vortices along its path. The undisturbed surface was given an initial disturbance and thereafter the paths of each vortex was determined by a numerical step-by-step method based on the velocity field due to every other vortex. The line joining these vortices, at any time, was assumed to be an approximation to the actual shape of the surface of discontinuity.

Zarodny and Greenberg (1973), used the vortex-point method to describe large amplitude but nonbreaking, two-dimensional, inviscid, incompressible, irrotational water waves. In their model, both the free surface, and the flat bottom boundary were represented by vortex sheets. The governing integro-differential equations of motion were in Lagrangian form and were integrated at finite steps using the Runge-Kutta-Merson technique which required five time derivatives at each step. The velocities of the vortex sheet and their complex images were calculated using the Biot-Savart integral.

More recently, Baker, et al (1980), have used this method to treat the problem of the growth of Taylor-Rayleigh instability to large amplitude. The governing equations of motion were solved iteratively. The time stepping was done using the implicit fourth-order Adams-Moulton method that requires only two time derivatives at each time step. The numerical results were in agreement with the results obtained by a conformal-mapping method and in qualitative agreement with experiments.

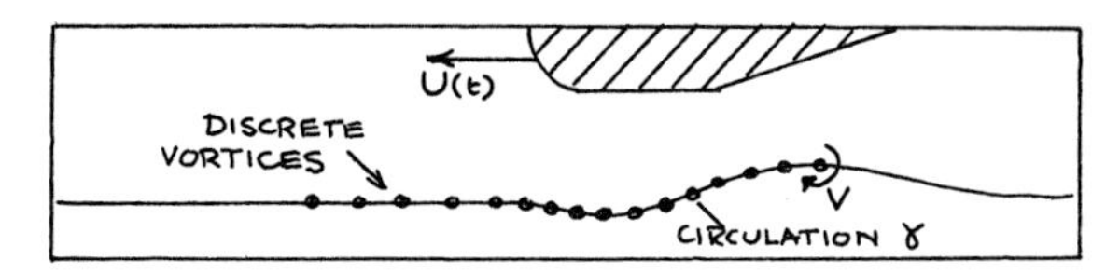

Fig. 6. Numerical model in which the interface is replaced by a series of vortices. Flow over the moving obstacle is represented by a distribution of sources.

The present flow is modelled by assuming it to be two dimensional, inviscid and incompressible while vorticity is generated at the interface due to the existence of a tilted surface across which there is a density jump and shear.

The density interface between the two fluid layers can then be represented by a vortex sheet, as shown in Figure 6. The normal velocity components are continuous across this curve, while there is a jump in tangential velocity components. This jump is the strength of the vortex sheet. The velocity of the sheet, defined as the average fluid velocity of a given point, is given by the Biot-Savart integral as a function of the vortex strengths (γ) and the relative positions of all of the other vortices. The equation of motion for each layer are used to derive an expression for γ

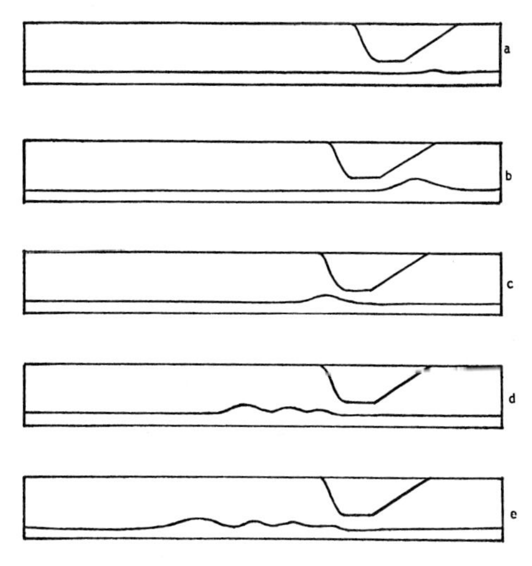

Fig. 7. Theoretical interface shapes for R = 0.772, maximum F_r = 0.672 and one half of a full sinusoidal path of period 12.9 secs. The times for each figure are a) 1.0 secs., b) 4.5 secs., c) 9.0 secs., d) 15.0 secs., e) 17.0 secs.

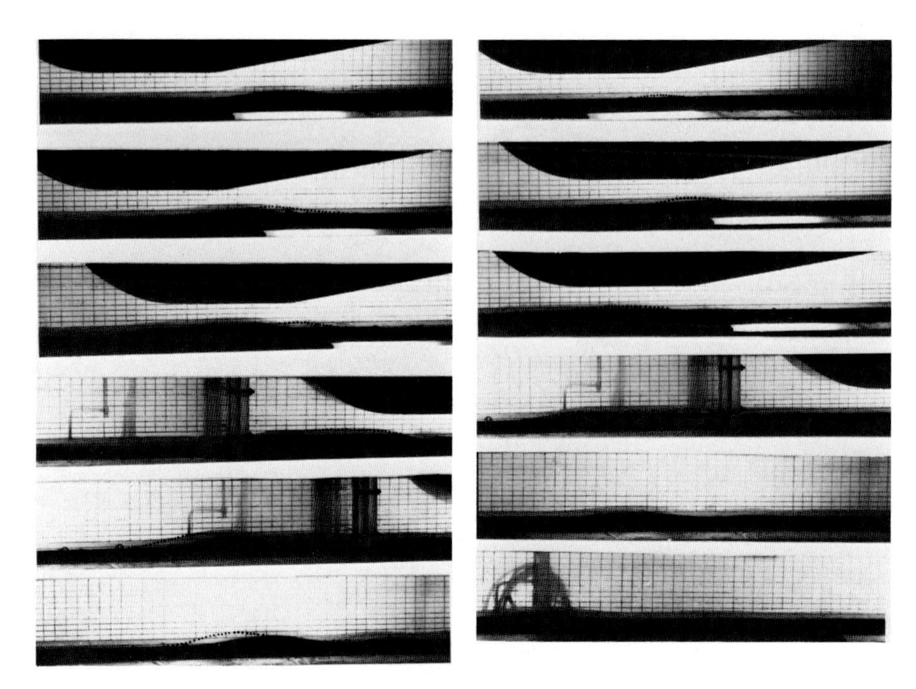

Fig. 8. (left) Experimental results for R = 0.814, maximum F_r = 0.665 and one half of the sinusoidal motion of the obstacle of full period 12.9 secs. From top to bottom, the photographs were taken 4.1, 6.6, 8.8, 10.2, 14.6 and 18.9 secs. from the start of the motion. The dotted lines in this and the figure which follow represent the theoretically determined surface shape as discussed in detail in section 4.4.

Fig. 9. (right) Experimental results for R = 0.814, maximum F_r = 0.327 and one half sinusoidal motion of full period 25.9 secs. Photographs taken 6.3, 8.8, 12.0, 17.4, 23.1 and 29.7 secs. from the start of the motion.

in terms of the slope of the interface and the relative accelerations of the fluid on each side.

The boundaries, including the complex, moving shape of the topography was modelled by a source distribution which like γ had to change at each time step to reproduce the rigid boundaries faithfully. The resulting equations were iterated numerically at each time step and advanced in time using the implicit fourth order Adams-Moulton method. The reader is referred to Lansing (1981) for the complex details that result from the use of this method.

In figure 7 we show results for one such integration where the evolutionary process described earlier in this section can be clearly seen for a case which corresponds to one of the experiments we have performed. The agreement, not shown for this case, is good except that the leading wave that is formed is about one wavewidth ahead of the experimentally observed wave front.

Detailed comparisons are shown in figures 8 and 9 for cases of the worst (figure 8) and best agreement (figure 9) between the numerical solution, shown dotted, and the experiment.

There are several reasons for such a discrepancy, in particular the experimental obstacle motion was not exactly sinusoidal while boundary layer separation from the backside of the obstacle presented an effective time-changing shape to the interface which was not the same as that modelled theoretically.

3. THE PROPOGATION OF INTERNAL, SOLITARY WAVES:

a) Introduction

Much of the original experimental and theoretical work on this subject was performed by Benjamin (1967), Davis and Acrivos (1967), and Maxworthy (1980) for the case of a nonrotating system. In the latter paper in particular

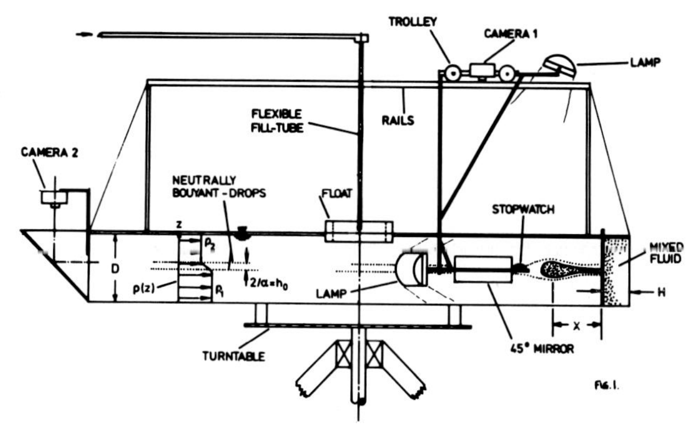

Fig. 10. Apparatus. We attempt to show all of the experimental apparatus and procedure on one drawing. The fill tube and float were removed before the experiment started. The barrier, shown in place here, was removed to create the initial internal gravity which is shown.

the evolution of an initial compact disturbance into an ordered sequence of solitary waves in both a plane and an axisymmetric geometry was studied in some detail.

More recently Maxworthy (1983) has considered a similar problem in which the experimental tank was also rotated, so that any initial disturbance produced solitary Kelvin waves, in which the Coriolis force due to the forward motion of the waves was ultimately balanced by a pressure gradient and hence change in wave amplitude along the wave crest. Elementary calculations show that the Rossby Number ($R_o = \frac{C}{2\Omega\lambda}$, where C is the internal wave speed, Ω the rotation rate and λ the wavelength) is of order one for most internal waves in natural systems; while the internal Rossby Radius of Deformation ($L_c = \frac{C}{2\Omega}$) is in the range of 3 to 5 kms. Since this latter scale is the one that determines the spatial scale of change of wave amplitude along the wave crest it is clear that rotation ultimately will be an important dynamical effect in the propagation of such internal waves, even though the approach to geostrophic adjustment may be quite slow at values of R_o of order one. Most of the interesting examples come from limnological studies e.g. Mortimer (1955), Thorpe (1974), Farmer (1978), Hunkins and Fliegel (1978), Farmer and Smith (1980) etc. where L_c is of the same order as the width (W) of the channel within which the waves propagate. In the experiments which follow we simulate this physical situation and produce results which suggest some interesting, future field measurements.

b) Apparatus:

A section of rectangular, plexiglass channel 360 cms long x 20 cms wide x 30 cms high was mounted on an existing rotating table (Figure 10). Before rotation the lower 15 cms was filled with a salt solution of known density (ρ_1 = 1.04 or 1.08 grm/cm^3). The table was then set into rotation at a known rate and the tank was filled slowly, through a floating diffuser, with fresh water of density ρ_2 = 1.00 grm/cm^3. The float and supply tube were then removed. After one hour, when diffusion had smoothed out all irregularities, two layers of dyed freon-kerosene droplets were introduced to settle at two known levels. The required drop densities were found by noting that the density profile was closely represented by

$$\rho(\eta)=\bar{\rho}(1 - \tilde{\omega} \tanh \alpha\eta)$$

(see Benjamin 1967, Faust 1981 and Maxworthy 1980) where η is the height from the mid-plane of the interface; $\bar{\rho}$ is the mean density $(\frac{\rho_1+\rho_2}{2})$; $\tilde{\omega}$ is a measure of the density difference and equals $(\rho_2-\rho_1)/(\rho_2 + \rho_1)$ and α is the inverse scale height of the density distribution. The drop densities were calculated so that they were in equilibrium at heights $1/\alpha$ from the mid level. Thus the distance between the layers was equal to $h_o = 2/\alpha$, while the distortions of the layers, as the waves propagated, were a convenient and consistent measure of the wave amplitude, which could then be related to the theoretical developments of Benjamin (1967) and Joseph (1977), for example.

The waves themselves were produced as in Maxworthy (1980) by trapping fluid behind a barrier (see Fig. 10) then mixing it up completely. Upon pulling out the barrier and coincidentally starting a stopwatch in the field of view of a recording camera, the mixed fluid collapsed along the mid-plane of the interface.

Because of their three-dimensional structure, these waves had to be viewed from three mutually orthogonal directions. A motorized camera was mounted at one end of the tank so that we could photograph a front view of the wave as it approached the end wall of the tank.

c) Results:

It is useful to consider a typical wave evolution so that the quantitative results can be presented in a logical way later. We consider initially the limit, $L_c \lesssim W$, for which the effects of rotation are most dramatic. Under such circumstances the removal of the barrier created an internal, gravity current, which because of the clockwise rotation, started to move along the left-hand wall of the tank (looking in the direction of motion). This mass of constant density fluid distorted the isopycnal surfaces symmetrically about the mid-plane of the interface, however the thickness of the region of mixed fluid varied across the channel (figures 11a). Significantly the wave front when viewed from above was

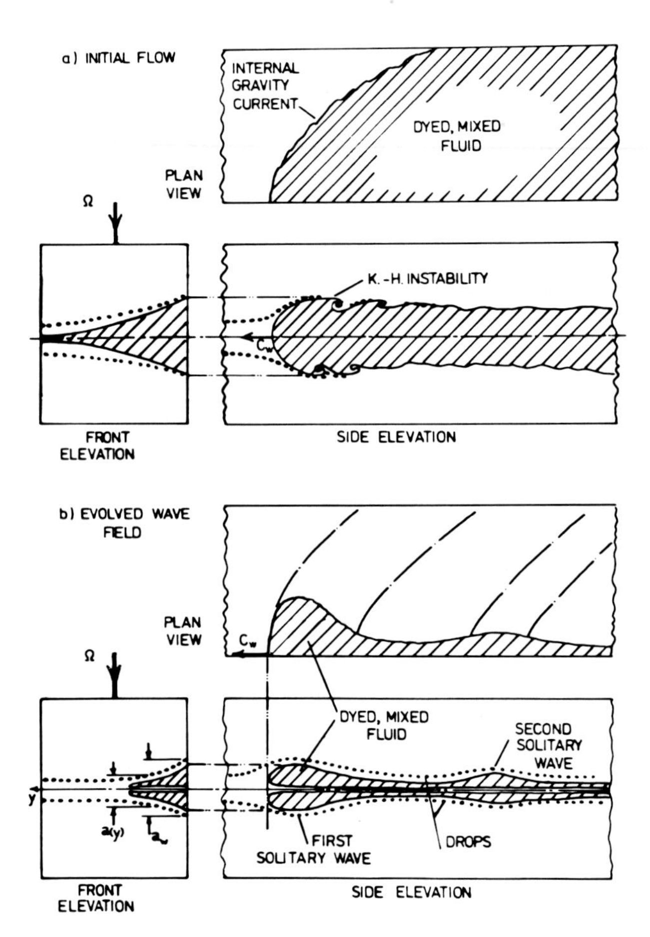

Fig. 11. a) Initial form taken by the internal gravity current shortly after barrier removal. Three views are shown to indicate the wedge shaped, three dimensional nature of this feature. For large amplitudes i.e. large H/h_o the front was turbulent with a Kelvin-Helmholtz (k-H) type of shear flow instability very evident.
b) Evolved wave field. The state shown in (a) has evolved into a sequence of solitary waves, two of which are drawn here. The wave amplitude decreases away from the left-hand wall and "closed stream-lines" only existed part-way across the tank.

inclined backwards while the side view was indistinguishable from the wave forms found in Maxworthy (1980). As the forward wave evolved further more waves began to emerge behind it, although the tank was too short to allow their full evolution. By the time the wave had almost reached the end of the tank it had the form shown in figures 11b.

The wave amplitude decreased as the wave propagated, mainly due to inertial wave drag, and the mixed fluid slowly leaked from the rear of the wave until, eventually, no closed stream surfaces could exist and the wave left the mixed fluid behind completely. In figure 11b we have also tried to indicate, by the chaindotted lines, that the lines of constant phase across the wave were curved backwards as indicated by the distortion of the sheet of marker drops. This curvature has no equivalent in the theory of linear Kelvin Waves since it is a reflection of non-linear effects, in particular, the dependence of wave speed upon wave amplitude, as is discussed in detail later.

In cases where L_c was larger than in the one described above the region of closed streamlines could extend complete across the tank. Even for the lowest rotation rate possible in the present equipment, for which $L_c \approx 210$ cms i.e. 10 times the tank width, the wave front was still noticeably curved.

The qualitative observations just presented can be expanded upon and quantitatively evaluated in several instances. We start by discussing the overall motion of the evolving first wave by measuring and plotting both its displacement (x) and the total wave height at the wall (a_w) as a function time (t). By taking the local slope through four points at three different locations along such x vs t curves, local wave speeds (C_w) were measured, these were made dimensionless by dividing by $(g \frac{(\rho_2 - \rho_1)h_o}{\rho})^{\frac{1}{2}}$. While the dimensionless amplitudes ($A_w = (\frac{a_w}{h}) - 1$), corresponding to the locations at which C_w was measured, were also found. These results are shown on figure 12 and are well represented by the curve

$$\frac{C_w}{(g'h_o)^{\frac{1}{2}}} = \{0.33 \pm 0.22\} \{1 + 0.51 A_w - 0.024 A_w^2\}$$

which is to be compared to Benjamin's (1967) expression for an internal wave in an infinitely deep fluid; in our notation,

$$\frac{C_w}{(g'h_o)^{\frac{1}{2}}} = 0.35 \{1 + 0.3 A_w\}$$

Curves representing both of these expressions are drawn in

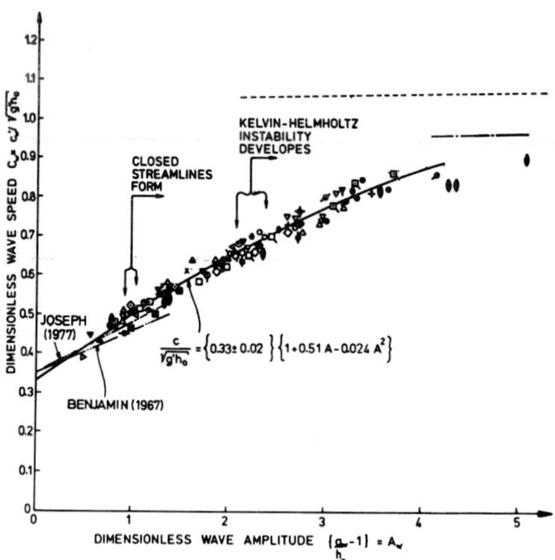

Fig. 12. Non-dimensional local wave velocity $C_w/(g'h_o)^{\frac{1}{2}}$ versus non-dimensional wave amplitude $A_w = \frac{a_w}{h_o} - 1$.

These results represent values taken over a twenty-fold range of L, no dependence on this parameter can be distinguished. —◆— from Faust (1981). Gravity current data from Britter and Simpson (1978)----; and (1981)—.—— . Benjamin (1967), ——..——, $C_w = 0.35(1 + 0.3A_w)$. Joseph (1977), ——...——, $C_w = 0.35\ (1 + 0.3\ A_w(1 + \frac{h_o}{D})$

figure 12. We noted especially that no dependence on L_c can be distinguished even though it varied by a factor of twenty over the whole range of the experiments.

In this figure, for low amplitudes, the difference between our results and those of Benjamin (1967) can be explained to some extent by the finite depth (D) of our wave guide. Using Joseph (1977) we can correct Benjamin's infinite depth result and also show this on figure 12 for a depth ratio (D/h_o) of 15, typical of our experimental range. It is also possible that some of the small scatter of figure 12 is due to the different values of D/h_o used but no consistent trend with this parameter can be determined from the data.

We also found that the wave amplitudes decay very rapidly at the higher rotation rates for the same initial conditions. This we attribute mainly to a wave drag experienced by the internal waves as they propagate. To the homogeneous

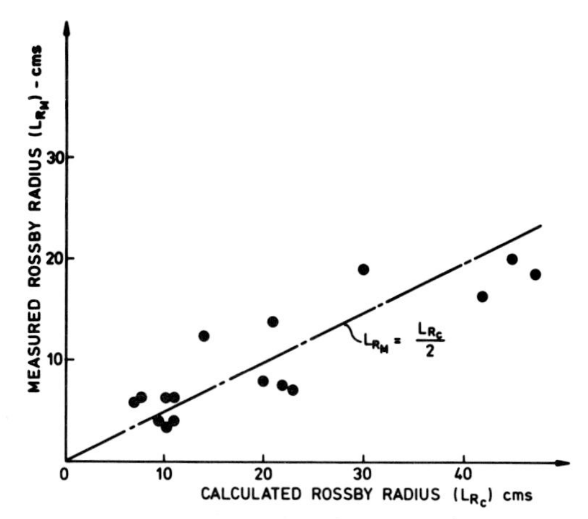

Fig. 13. Rossby Radius of deformation (L_M) measured from photographs of the transverse wave shape versus values calculated (L_C) from $C_W/2$.

fluid above and below the wave guide the internal wave appears as a moving topographic disturbance. This in turn generates inertial waves which drain energy from the internal wave and cause a rapid decay in amplitude. The reader is referred to Maxworthy (1983) for fuller details.

By photographing the wave as it approached the end wall of the tank we have been able to determine the cross-stream, exponentially varying structure of these waves. Figure 13 shows measurements of the Rossby Radius (L_M) determined from these photographs, as the tranverse distance from the wall at which the wave amplitude A(y) is 1/e of the wall value. The accuracy is not high but shows quite clearly and unexpectedly that the measured value (L_M) and the calculated value (L_C) of the Rossby Radius differ by at least a factor of two. The validity of this observation is reinforced in our final quantitative results in which we calculate the shape of the wave front based on rather elementary, but nonetheless satisfying, considerations.

The speed at which the whole wave propagates is determined by the wave amplitude at the wall (A_W). Points away from the wall have a smaller amplitude and hence lower wave speed. In order for them to propagate at the maximum speed the wave must be sloped backwards at an angle θ to the

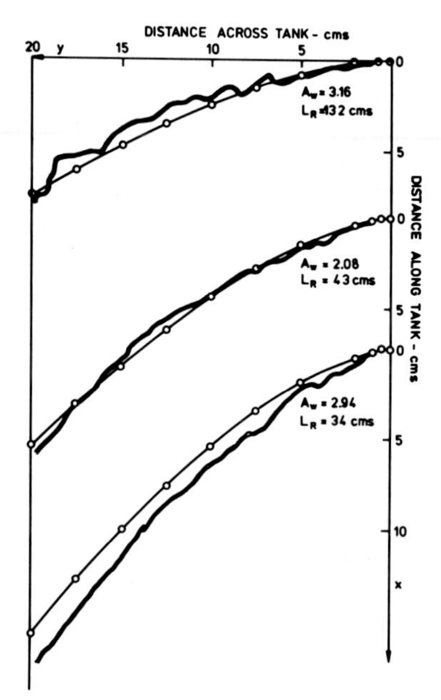

Fig. 14. Calculated versus observed wave shape from plane view of the wave as in figure 2 and 3. The basis for this calculation is shown in figure 4 and explained in the text.

direction of propagation given by Cos $\theta = C/C_w$. We make use of our experimentally determined values of both velocity (figure 12) and Rossby Radius (figure 13) to determine the local wave angle (θ). Then

$$\cos\theta\,(y) = \frac{1 + 0.51\,A_w e^{-2y/L_c} - 0.024\,A_w{}^2 e^{-4y^2/L_c}}{1 + 0.51\,A_w - 0.024\,A_w{}^2}$$

Clearly at $y = 0$, $\theta = 0^o$ and we can use a trivial "shooting" technique to obtain the wave shape knowing the value of θ at each point. The final comparison is then shown in figure 14 where the agreement between the observed and calculated front is seen to be quite satisfactory.

d) Discussion:

Based on available field data one can discuss the wave shapes that might be observable in natural systems. Unfortunately we have been able to find only two examples that appear to indicate the presence of curvature in naturally occurring internal waves.

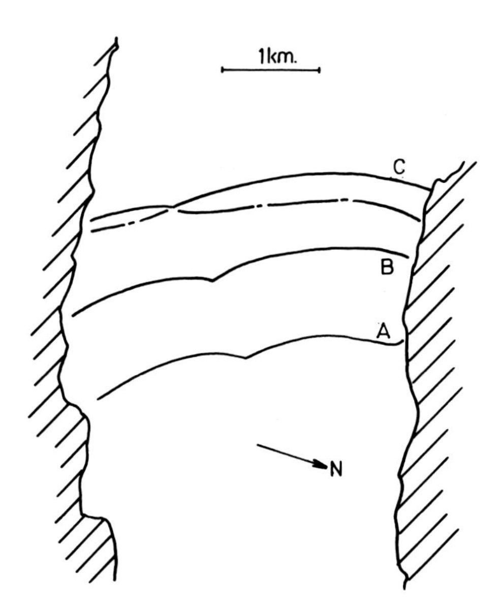

Fig. 15. Sketch of the shape of the first of a group of internal wave propagation westward in Knight Inlet, transferred from photographs supplied by Dr. D. M. Farmer of the Institute of Ocean Sciences, Sidney, B.C. The photographs were taken on August 16, 1977 at 16.11 hrs. 30 sec. for A, 16.27 hrs. 20 sec. for B and 16.41 hrs. 35 sec. for C. In the latter case we have also drawn the second wave, chain-dotted, since it appears to interact with the first wave to the west. The curved shape into which this wave evolves is clearly related to the curved shapes we have observed in our experiments, although exact correspondence is probably confused by changes in wave speed due to changes in fluid depth, shoreline curvature and mean flow variations.

Dr. David Farmer has kindly sent me three unpublished photographs of internal waves propagating westward on August 16, 1977 about 5-7 kms from the location of generation in Knight Inlet, taken with the camera looking almost vertically downwards. The shape and location of the leading wave is reproduced in figure 15 and it has begun to show the curved shape that one might expect of a Kelvin wave.

One further example is contained in Baines (1980) where he developed a theory of the "Southerly Buster", a coastal low pressure front which appears to be strongly influenced by the presence of the Great Dividing Range in southeastern Australia. As a result of the interaction of a cold front

Fig. 16. Satellite image of a Southerly Buster showing the backward curved shape as it propagates northwards, (from Baines 1980).

with this mountain barrier a large amplitude disturbance propagates rapidly northwards with the barrier to its left. A satellite photograph of the low level cloud associated with such a front is reproduced in figure 16. Note particularly the curved shape of the front, a phenomenon which was commented on by Baines and for which he anticipated the explanation given in this paper!

REFERENCES

1. Apel, J.R., and Holbrook, J.R., 1980, "The Sulu Sea Internal Soliton Experiment, Part A; Background and Overview", EOS. 61, No. 46, 1009.
2. Baines, P.G., 1980, The dynamics of the Southerly Buster, Australian Met. Mag., 28, 175.
3. Baker, G.R., Merion, D.I., and Orszag, S.A., 1980, "Vortex Simulation of the Rayleigh-Taylor Instability", Phys. Fluids, 25, 1485.
4. Benjamin, T.B., 1967, Internal waves of permanent form in fluids of great depth. J. Fluid Mechs., 29, 559.
5. Britter, R.E. and Simpson, J.E., 1978, Experiments on the dynamics of a gravity current head. J. Fluid Mechs., 88, 223.
6. Britter, R.E. and Simpson, J.E., 1981, A note on the structure of the head of an intrusive gravity current. J. Fluid Mechs., 112, 459.

7. Davis, R.A. and Acrivos, A., 1967, Solitary internal waves in deep water, J. Fluid Mechs., 29, 593.
8. Farmer, D.M., 1978, Observations of long nonlinear internal waves in a lake, J. Phys. Ocean., 8, 63.
9. Farmer, D.M. and Smith, J.D., 1980, Tidal interaction of stratified flow with a sill in Knight Inlet. Deep-Sea Research, 27A, 239.
10. Faust, K.M., 1981, Intrusion of a density front into a stratified environment. Report III, Univ. Karlsruhe, Inst. F. Wasserbau III.
11. Hunkins, D. and Fliegel, M., 1973, Internal undular surges in Seneca Lake: A natural occurence of solitons. J. Geophys. Res., 78, 539.
12. Joseph, R.I., 1977, Solitary waves in a finite depth fluid. J. Phys. (A: Math. Gen.) 10, L255
13. Lansing, F.S., 1981, The generation and propagation of internal solitary waves. Ph.D. Thesis University of Southern California.
14. Lansing, F.S. and Maxworthy, T., 1983, The generation and propagation of internal solitary waves. J. Fluid Mechs., subjudice.
15. Lee, C.Y. and Beardsley, R.C., 1974, "The generation of long internal waves in a weakly stratified shear flow," J. Geophys. Res., 79, 453-462.
16. Liu, A.K., 1981, "Internal solitons in the Sulu Sea: Comparison of wave theory with field data", Dynatech Report DT-8204-01.
17. Long, R.R., 1954, "Some Aspects of the Flow of Stratified Fluids 2: Experiments with a Two-Fluid System", Tellus, 6, 97.
18. Maxworthy, T., 1979, "A Note on the Internal Solitary Waves Produced by Tidal Flow Over a Three Dimensional Ridge", J. Geophys. Res., 84, 338.
19. Maxworthy, T., 1980, On the formation of nonlinear, internal waves from the gravitational collapse of mixed regions in two and three dimensions. J. Fluid Mechs. 96, 47.
20. Maxworthy, T., 1983, Experiments on solitary, internal Kelvin waves, J. Fluid Mechs., In press.

21. Mortimer, C.H., 1955, Some effects of the earth's rotation on water movements in stratified lakes. Proc. Intern. Assoc. Appl. Limnol. 12, 66.

22. Rosenhead, L., 1931, "The Formation of Vortices from Surface Discontinuity", Proc. Roy. Soc., A134, 170.

23. Thorpe, S.A., 1974, Near-resonant forcing in a shallow two-layer fluid: a model for the internal surge in Loch Ness? J. Fluid. Mechs., 63, 509.

24. Zarodny, S.J., and Greenberg, M.D., 1973, "On a Vortex Sheet Approach to the Numerical Calculation of Water Waves", J. Comp. Physics, 11, 440.

Supported by the Office of Naval Research under Contract No. N0004 - 82 - K - 0084 and by the Jet Propulsion Laboratories under Contract No. NASW7 - 100.

Departments of Mechanical and
Aerospace Engineering
University of Southern Calif.
Los Angeles, CA. 90089-1453
and, Space Sciences Division
Jet Propulsion Laboratory,
Pasadena, California 91109

INTERFACIAL INSTABILITIES CAUSED BY AIR FLOW OVER A THIN LIQUID LAYER

T. J. Hanratty

1. INTRODUCTION.

When air flows concurrently with a thin liquid film in an enclosed channel, a number of different wave forms are generated depending, principally, on the velocity of the air and the flow rate of the liquid. This paper summarizes progress that has been made in explaining these transitions.

The approach taken is to solve the linear momentum equations to determine whether small amplitude wavelike disturbances at the interface will grow or decay. In carrying out this analysis it is convenient to consider separately the gas and liquid flows. In this framework the gas flow is found to affect the stability of the liquid film through the imposition of pressure and shear forces at the interface. The prediction of the pressure and shear stress variation along a wavy surface over which a turbulent gas is flowing then becomes the central problem in predicting the stability of the liquid film.

The analysis of the stability of the liquid film is carried out by imposing a disturbance of the form

$$h' = a \exp i\alpha(x-Ct) \quad (1.1)$$

on the interface of the fully-developed gas-liquid flow depicted in Fig. 1. Here h' is the displacement of the interface from its time averaged location, x, the distance in the direction of flow and t, the time. The amplitude of the disturbance, a, and the wave number, $\alpha=2\pi/\lambda$, are positive quantities. The wave velocity, C, is complex,

ISBN 0-12-493220-7

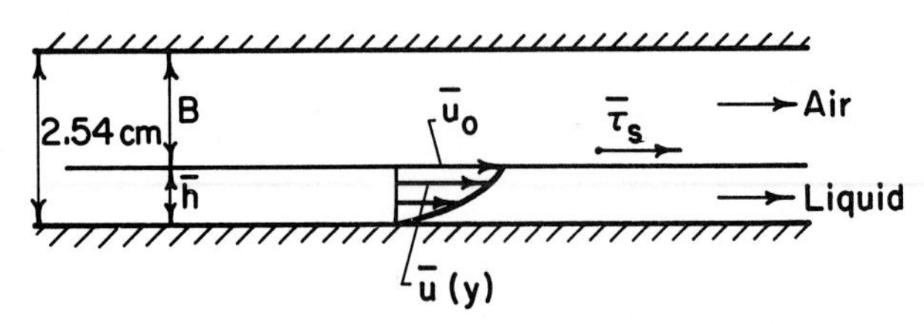

Fig. 1. System to which stability analysis is applied.

$$C = C_R + iC_I \tag{1.2}$$

The condition $C_I=0$ defines the transition from a stable to an unstable film. For $C_I>0$ waves will grow. The most rapidly growing wave is the one for which αC_I is a maximum.

The interfacial wave is accompanied by two-dimensional disturbances in the fully-developed velocity fields and in the gas phase pressure and shear stresses at the interface, designated as follows:

$$U = \overline{U}(y) + U', \quad V = V'$$
$$u = \overline{u}(y) + u', \quad v = v' \tag{1.3}$$

$$\tau_s = \overline{\tau}_s + \tau_s'$$
$$P = \overline{P} + P'. \tag{1.4}$$

The amplitude of the wave is assumed small enough that it induces a linear response in the velocity and stress fields, represented as

$$\frac{u'}{\hat{u}(y)} = \frac{v'}{\hat{v}(y)} = \frac{U'}{\hat{U}(y)} = \frac{V'}{\hat{V}(y)} = \frac{P_s'}{\hat{P}_s} = \frac{\tau_s'}{\hat{\tau}_s}$$
$$= a \exp i\alpha(x-Ct) \tag{1.5}$$

The amplitudes $a\hat{u}(y)$, $a\hat{v}(y)$, $a\hat{U}(y)$, $a\hat{V}(y)$, $a\hat{P}_s$ and $a\hat{\tau}_s$ are complex. Thus, if only the real parts of (1.1) and (1.5) are considered

$$h' = a \exp(\alpha C_I t) \cos\alpha(x-C_R t) \tag{1.6}$$

$$P_S' = a\ \exp(\alpha C_I t)[\hat{P}_{SR}\cos\alpha(x - C_R t) - \hat{P}_{SI}\sin\alpha(x - C_R t)] \quad (1.7)$$

$$\tau_S' = a\ \exp(\alpha C_I t)[\hat{\tau}_{SR}\cos\alpha(x - C_R t) - \hat{\tau}_{SI}\sin\alpha(x - C_R t)] \quad (1.8)$$

The amplitudes $a\hat{P}_{SR}$ and $a\hat{\tau}_{SR}$ are the components of P_S' and τ_S' in phase with the wave amplitude; the amplitudes $a\hat{P}_{SI}$ and $a\hat{\tau}_{SI}$ are the components, in phase with the wave slope.

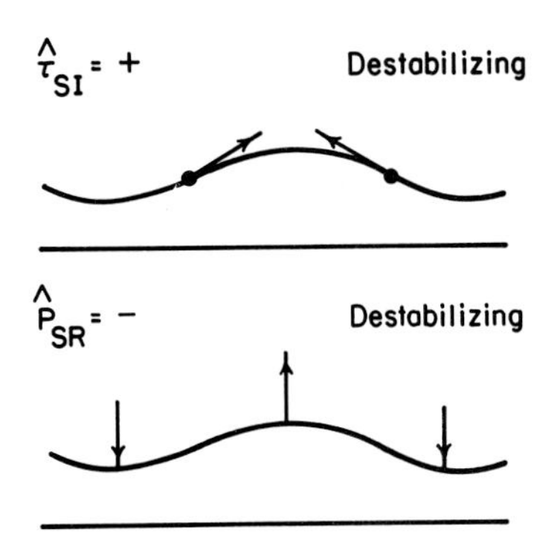

Fig. 2. Influence of surface stress on wave stability.

The amplitudes of the wave-induced variation of the velocity components in the liquid, $a\hat{u}(y)$ and $a\hat{v}(y)$, are determined from a solution of the linear momentum equations. This solution satisfies the no slip condition at the solid boundary and the kinematic condition at the interface. In addition, the perturbation of the shear stress in the liquid at the interface must equal τ_S' and the perturbation of the pressure at the interface must meet the condition

$$p_S' - P_S' = -\sigma \frac{\partial^2 h'}{\partial x^2}, \quad (1.9)$$

where σ is the surface tension. The solution can satisfy these five conditions for a given α only for a certain value of the complex wave velocity C. The determination of this eigenvalue then reveals whether the disturbance will grow or decay.

The complex amplitudes $a\hat{\tau}_S$ and $a\hat{P}_S$, that enter the eigenvalue problem through the boundary conditions at the interface, are determined from a solution of linearized momentum equations for the gas. The gas streamlines are compressed in the crest region and are expanded in the trough region. Thus, according to the Bernoulli equation, one expects $\hat{P}_{SR}$ to be negative and destabilizing; i.e., it will supply a suction at the crest. Similarly, because of the compression of the streamlines at the crest, $\hat{\tau}_{SR}$ will be positive. Calculations by Benjamin [5] indicate that the maximum in the shear stress occurs upstream of the crest, $\hat{\tau}_{SI}$ = plus value, and that the minimum pressure occurs downstream of the crest, $\hat{P}_{SI}$ = plus value. As shown in Fig. 2, a positive $\hat{\tau}_{SI}$ would be destabilizing. For a wave propagating in the positive x-direction, a positive $\hat{P}_{SI}$ is destabilizing in that it is accompanied by a transmission of energy from the gas phase to the disturbance in the liquid.

2. OBSERVED INSTABILITIES.

The type instabilities observed for concurrent gas-liquid flow are illustrated in Fig. 3, which summarizes studies by Engen [12], Hershman [13], Woodmansee [24], Miya [19], Cohen [6] and Craik [8] for the flow of air and water in a 2.54 x 30.5 cm horizontal enclosed channel.

Consider first observations for a thick liquid layer flowing along the bottom of the channel (large $\mathbf{R}_L$). At very low air velocities the interface is smooth. Long crested two-dimensional waves with wavelengths of 2.2-3.0 cm and with wave velocities greater than the liquid velocity appear as the air velocity is increased [6], keeping the liquid flow rate constant. A slight increase of the gas velocity above that required to produce two-dimensional waves causes the interface to break into a pebbled structure with a distance between crests of 0.5-1.0 cm [6]. The two-dimensional waves are found to be more stable on liquids of larger viscosity than water; i.e., they exist over a wider range of gas velocities [13].

These waves are generated because wave-induced pressure variations in the gas that are in phase with the wave slope ($\hat{P}_{SI}$ = plus value) transfer energy from the air to the liquid through velocity fluctuations normal to the interface [6,8]. When the rate of this energy input exceeds the rate of dissipation in the liquid, an instability is predicted.

As indicated in the previous section, energy can also be transmitted from the air to the liquid by shear stress variations in phase with the wave height ($\hat{\tau}_{SR}$ = plus value) through liquid velocity fluctuations tangential to the interface. The shear stress amplitude is much smaller than the pressure amplitude for air flows, so the influence of $\hat{\tau}_S$ is usually of secondary importance. However for very thin films, $\tanh \alpha \bar{h} \to 0$, wave-induced velocity fluctuations in the liquid have much larger components in the tangential direction than in the normal direction. Consequently, the shear stress amplitude can be an important, and even a dominant, mechanism for transferring energy from the air to the liquid [6,14]. Waves which are dominated by $\hat{\tau}_S$ usually have a different appearance from the waves observed on thicker films, in that they have a steep front and gradually sloping back. Examples of these are the capillary ripples that appear on

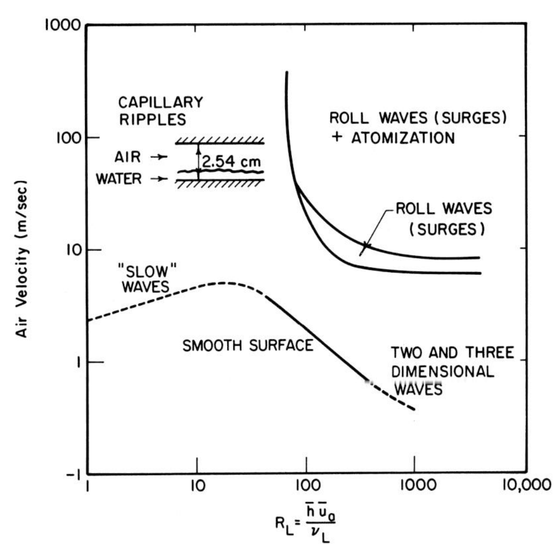

Fig. 3. Wave patterns observed in gas-liquid flows.

thin liquid films at high gas velocities [25] and the "slow waves" that appear at low gas velocities when the liquid velocity is decreased [8].

At high gas velocities wave-induced pressure variations in phase with the wave amplitude, $\hat{P}_{SR}$, become a dominant factor. For air-water flow in a horizontal channel, this is first demonstrated with the appearance of surges in the flow rate of the liquid film, called roll waves [13,12,14,3]. These may be considered to originate from large wavelength disturbances for which the destabilizing influences of inertia and of pressure forces in phase with the wave height are larger than the stabilizing influence of gravity.

For very high gas velocities wave-induced pressure variations in phase with the wave height can be so large that they overcome the stabilizing influence of surface tension. Under these circumstances small wavelets riding on top of the roll waves are atomized [24].

For thin liquid films the transition to the roll wave regime is more sensitive to changes in the liquid flow rate than to changes in the gas velocity [24]. In fact, it is found that no matter how high the gas velocity there is always a critical liquid flow rate below which roll waves will not be present. Andreussi, Asali and Hanratty [3] have recently shown that for very large wavelength disturbances the component of the interfacial stress in phase with the wave slope, $\hat{\tau}_{SI}$, can become stabilizing. This could be the principal factor giving rise to a critical liquid Reynolds number for the appearance of roll waves at large gas velocities.

3. PREDICTION OF $\hat{\tau}_S$ and $\hat{P}_S$.

(a) Small wavelength disturbances

Most of the theoretical analyses of interfacial instabilities have used the solution by Benjamin [5], which neglects the effect of turbulence on the wave-induced flow, to predict $\hat{P}_S$ and $\hat{\tau}_S$. Benjamin formulated the linear momentum equations for the gas flow in a curvilinear coordinate system which conforms to a small amplitude sinusoidal wave at $y=0$. However, the nonhomogeneous terms in the momentum equations introduced because of the use of curvilinear coordinates were neglected, so that the expressions derived for $\hat{\tau}_S$ and $\hat{P}_S$ are the same as would be obtained if the problem were formulated in a Cartesian coordinate system. Thorsness, Morrisroe and Hanratty [23,26] have recently examined possible errors caused by the neglect of turbulence effects and by the use of a Cartesian coordinate system.

They formulated the problem in boundary layer coordinates for which the x-axis is tangent to the wave surface and the y-axis perpendicular to it. The turbulent Reynolds stresses in this coordinate system are designated as $R_{ij}\rho_G$. The presence of the wave introduces a disturbance in the turbulence properties so that $R_{ij}=\overline{R}_{ij}(y)+R'_{ij}$, with

$$R'_{ij} = a\hat{R}_{ij}(y) \exp i\alpha(x-Ct) \tag{3.1}$$

A stream function for the gas flow

$$\Psi = \int_0^y \overline{U}(y)\, dy + aF(y) \exp i\alpha(x-Ct) \tag{3.2}$$

is defined such that wave-induced velocity components in the x and y directions are given by

$$U = h_y^{-1}\, \partial\Psi/\partial y, \quad V = -h_x^{-1}\, \partial\Psi/\partial x, \tag{3.3}$$

where h_x and h_y are the linearized metric functions for the boundary layer coordinate system:

$$h_x = 1 + a\alpha^2 y \exp i\alpha(x-Ct), \quad h_y = 1. \tag{3.4}$$

From the linearized x and y momentum balances the following equation is obtained for F:

$$i\alpha\left[(\overline{U}-C)\left(F^{II} - \alpha^2 F\right) - \overline{U}^{II}F + \alpha^2\overline{U}^2\right]$$
$$= \nu_G\left[F^{IV} - 2\alpha^2 F^{II} + \alpha^4 F + 2\alpha^2\overline{U}^{II} - \alpha^4\overline{U}\right] + \mathcal{R}, \tag{3.5}$$

where ν_G is the kinematic viscosity of the gas. The terms on the left-hand side are the inertia terms associated with the wave-induced flow, with $\alpha^2\overline{U}^2$ representing a centripetal acceleration associated with the use of a curvilinear coordinate system. The term $\mathcal{R}$ contains the amplitudes of the wave-induced variation of the Reynolds stresses $\hat{R}_{xx}$, $\hat{R}_{yy}$

and $\hat{R}_{xy}$:

$$\mathcal{R} = i\alpha^3 \overline{R}_{xx} + 3\alpha^2 \overline{R}_{xy}^{I} + i\alpha(\hat{R}_{xx}^{I} - \hat{R}_{yy}^{I}) + \alpha^2 \hat{R}_{xy} + \hat{R}_{xy}^{II} \tag{3.6}$$

Equation (3.5) is solved subject to the boundary condition that the velocity components in the gas are equal to those in the liquid at the wave surface,

$$F(0) = C - \overline{u}_0, \quad F^I(0) = \hat{u}(\overline{h}) \tag{3.7}$$

and the condition of parallel flow far from the surface,

$$F = (\overline{U} - C), \quad F^I = \overline{U}^I \text{ at large } y. \tag{3.8}$$

Equation (3.8) implies that the wavelength is small enough that the upper boundary does not affect the wave-induced flow.

From the solution of (3.6), (3.7) and (3.8) the shear stress and pressure at the wave interface can be calculated since, at y=0

$$\hat{P}_S = (-i\mu_G/\alpha)\left[F^{III}(0) + \alpha^2 \overline{U}^I(0)\right], \tag{3.9}$$

$$\hat{\tau}_S = \mu_G F^{II}(0), \tag{3.10}$$

where μ_G is the dynamic viscosity of the gas.

The principal theoretical problem in calculating $\hat{P}_S$ and $\hat{\tau}_S$ is the specification of $\mathcal{R}$. The quasi-laminar assumption, $\mathcal{R}=0$, implies that the only effect of the turbulence on the wave-induced flow is in the specification of $\overline{U}(y)$. By considering measurements for flow over solid wavy surfaces Abrams, Frederick and Hanratty [1,11] concluded that the quasi-laminar assumption is valid only for very large values of the dimensionless wave number, $\alpha\nu_G/v_G^*$, where v_G^* is the gas-phase friction velocity, $(\overline{\tau}_S/\rho_G)^{1/2}$.

Thorsness, et al. [23], found that good agreement with the measured shear stress variation along a small amplitude solid wavy surface could be obtained by modeling the Reynolds stresses with a zero equation eddy viscosity model developed by Loyd, Moffat and Kays [17].

The Reynolds stress terms in (3.6) are represented as

$$R_{ij} = -\delta_{ij} \frac{q^2}{3} + 2\nu_T E_{ij}, \qquad (3.11)$$

where $q^2/2$ is the turbulent kinetic energy per unit mass, ν_T, the turbulent eddy viscosity, E_{ij}, the rate of strain tensor. The Prandtl mixing-length equation gives

$$\nu_T = \ell^2 |2E_{xy}| \qquad (3.12)$$

Good agreement with velocity measurements over a flat plate can be obtained if the mixing-length is given as

$$\ell_o = 0.41y \left[1 - \exp\left(\frac{-y\ (\tau_S)^{1/2}}{\rho^{1/2}\nu A}\right)\right]. \qquad (3.13)$$

The argument of the exponential term is the damping factor, D, which represents the influence of viscosity on the scale of the eddies. The van Driest parameter, A=25, is therefore a measure of the thickness of the viscous region close to a wall.

The straightforward application of (3.13) to flow over a wavy surface would indicate that the viscous region, where dampening of the mixing-length occurs, would thin out in the regions of high τ_S and would thicken in regions of low τ_S. However, this is not expected to be correct because the presence of pressure gradients causes a large variation of τ with distance from the boundary and a drastic change in the production of turbulence in the viscous wall region.

This can be taken into account by redefining the damping factor as

$$D = \frac{y\tau^{1/2}(y)}{\rho^{1/2}\nu A}, \qquad (3.14)$$

where the local shear stress, $\tau(y)$, is used instead of the value at the surface, τ_S. In favorable pressure gradients $\tau(y)$ is less than the value at the wall, so this formulation will predict a thickening of the viscous wall region.

However, Loyd, Moffat and Kays [17] and Reynolds [21,22] found that equation (3.14) with A=25 does not correctly predict the effect of pressure gradient in equilibrium boundary layers. They, therefore, allowed A to be a function of a dimensionless pressure gradient $\frac{dP}{dx}\nu / \rho v^{*3}$, defined in Fig. VI-2 of their report [17]. For small values of this parameter, their results are described by the function

$$A = 25 \left[1 + k_1\left(\frac{dP}{dx}\nu / \rho v^{*3}\right) + k_2\left(\frac{dP}{dx}\nu / \rho v^{*3}\right)^2 + \ldots\right] \quad (3.15)$$

with k_1=-30 and k_2=1.54x10^3.

In situations for which the pressure gradient is varying rapidly, Loyd, Moffat and Kays [17] suggest that the flow close to the wall sees an effective pressure gradient given by the equation

$$\frac{d\left(\frac{dP}{dx}\right)_{eff}}{d\left(\frac{xv^*}{\nu}\right)} = \frac{\left(\frac{dP}{dx}\right) - \left(\frac{dP}{dx}\right)_{eff}}{k_L}, \quad (3.16)$$

where k_L is the relaxation constant approximately equal to 3000. Equation (3.15) now gives the value of A with $\left(\frac{dP}{dx}\right)_{eff}$ substituted for $\frac{dP}{dx}$. Because of the assumption of small amplitude waves we need consider only the first correction term in (3.15). The combination of (3.15) and (3.16) gives

$$A = 25 + \frac{25i\left(\frac{a\hat{P}_S}{\rho v^{*2}}\right)\left(\frac{\alpha\nu}{v^*}\right)k_1}{1 + i\left(\frac{\alpha\nu}{v^*}\right)k_L} \exp i\alpha(x-Ct) \quad (3.17)$$

Abrams, Frederick and Hanratty [1] indicate that reasonable agreement with measurements for solid surfaces can be obtained using k_1=-55 and k_L=3000. They called these closure equations Model D^* to differentiate them from the Model D of Thorsness [23,26], which used a damping factor

$$D = \frac{y\tau_S^{1/2}}{\rho^{1/2}\nu A}, \quad (3.18)$$

rather than (3.14).

As pointed out by Thorsness, et al. [23], the influence of turbulent stresses on the wave induced flow is important only close to the surface. Consequently, the evaluation of R_{ij} using (3.11) and (3.12) can be simplified because a boundary layer assumption can be made whereby the normal stresses $\overline{R}_{xx}$, $\hat{R}_{xx}$ and $\hat{R}_{yy}$ can be ignored.

(b) Large wavelength disturbances

The roll waves that are observed for concurrent gas-liquid flows in an enclosed channel are of such large wavelength that the assumption (used in the previous section) that the top boundary is not affecting the wave-induced disturbances in the gas flow is not correct. The situation is further complicated because the long wavelength disturbances can appear on a liquid surface which is already roughened with waves of smaller length.

Under these circumstances it is convenient to use a shallow gas assumption to solve the disturbance equations for the gas. According to this assumption the pressure in the gas phase is varying in the y-direction only because of changes of hydrostatic head and the velocity profile is closely approximated by that for flow without waves being present.

If the shear at the interface in the absence of a long wave disturbance is given as

$$\tau_S = \rho_G U_a^2 f_S \frac{1}{2} , \tag{3.19}$$

where

$$U_a(B) = \int_h^{h+B} U \, dy, \tag{3.20}$$

then the same relation is assumed to hold when long waves are present, i.e.,

$$\tau_S = \rho_G (U_a - C_R)^2 f_S \frac{1}{2} . \tag{3.21}$$

If the amplitude of the waves is small enough that second order terms can be ignored, then the following equation for the wave-induced variation of the shear stress, τ_S', is obtained:

$$\tau_S' = \bar{\tau}_S \left[\frac{f_S'}{\bar{f}_S} + 2\frac{U_a'}{\bar{U}_a - C_R} \right] \tag{3.22}$$

For a smooth interface f_S would only be a function of the gas Reynolds number, $\mathbf{R_G}$. However, for wave roughened interfaces the f_S could change with a change of the film thickness because of a change in roughness of the film. Therefore, following Miya, et al. [19], we assume

$$f_S' = \left(\frac{\partial \bar{f}_S}{\partial \bar{\mathbf{R}}_\mathbf{L}}\right)\mathbf{R_L}' + \left(\frac{\partial \bar{f}_S}{\partial \bar{\mathbf{R}}_\mathbf{G}}\right)\mathbf{R_G}' \tag{3.23}$$

From conservation of mass for the gas $\mathbf{R_G}'=0$ and

$$U_a' = (\bar{U}_a - C_R)\frac{h'}{B} \ . \tag{3.24}$$

From conservation of mass for the liquid,

$$\mathbf{R_L}' = \bar{\mathbf{R}}_\mathbf{L} \frac{C_R}{\bar{u}_a} \frac{h'}{\bar{h}} \ , \tag{3.25}$$

where u_a is the average velocity of the liquid defined as

$$u_a = \frac{1}{h}\int_o^h u \, dy \tag{3.26}$$

The substitution of (3.23), (3.24) and (3.25) into (3.22) yields the following relation for $\hat{\tau}_S$:

$$\frac{a\hat{\tau}_{SR}}{\bar{\tau}_S} = \frac{2a}{B} + \frac{\bar{\mathbf{R}}_\mathbf{L}}{\bar{f}_S} \frac{\partial \bar{f}_S}{\partial \bar{\mathbf{R}}_\mathbf{L}} \frac{C_R}{\bar{u}_a} \frac{a}{\bar{h}} \tag{3.27}$$

$$\hat{\tau}_{SI} = 0 \tag{3.28}$$

Similar expressions can be derived for the amplitude of the shear stress on the smooth top wall of the enclosed channel:

$$\frac{a\hat{\tau}_{BR}}{\bar{\tau}_S} = \frac{\bar{\tau}_B}{\bar{\tau}_S} \frac{2a}{B} \tag{3.29}$$

$$\hat{\tau}_{BI} = 0 \tag{3.30}$$

The pressure amplitude is derived from the integral forms of the conservation of mass and momentum for the gas. If the shape factor

$$\Gamma_G = \int_h^{h+B} U^2 dy \Big/ B\overline{U}_a^2 \tag{3.31}$$

is taken to be approximately equal to unity for a turbulent gas flow, the following relations are obtained:

$$\hat{P}_{SR} = \frac{\rho_G}{B}\left[-(\overline{U}_a - C_R)^2 - \frac{1}{\alpha\rho_G}(\hat{\tau}_{SI} + \hat{\tau}_{BI})\right] \tag{3.32}$$

$$\hat{P}_{SI} = \frac{1}{\alpha B}\left[(\hat{\tau}_{SR} + \hat{\tau}_{BR}) - \frac{\partial \overline{P}}{\partial x}\right]. \tag{3.33}$$

4. STABILITY ANALYSIS OF THE LIQUID FILM.

(a) Large $(\alpha\overline{h})(hC_R/\nu_L)$

For thick liquid films the stability analysis can be carried out using the condition that $(\alpha\overline{h})(\overline{h}C_R/\nu_L)$ is a large number. As shown by Cohen and Hanratty [6], the following equation for C is then obtained if the eigenvalue problem outlined in Section 1 is solved, using the restrictions of $C_R/\overline{u}_o > 1$, $\frac{d^2\overline{u}}{dy^2}\Big/C_R\alpha^2 << 1$, $(\alpha\overline{h})^{1/2}(\overline{h}C_R/\nu_L)^{1/2}(C_R/\overline{u}_o) >> 1$, and $(\alpha\overline{h})^{1/2}(\overline{h}\overline{u}_o/\nu_L)^{1/2}(\overline{u}_o - C_R)^{3/2}\Big/\overline{u}_o^{\,3/2} >> 1$:

$$\begin{aligned}
\alpha(\overline{u}_o - C)^2\coth(\alpha\overline{h}) &- \left(\frac{d\overline{u}}{dy}\right)_o(\overline{u}_o - C) \\
&= G + \frac{i\hat{\tau}_S}{\rho_L}\left[\coth(\alpha\overline{h}) - \frac{1}{\alpha(\overline{u}_o - C)}\left(\frac{d\overline{u}}{dy}\right)_o\right] \\
&- (\alpha\overline{h})^{1/2}\left(-i\,\frac{\overline{h}C}{\nu_L}\right)^{-1/2}\left[\alpha(\overline{u}_o - C)^2[1 - \coth^2(\alpha\overline{h})]\right] \\
&+ 4i(\alpha\overline{h})\left(\frac{\overline{h}(\overline{u}_o - C)}{\nu_L}\right)^{-1}\alpha(\overline{u}_o - C)^2\left[\coth(\alpha\overline{h}) - \frac{1}{\alpha(\overline{u}_o - C)}\left(\frac{d\overline{u}}{dy}\right)_o\right]
\end{aligned} \tag{4.1}$$

$$G = \frac{\hat{P}_S}{\rho_L} + \frac{\sigma\alpha^2}{\rho_L} + g, \tag{4.2}$$

where only the highest order terms in $(\bar{h}C_R/\nu_L)$ have been retained and $\bar{u}_o$, $\left(\frac{d\bar{u}}{dy}\right)_o$ are the velocity and the velocity gradient in the liquid evaluated at the average location of the interface. This is the same as the result obtained by Craik [7] except for a minor difference in the formulation of the normal stress condition at the interface. (Craik does not include the first term on the right side of Equation (36) in the paper by Cohen and Hanratty.)

The terms on the left side of (4.1) give the destabilizing effect of inertia. The term G represents the stabilizing effect of surface tension and gravity and the destabilizing effect of wave-induced pressure variations in the gas. For large $(\alpha\bar{h})(\bar{h}C_R/\nu_L)$ viscous dissipation occurs in thin boundary layers at the interface and at the wall, represented by the fourth and third terms on the right side of (4.1). Viscous dissipation at the wall is stronger than at the interface since it is of higher order in the Reynolds number. However, for wavelengths small compared to the film height $[\coth^2(\alpha\bar{h})-1]\rightarrow 0$ and viscous dissipation at the wall is not significant.

The $\hat{\tau}_S$ term represents the destabilizing influence of the wave-induced variation of the shear stress in the gas. The quantities in the brackets multiplying $\hat{\tau}_S$ equal the ratio amplitudes of the wave-induced variation of the streamwise and normal velocity components at the interface, $\hat{u}(0)/\hat{v}(0)$. If (4.1) is multiplied by $\hat{v}(0)$ the resulting terms containing $\hat{P}_S$ and $\hat{\tau}_S$, respectively, represent the energy transferred from the gas to the liquid by wave-induced pressure and shear variations in the gas. Calculations carried out by Thorsness, et al. [23] show that $\hat{\tau}_S$ is at least an order of magnitude smaller than $\hat{P}_S$. On this basis, one would expect energy transport by pressure variations to be far more important than by shear stress variations. However, as pointed out by Cohen and Hanratty [6] $\hat{u}_o \gg \hat{v}_o$ for small values $\alpha\bar{h}$. Consequently, one can expect the $\hat{\tau}_S$ term to play a dominant role as $\alpha\bar{h}\rightarrow 0$.

In order to do a calculation on the stability of the film for a given flow condition one needs to determing C_R and

C_I. The two equations defining these quantities are obtained by separately equating the real and imaginary parts of (4.1), keeping only terms of the highest order in $(\bar{h}C_R/\nu_L)$:

$$0 = (C_R-\bar{u}_o)^2 + \left(\frac{d\bar{u}}{dy}\right)_o (C_R-\bar{u}_o)\,\frac{\tanh(\alpha\bar{h})}{\alpha} - \left(\frac{\hat{P}_{SR}}{\rho_L\alpha} + \frac{\sigma\alpha}{\rho_L} + \frac{g}{\alpha}\right)\tanh(\alpha\bar{h}) + \frac{\hat{\tau}_{SI}}{\rho_L\alpha} - C_I^2$$
$$+ \frac{\tanh(\alpha\bar{h})}{\alpha}\left(\frac{d\bar{u}}{dy}\right)_o \frac{\hat{\tau}_{SI}}{\rho_L\alpha}\left[\frac{(C_R-\bar{u}_o)}{[(C_R-\bar{u}_o)^2+C_I^2]} - \frac{\hat{\tau}_{SR}}{\rho_L\alpha}\,\frac{C_I}{[(C_R-\bar{u}_o)^2+C_I^2]}\right] \quad (4.3)$$

$$\frac{\hat{P}_{SI}}{\rho_L} + \frac{\hat{\tau}_{SR}}{\rho_L\alpha}\left[\alpha\coth(\alpha\bar{h}) + \frac{(C_R-\bar{u}_o)}{[(\bar{u}_o-C_R)^2+C_I^2]}\left(\frac{d\bar{u}}{dy}\right)_o\right] + \frac{\hat{\tau}_{SI}}{\rho_L\alpha}\,\frac{C_I}{[(C_R-\bar{u}_o)^2+C_I^2]}\left(\frac{d\bar{u}}{dy}\right)_o = 4\alpha^2\nu_L(C_R-\bar{u}_o)\coth(\alpha\bar{h})$$
$$+ 4\alpha\nu_L\left(\frac{d\bar{u}}{dy}\right)_o + 2(C_R-\bar{u}_o)C_I\alpha\coth(\alpha\bar{h}) + C_I\left(\frac{d\bar{u}}{dy}\right)_o$$
$$+ \frac{\nu_L^{1/2}\alpha^{3/2}[\coth^2(\alpha\bar{h})-1]C_R^{1/2}}{(2)^{1/2}(C_R^2+C_I^2)^{1/2}}\Big[[(C_R-\bar{u}_o)-C_I^2](\cos\theta-\sin\theta) + (C_R-\bar{u}_o)2C_I(\cos\theta+\sin\theta)\Big] \quad (4.4)$$

$$\theta = \frac{1}{2}\tan^{-1}\frac{C_I}{C_R}\,. \quad (4.5)$$

(b) Solution for $(\alpha\bar{h})\to 0$

As indicated in the previous section, the wave-induced shear stresses in the gas assume a more important role than wave-induced pressure variations for $(\alpha\bar{h})\to 0$. However, in this limit (4.1) may not be a proper solution of the eigenvalue problem since the assumptions of $(\alpha\bar{h})(\bar{h}C_R/\nu_L)$ large and $(d^2\bar{u}/dy^2)\big/C_R\alpha^2 \ll 1$ would no longer be valid.

An approximate solution in this limit can be obtained by using integral forms of the mass and momentum balances. Define an average velocity

$$u_a = \frac{1}{h}\int_o^h u\,dy. \tag{4.6}$$

Then the mass balance equation is

$$h\frac{\partial u_a}{\partial x} + u_a\frac{\partial h}{\partial x} + \frac{\partial h}{\partial t} = 0. \tag{4.7}$$

For $(\alpha\bar{h})\to 0$ a shallow water assumption can be made whereby the pressure in the y-direction varies only because of variation of hydrostatic head:

$$p = P_S + (h-y)\rho_L g\,\sin\beta - \sigma\frac{\partial^2 h}{\partial x^2} - 2\mu\left(\frac{\partial u}{\partial x}\right)_S - 2\tau_S\frac{\partial h}{\partial x}, \tag{4.8}$$

where β is the angle the channel makes with the vertical. The integration of the x-momentum equation using (4.8) to define $\frac{\partial p}{\partial x}$ gives the following momentum balance:

$$\frac{\partial u_a}{\partial t} + (2\Gamma-1)u_a\frac{\partial u_a}{\partial x} + (\Gamma-1)\frac{u_a^2}{h}\frac{\partial h}{\partial x} + u_a^2\frac{\partial \Gamma}{\partial x} = \frac{\tau_S}{\rho_L h} - \frac{\tau_w}{\rho_L h}$$

$$+ g\cos\beta - \frac{1}{\rho_L}\frac{\partial P_S}{\partial x} - g\sin\beta\frac{\partial h}{\partial x} + \frac{\sigma}{\rho_L}\frac{\partial^3 h}{\partial x^3}$$

$$+ \frac{1}{\rho_L h}\int_o^h \frac{\partial \tau_{xx}}{\partial x}\,dy + \frac{2\tau_S}{\rho_L}\frac{\partial^2 h}{\partial x^2} + 2\nu_L\frac{\partial^2 u_S}{\partial x^2}, \tag{4.9}$$

where Γ is a shape factor characterizing the velocity profile,

$$\Gamma = \frac{1}{hu_a^2}\int_o^h u^2\,dy. \tag{4.10}$$

Equations for small disturbances can be obtained by equating each of the variables in (4.7) and (4.9) to the sum of the time averaged and the wave-induced variation and by neglecting terms which are quadratic in the disturbance. The following equation for the amplitude of the disturbance velocity is thus obtained from the linearized mass balance equation:

$$\overline{h}\hat{u}_a = C - \overline{u}_a \,. \tag{4.11}$$

In the limit being considered, $\alpha\overline{h}\to 0$, the last three terms in (4.9) can be neglected. The linearized momentum balance equation then gives the following relation if (4.11) is used to eliminate $\hat{u}_a$:

$$C^2 + \overline{u}_a^2\Gamma - 2\overline{u}_a\overline{\Gamma}\,C - \overline{h}\overline{u}_a^2\hat{\Gamma} = i\frac{\hat{\tau}_S}{\rho_L\alpha} - i\frac{\hat{\tau}_W}{\rho_L\alpha} + \frac{ig\cos\beta}{\alpha} - \frac{i}{\rho_L\alpha}\frac{d\overline{P}_S}{dx} + \overline{h}\,\frac{\hat{P}_S}{\rho_L} + g\overline{h}\,\sin\beta + \frac{\alpha^2\overline{\sigma}h}{\rho_L}. \tag{4.12}$$

Here the amplitudes are defined in the same manner as in (1.5):

$$h' = \frac{u_a'}{\hat{u}_a(y)} = \frac{\Gamma'}{\hat{\Gamma}} = \frac{\tau_W'}{\hat{\tau}_W} = a\;\exp i\alpha(x-Ct). \tag{4.13}$$

The similarity between (4.12) and (4.1) in the limit $\alpha\overline{h}\to 0$ is to be noted. The stabilizing influences of gravity and surface tension and the destabilizing influence of the wave-induced pressure variations are the same in the two equations. The $\hat{\tau}_W$ term in (4.12) replaces the viscous terms in (4.1). The term containing $g\cos\beta - \frac{1}{\rho_L}\frac{d\overline{P}_S}{dx}$ does not have an analog in (4.1) because of the assumption of $\frac{d^2\overline{u}}{dy^2}\Big/C_R\alpha^2 \ll 1$ used in its derivation.

The quantities on the left side of (4.12) represent the inertia terms. It can be shown that these are positive, and therefore destabilizing, provided $C_R > \overline{u}_o$, where $\overline{u}_o$ is the liquid velocity at the interface [3]. For example, consider the case of a plug flow for which $\overline{u}_o = \overline{u}_a$, $\overline{\Gamma}=1$ and $\hat{\Gamma}=0$. The inertia terms are then equal to $(\overline{u}_a - C)^2$ and, consequently always destabilizing. For $\overline{u}_a = \overline{u}_o = C$ the effects of inertia vanish.

However, in general, inertia can be stabilizing or destabilizing. For $\hat{\Gamma}=0$, it is seen that inertia effects vanish for

$$\frac{C}{\bar{u}_a} = \bar{\Gamma} \pm [\bar{\Gamma}(\bar{\Gamma}-1)]^{1/2}. \qquad (4.14)$$

For very thin films for which the velocity of the liquid varies linearly with distance from the wall, $\Gamma=4/3$ and $\bar{u}_o/\bar{u}_a=2$. In this case, inertia vanishes at $C/\bar{u}_a=2$ and at $C/\bar{u}_a=2/3$. The inertia terms are negative for $2/3<C/\bar{u}_a<2$ and are positive outside this range.

For disturbances of very long wavelength the change of height of the liquid with distance is gradual enough that the shape of the velocity profile would be closely approximated by that which would exist for a flow without a disturbance. Therefore, $\hat{\tau}_w$ and $\hat{\Gamma}$ are evaluated by assuming that the relationships between τ_w and Γ with u_a and h are the same as would be determined for an undisturbed flow. Depending on the situation, the film flow, prior to the appearance of a long wavelength disturbance, may be either a steady laminar flow or unsteady with small wavelength waves at the interface.

For laminar flow,

$$\tau_w = \frac{2\mu u_a}{h} - \frac{h\tilde{P}}{3}, \qquad (4.15)$$

$$\Gamma = \frac{4}{3} + \frac{1}{270}\frac{\tilde{P}^2 h^4}{u_a^2 \mu^2} + \frac{1}{18}\frac{h^2\tilde{P}}{\mu u_a}, \qquad (4.16)$$

$$\tilde{P} = \frac{dP}{dx} - \rho_L g \cos\beta . \qquad (4.17a)$$

$$= (\tau_S - \tau_w)/h \qquad (4.17b)$$

If the film is thin and highly sheared $h\tilde{P}$ can be neglected so that

$$\tau_w = \frac{2\mu u_a}{h}, \quad \Gamma=\frac{4}{3} . \qquad (4.18)$$

For the case of a free falling film with no gas flow, $\tilde{P}=-\tau_w/h$ so that

$$\tau_w = \frac{3\mu u_a}{h}, \quad \Gamma = \frac{6}{5} . \qquad (4.19)$$

In many instances the flow in the undisturbed film may be

approximated by a turbulent flow relation [15],

$$\frac{u}{v_C^*} = f\left(\frac{y\, v_C^*}{\nu_L}\right), \tag{4.20}$$

with

$$v_C^* = (\tau_C/\rho_L)^{1/2} \tag{4.21}$$

$$\tau_C = \frac{2}{3}\tau_w + \frac{1}{3}\tau_S = \tau_w + \frac{1}{3}h\tilde{P} \tag{4.22}$$

The integration of (4.20) from 0 to h^+ yields the following result [15]:

$$\frac{hv_C^*}{\nu_L} = g(\mathbf{R_L}) \tag{4.23}$$

$$g(\mathbf{R_L}) = \left[(1.414\ \mathbf{R_L}^{0.5})^{2.5} + (0.132\ \mathbf{R_L}^{0.9})^{2.5}\right]^{0.4} \tag{4.24}$$

As shown by Hanratty and Hershman [13] this can be rewritten in the form

$$\tau_w = \frac{2\mu u_a}{h} A(\mathbf{R_L}) - \frac{1}{3}h\tilde{P} \tag{4.25}$$

$$A(\mathbf{R_L}) = g^2(\mathbf{R_L})\Big/2\mathbf{R_L} \tag{4.26}$$

For the case of a laminar film $A(\mathbf{R_L})=1$ and (4.25) is equivalent to (4.15).

Similarly, from (4.18),

$$\Gamma = \frac{h^+}{\mathbf{R_L}^2} \int_o^{h^+} f^2(y^+)\, dy^+ \tag{4.27}$$

Since h^+ is a function of $\mathbf{R_L}$, it follows that Γ is also a function of $\mathbf{R_L}$.

It is assumed that (4.25) holds both for the disturbed and undisturbed flow, with $\tilde{P}$ given by (4.17b). Consequently the variables can be replaced by the sum of a time averaged and a fluctuating component. The subtraction of the time averaged equation and the neglect of quadratic terms in the fluctuating quantities give an equation for τ_w' or $\hat{\tau}_w$. If

(4.11) is used to eliminate $\hat{u}_a$ from this equation the following results:

$$\hat{\tau}_W = \frac{3\mu\bar{u}_a A(\bar{\mathbf{R}}_L)}{\bar{h}^2}\left[\frac{C}{\bar{u}_a} - 2 + \frac{C}{\bar{u}_a}\frac{\bar{\mathbf{R}}_L}{\bar{A}}\frac{d\bar{A}}{d\bar{\mathbf{R}}_L}\right] - \frac{1}{2}\hat{\tau}_S \tag{4.28}$$

Similarly, from (69),

$$\hat{\Gamma} = \frac{d\bar{\Gamma}}{d\bar{\mathbf{R}}_L}\bar{\mathbf{R}}_L\frac{C}{\bar{u}_a\bar{h}} \tag{4.29}$$

For the case of a laminar flow the equation for $\hat{\tau}_W$ is the same as (4.28) with $A(\bar{\mathbf{R}}_L)=1$. From (58) and (59b), the following equations for $\bar{\Gamma}$ and $\hat{\Gamma}$ are obtained:

$$\bar{\Gamma} = \frac{4}{3} + \frac{1}{270}\mathbf{P}^2 + \frac{1}{18}\mathbf{P}\ , \tag{4.30}$$

with $\mathbf{P}=\bar{\tilde{P}}\,\bar{h}^2/\mu\bar{u}_a$,

$$\bar{h}\hat{\Gamma} = \frac{1}{270}\mathbf{P}^2\left(4 - \frac{2C}{\bar{u}_a}\right) + \frac{1}{270}\mathbf{P}\left(42 - \frac{21C}{\bar{u}_a}\right) + \frac{1}{270}\left(90 - \frac{45C}{\bar{u}_a}\right)$$
$$+ \frac{3}{2}\frac{\hat{\tau}_S}{\bar{\tilde{P}}}\left(\frac{2}{270}\mathbf{P}^2 + \frac{1}{18}\mathbf{P}\right) \tag{4.31}$$

The two equations defining C_R and C_I are obtained by equating the real and imaginary parts of (4.12) to zero:

$$-C_I^2 + C_R^2 - 2\bar{\Gamma}\bar{u}_a C_R + \bar{\Gamma}\bar{u}_a^2 - \bar{h}\bar{u}_a^2\hat{\Gamma}_R = -\frac{\hat{\tau}_{SI}}{\rho_L\alpha} + \frac{\hat{\tau}_{WI}}{\rho_L\alpha}$$
$$+ \frac{\alpha\bar{h}\hat{P}_{SR}}{\rho_L\alpha} + g\bar{h}\sin\beta + \frac{\alpha^2\sigma\bar{h}}{\rho_L} \tag{4.32}$$

$$-\bar{h}\bar{u}_a^2\hat{\Gamma}_I + 2C_I(C_R - \bar{\Gamma}\bar{u}_a) = \frac{\hat{\tau}_{SR}}{\alpha\rho_L} - \frac{\hat{\tau}_{WR}}{\alpha\rho_L} + \frac{g\cos\beta}{\alpha}$$
$$- \frac{1}{\alpha\rho_L}\frac{d\bar{P}_S}{dx} + \frac{\alpha\bar{h}\hat{P}_{SI}}{\rho_L\alpha} \tag{4.33}$$

From (4.28) it is noted that $\hat{\tau}_{WR}$ is strongly related to C_R. Thus (4.33) may be viewed as the equation defining C_R under neutral stability conditions. This velocity is the kinematic wave velocity defined by Lighthill and Whitham [16]. In this context, (4.34) then defines the dynamic conditions necessary for neutral stability.

Different results from (4.28), (4.30), (4.32), (4.33) and (5.4), (5.5), to be developed in the next section, would be obtained if (4.17a) were used instead of (4.17b) when evaluating wave-induced variations in $\tilde{P}$. This accounts for differences in (4.28), (5.3), (5.4) from analogous equations developed by Hanratty and Hershman [13] and by Hanratty and Woodmansee [14]. The principal difference is the appearance of $\hat{\tau}_S$ in (4.28) and the 3/2 factor multiplying $\hat{\tau}_{SI}$ in (5.5). We have chosen to use (4.17b) instead of (4.17a) to evaluate disturbances in $\tilde{P}$ since (4.17b) appears more consistent with the pseudosteady state assumption and since the final results are more consistent with results derived by Craik [8] in a very different way for a laminar film.

5. INTERPRETATION OF FILM INSTABILITIES.

(a) Two and Three-Dimensional Waves

Cohen and Hanratty [6] considered three mechanisms for the generation of the two-dimensional waves observed to appear on a smooth liquid film with an increase in gas velocity. They argued that, since the wave velocity is greater than the liquid velocity, energy is not transmitted from the mean flow in the film to the disturbances. This means that the waves are not the result of the Tollmien-Schlichting instability in the liquid described by Miles [18] and by Feldman [9]. They also argued that the waves are not caused by turbulent pressure fluctuations in the gas flow above the liquid surface. Smooth interfaces were observed in the presence of turbulent gas flows. Since the wave velocities are much smaller than the average velocity, a resonance mechanism of the type described by Phillips [20] could not be reasonable for the generation of waves. Components of the turbulent pressure fluctuations with wavelengths comparable to those observed at the liquid interface would have velocities much larger than the wave velocities, since they would be comparable in magnitude to the average gas velocity. Consequently Cohen and Hanratty concluded that the waves receive their energy from the mean gas flow through pressure and shear stress variations induced by the waves along the interface.

For water and liquids of moderate viscosity, $(\alpha\bar{h})(\bar{h}C_R/\nu_L)$ is large at the transition. Consequently (4.3) and (4.4) can be used to predict conditions for the initiation of waves. These equations can be simplified since the values of $\alpha\bar{h}$ are large enough at transition that the influence of $\hat{\tau}_S$ can be neglected. The following conditions for neutral stability are therefore obtained:

$$0 = (C_R-\bar{u}_o)^2 + (C_R-\bar{u}_o)\,\frac{\tanh(\alpha\bar{h})}{\alpha}\left(\frac{d\bar{u}}{dy}\right)_o - \left(\frac{\hat{P}_{SR}}{\rho_L\alpha} + \frac{\sigma\alpha}{\rho_L} + \frac{g}{\alpha}\right)\tanh(\alpha\bar{h}), \tag{5.1}$$

$$\frac{\hat{P}_{SI}}{\rho_L} = 4\alpha^2\nu_L(C_R-\bar{u}_o)\coth(\alpha\bar{h}) + 4\alpha\nu_L\left(\frac{d\bar{u}}{dy}\right)_o + \frac{\nu_L^{1/2}\alpha^{3/2}(C_R-\bar{u}_o)^2}{(2\ C_R)^{1/2}}\,[\coth^2(\alpha\bar{h})-1] \tag{5.2}$$

At transition, the gas velocity is small enough that $\hat{P}_{SR}/\rho_L\alpha$ can be neglected compared to $(\sigma\alpha/\rho_L+g/\alpha)$. Consequently (5.1) indicates that the wave velocity is given by the classical dispersion relation. Equation (5.2) indicates that transition occurs when energy fed to the waves by pressure variations in phase with the wave slope is larger than the dissipative effects of viscosity. Since α is inversely proportional to the wavelength the dissipation in the viscous layer at the interface, described by the first two terms on the right side of (5.2), becomes quite large for small wavelengths. For large wavelengths, where $[\coth^2(\alpha\bar{h})-1]$ is not small, dissipation in the viscous layer at the wall is much larger than in the viscous layer at the interface since it is of higher order in the liquid Reynolds number. Therefore, the waves first observed on the liquid film have a length of the order of the liquid height. The wavelengths are small enough so viscous dissipation in the wall layer is relatively unimportant; yet, they are large enough that dissipation in the viscous layer at the interface is not too great. For lower liquid Reynolds numbers the film is thinner and the wave number is larger. This gives rise to larger viscous dissipation. Consequently a larger $\hat{P}_{SI}$ or larger gas velocity is needed to generate these waves, as indicated in Fig. 3.

Cohen and Hanratty [6] used (5.1) and (5.2) to determine neutral stability conditions. They evaluated $\hat{P}_{SI}$ from a solution of the linearized momentum equations for the gas which used a quasi-laminar model and a Cartesian coordinate system. The wave velocities observed by Cohen, although greater than the liquid velocity, were small compared to the gas velocity; i.e., $C_R/v_G^*=1-3$. Consequently, one would expect turbulence models developed from measurements of flow over solid wavy surfaces to be applicable. The analysis of such measurements presented by Thorsness, et al. [23,26], indicate that for the range of wave numbers covered by Cohen's experiments, $\alpha\nu_G/v_G^*=0.03-0.045$, the assumptions used by Cohen could cause errors in the calculation of $\hat{P}_S$ and $\hat{\tau}_S$.

Consequently Frederick [1,11] recently reanalyzed the results of Cohen. He evaluated $\hat{P}_{SI}$ by solving (3.5) using Model D^*, discussed in section 3(a), to calculate the wave-induced variation of the Reynolds stress. This calculation showed an interesting influence of C_R/v_G^*. The amplitude of the wave-induced pressure variations decreased with increasing C_R/v_G^*. However, the phase angle decreased and, since $\hat{P}_{SI}=|\hat{P}|\sin\theta_P$, the values of $\hat{P}_{SI}$ calculated for finite values of C_R/v_G^* are very close to what is calculated for a solid wave ($C_R=0$).

Figures 4 and 5 compare Frederick's calculations with the observations of Cohen. The solid curve marked Model D^* encloses a region in which $C_I>0$. It is noted that the predicted critical gas Reynolds numbers agree closely with the measured values of 1870 and 3600. Figs. 4 and 5 also give values of the wave numbers for waves observed at gas Reynolds numbers above the critical. These are in approximate agreement with the line indicating the most rapidly growing wave number; i.e., the locus of maxima in αC_I calculated from (4.3) and (4.4).

The neutral stability curve marked Model A in these figures was determined using values of $\hat{P}_{SI}$ calculated from (3.5) using the quasi-laminar assumption, $\mathcal{R}=0$. It is noted that calculations that neglect turbulence effects give critical gas Reynolds numbers which are good approximations to experimental measurements, but not as good as Model D*.

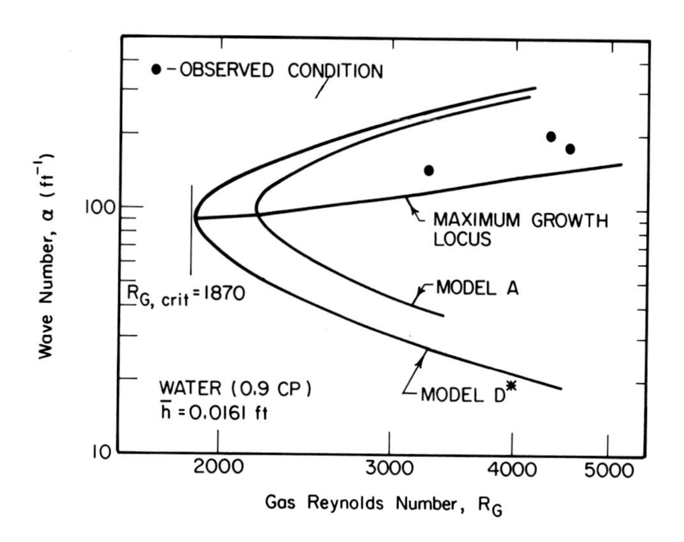

Fig. 4. Comparison of observed waves with stability theory.

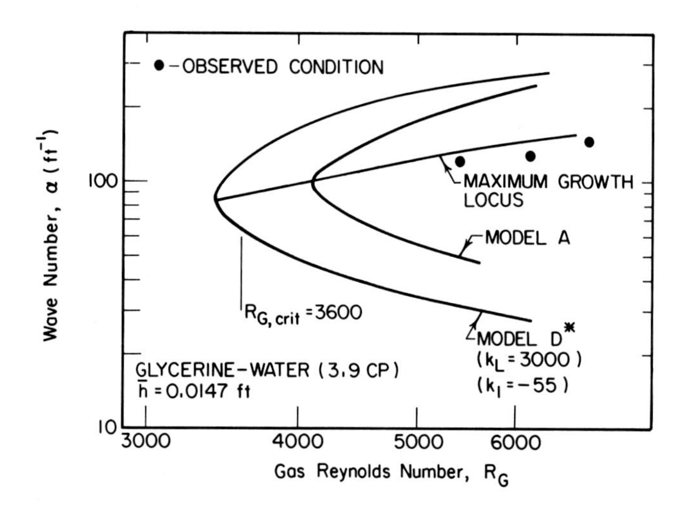

Fig. 5. Comparison of observed waves with stability theory.

At gas Reynolds numbers above those indicated in Fig. 4 for a water film the two-dimensional waves break into a three-dimensional pattern. This probably results from an instability of the two-dimensional waves. This instability appears to be inhibited by an increase in liquid viscosity or by a decrease in the channel width.

(b) Slow Waves

If one considers the stability of a very thin laminar film to disturbances with wavelength much greater than the film thickness, equations (4.32) and (4.33) must be used. Since laminar flow is assumed, $\hat{\tau}_W$ is given by (4.28) with A=1 and $dA/d\mathbf{R}_L=0$. The shape factors $\overline{\Gamma}$ and $\hat{\Gamma}$, are given by (4.30) and (4.31).

It is to be noted that for $\alpha\bar{h}$ very small the terms $\hat{\tau}_{SR}$ and $\hat{\tau}_{SI}$ can become much more important than the $\hat{P}_{SR}$ and the $\hat{P}_{SI}$ terms in (4.32) and (4.33). Therefore, the energy transfer from the gas to the liquid might no longer be dominated by gas phase pressure variations in phase with the wave slope as indicated by (5.2) for thick films.

From (4.33), (4.28), and (4.31) the following equations are obtained defining C_R and C_I:

$$\frac{C_R}{\bar{u}_a}\left[1+\frac{2}{3}\mathbf{R}_L(\alpha\bar{h})\frac{C_I}{\bar{u}_a}\right] = 2-\frac{1}{3}\mathbf{P}+\frac{1}{3}\mathbf{R}_L(\alpha\bar{h})\frac{C_I}{\bar{u}_a}\left(\frac{15}{6}+\frac{1}{30}\mathbf{P}\right)$$

$$+\,\mathbf{R}_L\boldsymbol{\tau}(\alpha\bar{h})\left[\frac{1}{2}\,\frac{\hat{\tau}_{SR}}{\alpha\bar{\tau}_S}+\frac{1}{3}(\alpha\bar{h})\frac{\hat{P}_{SI}}{\alpha\bar{\tau}_S}-\frac{1}{2}(\alpha\bar{h})\mathbf{R}_L\frac{\hat{\tau}_{SI}}{\alpha\bar{\tau}_S}\left(\frac{1}{18}+\frac{2}{270}\mathbf{P}\right)\right] \tag{5.3}$$

$$-\left(\frac{C_I}{\bar{u}_a}\right)^2+\left(\frac{C_R}{\bar{u}_a}\right)^2-2\overline{\Gamma}\left(\frac{C_R}{\bar{u}_a}\right)+\overline{\Gamma}-\frac{1}{270}(4\mathbf{P}^2+42\mathbf{P}+90)$$

$$+\,\frac{1}{270}\,\frac{C_R}{\bar{u}_a}(2\mathbf{P}^2+21\mathbf{P}+45)-\boldsymbol{\tau}\mathbf{R}_L(\alpha\bar{h})\frac{3}{2}\,\frac{\hat{\tau}_{SR}}{\alpha\bar{\tau}_S}\left(\frac{2}{270}\mathbf{P}+\frac{1}{18}\right)$$

$$=-\tau\left[\frac{3}{2}\,\frac{\hat{\tau}_{SI}}{\bar{\tau}_S\alpha}-(\alpha\bar{h})\frac{\hat{P}_{SR}}{\bar{\tau}_S\alpha}-\frac{\rho_L g\bar{h}}{\bar{\tau}_S}\sin\beta-\frac{(\alpha\bar{h})^2\sigma}{\bar{\tau}_S\bar{h}}\right]$$

$$+\,\frac{3}{\mathbf{R}_L(\alpha\bar{h})}\,\frac{C_I}{\bar{u}_a} \tag{5.4}$$

with

$$\tau = \frac{\bar{\tau}_S}{\rho_L \bar{u}_a^2} \tag{5.5}$$

$$\bar{\Gamma} = \frac{4}{3} + \frac{1}{270}\mathbf{P}^2 + \frac{1}{18}\mathbf{P}. \tag{4.16}$$

Equation (5.3) may be looked upon as defining $C_R/\bar{u}_a$. The terms on the left side of (5.4) represent the destabilizing effect of inertia. The film is stabilized by gravity and surface tension, but is destabilized by the wave-induced surface stresses, since $\hat{\tau}_{SI}$ is positive and $\hat{P}_{SR}$ is negative.

For a free flow, for which $\hat{\tau}_{SR}=\hat{\tau}_{SI}=\hat{P}_{SR}=0$ and $\mathbf{P}=-3$ the following conditions are obtained from (5.4) and (5.5) for neutral stability

$$\frac{C_R}{\bar{u}_a} = 3 \tag{5.6}$$

$$3\bar{u}_a^2 = \bar{h}g \sin\beta + \frac{\alpha^2 \sigma \bar{h}}{\rho_L} \tag{5.7}$$

For gas flows over very thin film $\mathbf{P}\cong 0$ and equations (5.4) and (5.5) may be written as follows for neutral stability:

$$\frac{C_R}{\bar{u}_a} = 2 + (\alpha\bar{h})\mathbf{R_L}\tau\left[\frac{1}{2}\frac{\hat{\tau}_{SR}}{\alpha\bar{\tau}_S} + \frac{1}{3}(\alpha\bar{h})\frac{\hat{P}_{SI}}{\alpha\bar{\tau}_S} - \frac{1}{36}(\alpha\bar{h})\mathbf{R_L}\frac{\hat{\tau}_{SI}}{\alpha\bar{\tau}_S}\right] \tag{5.8}$$

$$0 = +\frac{3}{2}\frac{\hat{\tau}_{SI}}{\alpha}\left[1-\frac{1}{36}(\alpha\bar{h})\mathbf{R_L}\right] + (\alpha\bar{h})\ \mathbf{R_L}\frac{2}{3}\frac{\hat{\tau}_{SR}}{\alpha}$$
$$+\frac{1}{2}(\alpha\bar{h})^2 R_L\frac{\hat{P}_{SI}}{\alpha} - (\alpha\bar{h})\left[\frac{\hat{P}_{SR}}{\alpha} + \frac{\rho_L g \sin\beta}{\alpha} + \sigma\alpha\right] \tag{5.9}$$

If the velocity is small enough that

$$\frac{\hat{P}_{SR}}{\alpha\rho_L} + \frac{g \sin\beta}{\alpha} + \frac{\sigma\alpha}{\rho_L} > 0, \tag{5.10}$$

equations (5.1) and (5.2) indicate that for thick liquid

films a wavy interface can be stabilized by decreasing the film height. However, (5.9) indicates that the net stabilizing influence of the terms in (5.10) decreases and the destabilizing influence of $\overline{\tau}_{SI}/\alpha$ remains constant as $\alpha\overline{h}$ becomes smaller. Consequently, for small enough $\alpha\overline{h}$, equation (5.9) predicts the film again becomes unstable.

This transition from a stable to a wavy film with decreasing liquid flow was first discovered by Craik [8]. For the conditions of his experiments $(\alpha\overline{h})\mathbf{R}_L$ was small so that (5.9) simplifies further to

$$0 = \frac{3}{2}\frac{\hat{\tau}_{SI}}{\alpha} - (\alpha\overline{h})\left[\frac{\hat{P}_{SR}}{\alpha} + \frac{\rho_L g \sin\beta}{\alpha} + \sigma\alpha\right]. \qquad (5.11)$$

Since $\hat{\tau}_{SI}$ varies directly with gas velocity, (5.11) is consistent with the observation by Craik that the film thickness at transition increases with increasing gas velocity. However, his calculated film thickness at transition is higher than is indicated by the experiments. (See Figure 11 in the Craik paper.) Craik used the Benjamin results to estimate $\hat{\tau}_{SI}$. For the reasons given in section 3a, these estimates could be in considerable error in the range of α for which this instability occurs. It is quite possible that better agreement could be obtained if a more accurate solution for the wave-induced flow in the gas were used. This is currently under investigation [10,2].

(c) Capillary Ripples

The ripples observed on very thin liquid films at high gas velocities for both vertical and horizontal flows [4,25] occur under conditions for which (5.3) and (5.4) should be applicable.

They are similar to the Craik waves in that $\hat{\tau}_S$ is exerting an important destablizing influence. However, they differ in that the stabilizing influence of surface tension is much more important than gravity. This is in evidence because their properties appear to be insensitive to the orientation of the duct and because the wavelengths are so small. They also differ from the Craik waves in that $\mathbf{R}_L$ can be large so that $(\alpha\overline{h})\mathbf{R}_L$ need not be a small number. This means that the only simplifications that can be made in (5.3) and (5.4) are to neglect the gravity term and to assume $\mathbf{P}\cong 0$, $\overline{\Gamma}=4/3$.

For neutral stability, $C_R/\bar{u}_a = 2$ if effects of $\hat{\tau}_{SR}$ and $\hat{P}_{SI}$ can be neglected and inertia effects vanish. However, because $(\alpha\bar{h})\mathbf{R_L}$ need not be small, the influence of surface stresses on $C_R/\bar{u}_a$ cannot be ignored. Consequently, the interpretation provided by (5.9) is that capillary ripples will occur for α small enough that the stabilizing influence of surface tension is not sufficient to overcome the destabilizing influences of the surface stresses $\hat{\tau}_{SI}$, $\hat{P}_{SR}$ and of inertia, represented by the $\hat{\tau}_{SR}$, $\hat{P}_{SI}$ terms in (5.9).

Asali [4] has recently examined such an interpretation by using (5.3) and (5.4) to calculate wave growth for conditions characterizing experiments for which these capillary ripples have been observed. For this analysis he used Model D^*, outlined in section 3a, to solve the gas flow equations for $\hat{\tau}_{SI}$ and $\hat{\tau}_{SR}$. Approximate agreement was noted between the observed wavelengths and the calculated wavelength of the fastest growing wave.

(d) Roll Waves - Thick Films

The surges in the flow rate of thick films, called roll waves, observed with an increase in gas velocity are also characterized by a long wavelength approximation, equations (4.32) and (4.33). However, they differ from the waves described in sections 5b and 5c in a number of important ways. The roll waves appear on a film which is highly roughened by waves. Surface tension affects these waves only insofar as it affects the surface roughness. Consequently its influence is felt through the τ_S term. In addition, these waves are not only long compared to the film thickness, but also compared to the thickness of the gas space. Because of this, none of the surface stress terms appearing in (4.32) and (4.33) vanish for $\alpha\bar{h}\to 0$.

If (3.27), (3.29), (3.30)', (3.32), (3.33) are used to approximate the surface stresses and (4.25), (4.26), (4.27), (4.28), and (4.29) are used to calculate τ_W and Γ the following equations for C_R and C_I are obtained from (4.32) and (4.33):

$$\frac{C_R}{\bar{u}_a} = \frac{2 - \frac{1}{3}\frac{P}{A} - \frac{\mathbf{H}}{3}\frac{\mathbf{P}_1}{A} + \frac{2}{3}\mathbf{T}\,\mathbf{H}\left[\frac{3}{2} + \mathrm{H}(1+\gamma)\right]}{1 + \frac{\mathbf{R_L}}{A}\frac{dA}{d\mathbf{R_L}} - \frac{1}{3}\mathbf{T}\frac{\mathbf{R}_L}{f_S}\frac{\partial f_S}{\partial \mathbf{R_L}}\left(\frac{3}{2} + \mathbf{H}\right)} , \quad (5.12)$$

$$-C_I^2 + C_R^2 - 2C_R\bar{u}_a\left(\bar{\Gamma} + \frac{d\bar{\Gamma}}{d\mathbf{R_L}}\mathbf{R_L}\right) + \bar{\Gamma}$$

$$= \frac{3\bar{u}_a^2}{\mathbf{R_L}(\alpha\bar{h})}\frac{C_I}{\bar{u}_a}\left(A + R_L\frac{dA}{d\mathbf{R_L}}\right) - \frac{\rho_G}{\rho_L}\mathbf{H}(\bar{U}_a - C_R)^2$$

$$+ g\bar{h}\sin\beta - \frac{3\hat{\tau}_{SI}}{2\alpha\rho_L}(1 - \tfrac{2}{3}\mathbf{H}), \tag{5.13}$$

with dimensionless groups defined as

$$\mathbf{H} = \frac{\bar{h}}{B} \qquad \mathbf{T} = \bar{\tau}_S \Big/ \frac{A\mu\bar{u}_a}{\bar{h}} \qquad \mathbf{P}_1 = \frac{\frac{d\bar{P}}{dx}\bar{h}^2}{\mu\bar{u}_a}. \tag{5.14}$$

It is to be noted that $\mathbf{T}$ is of order unity.

For the conditions of the experiments described in Fig. 3, the ratio of the film height to the height of the gas space is small. Therefore, all terms in (5.12) and (5.13) multiplied by $\mathbf{H}$ can be ignored with the exception of the term representing wave induced pressure variations in the gas, $\rho_G\mathbf{H}(\bar{U}_a - C_R)^2/\rho_L$. The ratio of the gas velocity to the liquid velocity $\bar{U}_a/\bar{u}_a$ is large so the quantity $\mathbf{H}\bar{U}_a^2/\bar{u}_a^2$ can be large even though $\mathbf{H}$ is small.

The neutral stability conditions for small $\mathbf{H}$ can therefore be expressed as

$$\frac{C_R}{u_a} = \frac{2 - \frac{1}{3}\frac{\mathbf{P}}{A}}{1 + \frac{\mathbf{R_L}}{A}\frac{dA}{d\mathbf{R_L}} - \frac{1}{2}\mathbf{T}\frac{\mathbf{R_L}}{f_s}\frac{\partial f_s}{\partial \mathbf{R_L}}} \tag{5.15}$$

$$C_R^2 - 2C_R\bar{u}_a(\bar{\Gamma} + \frac{d\bar{\Gamma}}{d\mathbf{R_L}}\mathbf{R_L}) + \bar{\Gamma}$$

$$= g\bar{h}\sin\beta - \frac{\rho_G}{\rho_L}\mathbf{H}(\bar{U}_a - C_R)^2 - \frac{3\hat{\tau}_{SI}}{2\alpha\rho_L} \tag{5.16}$$

For the pseudo-steady state assumption for the gas phase discussed in section 3b, the $\hat{\tau}_{SI}$ term in (5.16) can be neglected. From (5.16) instability is predicted when the gas velocity is large enough that the destabilizing influence of gas pressure variations and inertia (the left side of (5.12)) overcome the stabilizing influence of gravity. For very thin films, for which (4.18) is applicable, $\overline{\Gamma}=\frac{4}{3}$ and $\frac{d\overline{\Gamma}}{d\mathbf{R}_L}=0$. If it is also assumed that for thin enough films roughness effects are small, $\frac{\partial f_s}{\partial \mathbf{R}_L}=0$, inertia terms vanish and neutral stability is defined by the equations

$$(\overline{U}_a - C_R)^2 = \frac{\rho_L}{\rho_G}(B-\overline{h})g\ \sin\beta, \tag{5.17}$$

$$\frac{C_R}{\overline{u}_a} = 2 \tag{5.18}$$

This predicts a limiting gas velocity above which all films will be unstable to these large wavelength disturbances.

Hanratty and Hershman [13] used equations similar to (5.15) and (5.16) to explain the transition. Miya [19] and Asali [3] repeated these calculations because Hershman and Hanratty ignored the effects of surface roughness on τ_S. A comparison of the calculations by Asali and the experimental observations of Hershman and Hanratty are shown in Fig. 6.

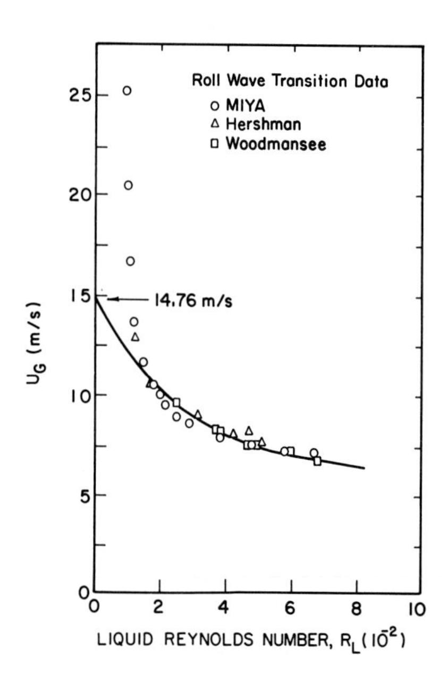

Fig. 6. Comparison of observed roll wave transition with calculations using $\hat{\tau}_{SI}=0$.

It is noted that good agreement between theory and experiment is obtained for thick films. However, for very thin films this is not the case. The transition appears to be not very sensitive to gas velocity and stable liquid films are found for gas velocities larger than the critical predicted by (5.17).

(e) Roll Waves - Thin Films

The comparison of the theory developed in the last section with experiment raises the question as to what stabilizes the film at low liquid Reynolds numbers. It is quite clear that for these very thin films the stabilizing influence of gravity is relatively unimportant since approximately the same critical liquid Reynolds number is found for horizontal and for vertical flows at large gas velocities.

Possible causes for the stabilization have recently been explored by Andreussi, Asali, and Hanratty [3]. From an examination of (4.32) they concluded that a negative value of $\hat{\tau}_{SI}$ or a positive value of $\hat{\tau}_{WI}$ in the neutral stability relation for very thin films is needed. These could arise because of relaxation effects in the liquid film. In deriving the relations for $\hat{\tau}_S$ and $\hat{\tau}_W$ presented in sections 3b and 4b, a pseudo-steady state assumption was made whereby the dependencies of τ_S and τ_W on h, u_a and U_a are the same as for a steady flow and that h^+ is a function of $\mathbf{R_L}$. Andreussi, et al., explored simple empirical methods to relax the pseudo-steady state assumption.

The dependency of f_s on $\mathbf{R_L}$ in (3.23) comes about because the wave roughness of the interface increases with increasing liquid flow rate or film height. It is possible that the roughness does not adjust immediately with changes in height and that there is a lag in the change of τ_i with a change in $\mathbf{R_L}$. Because of this, one could argue that an effective Reynolds number, $\mathbf{R_{Le}}$, at which τ_S should be calculated is defined by the rate equation

$$\frac{d}{dx}(\mathbf{R_{Le}}) = \frac{\mathbf{R_L} - \mathbf{R_{Le}}}{\gamma_i \bar{h}}, \qquad (5.19)$$

where $\gamma_i \bar{h}$ is a lag parameter proportional to the film height. For the case of long wavelength disturbances of the same form as (4.13)

$$\hat{\mathbf{R}}_{\mathbf{Le}} = \hat{\mathbf{R}}_{\mathbf{L}}\ (1 - i\alpha\gamma_i\bar{h}) \tag{5.20}$$

and $\hat{\tau}_{SI}$ is given by

$$\hat{\tau}_{SI} = -\bar{\tau}_S\alpha\gamma_i\ \frac{\bar{\mathbf{R}}_{\mathbf{L}}}{\bar{f}_s}\ \frac{\partial \bar{f}_s}{\partial \bar{\mathbf{R}}_{\mathbf{L}}}\ \frac{C_R}{\bar{u}_a}\ , \tag{5.21}$$

rather than the value of zero indicated by the pseudo-steady state assumption, equation (3.28). The real part of the stress amplitude, $\hat{\tau}_{SR}$, would still be given by (3.27).

A similar approach can be used to relax the expression for τ_W from that derived using a pseudo-steady state assumption. The disturbance waves have a velocity greater than the fluid velocity at the interface. This means that fluid of relatively low turbulence is entrained at the front of the waves and that this fluid becomes more turbulent as the height of the liquid above it increases. It could be argued that because the fluid does not adjust immediately to the surroundings there is a tendency for the increase in the intensity of turbulent transport to lag the increase in film height. This would influence the wall drag to be a maximum on the upstream side of the wave.

Andreussi et al. [3], found that relaxation of the expression for τ_S' gives a dependency of the critical liquid Reynolds number on fluid properties which is more consistent with measurements than does a relaxation of the expression for τ_W'. They used the relations,

$$f_S = f_o \left[1 + 0.0186\ \frac{h(\tau_S/\rho_G)^{1/2}}{\upsilon_G} \right] \tag{5.22}$$

$$f_o = 0.046\mathbf{R}_{\mathbf{G}}^{-0.2}\ , \tag{5.23}$$

derived from measurements, and (5.15), (5.16) and (5.21) to predict a transition liquid Reynolds number, $\mathbf{R}_{\mathbf{L}}$, for thin liquid films at high gas velocities that depends weakly on a dimensionless fluid property group, $\mu_L\rho_G^{1/2}/\mu_G\rho_L^{1/2}$. Good agreement with experimental data is obtained if γ_i is selected to be equal to 48. The wavelength of the ripples on the liquid film is about 25 times the film height. Therefore, a relaxation length, $\gamma_i\bar{h}$, equal to approximately two ripple lengths is indicated.

It should be pointed out that this is not the only explanation for this observed transition, and that more work needs to be done on this problem because it is of critical importance in the interpretation of entrainment results for annular gas-liquid flows [4]. Another possible explanation is that an instability to long wavelength disturbances does occur but that it remains as a slow moving swell that is not observed. This swell would not grow into an observed disturbance wave until it assumes a velocity much larger than the liquid film. According to (5.15) large values of $C_R/\overline{u}_a$ would be predicted if there were a sharp increase of f_S with increasing liquid Reynolds number. In fact, (5.15) predicts an infinite $C_R/\overline{u}_a$ if $\frac{\mathbf{R_L}}{f_S}\frac{\partial f_S}{\partial \mathbf{R_L}}$ is large enough. If f_S were described by an equation of the form of (5.22), the above explanation would indicate that transition would be observed for a fixed value of hv_G^*/υ_G. Such a result indicates a much stronger dependency of the critical liquid Reynolds number on $\mu_L\rho_G^{1/2}/\mu_G\rho_L^{1/2}$ than is observed.

(f) Atomization of Liquid Films

As discussed in sections 2, atomization of water films is observed for very high gas velocities for all film flow rates except those below which roll waves do not appear. An explanation for this occurrence can be obtained by examining (4.4), (4.5), (5.1), and (5.2). For situations for which

$$-\frac{\hat{P}_{SR}}{\rho_L\alpha} > \frac{\sigma\alpha}{\rho_L} + \frac{g}{\alpha}, \tag{5.24}$$

i.e., the destabilizing influence of wave-induced pressure variations is greater than the stabilizing influence of gravity and surface tension, very rapid wave growth occurs and it is not possible to find a solution to the equations for neutral stability, (5.1) and (5.2).

The Kelvin-Helmholtz solution for this instability condition is obtained by assuming an inviscid plug flow for the gas. Such an assumption leads to the result

$$\hat{P}_{SR} = -(\overline{U}_a - C_R)^2\alpha\rho_G. \tag{5.25}$$

The substitution of (5.25) into (5.24) gives the classical result for air-water that instability occurs for a wavelength of 1.7 cm at a gas velocity of 6.95 m/sec.

The experiments described in section 2 differ from the Kelvin-Helmholtz analysis in an important way: Because the actual velocity is not uniform the Kelvin-Helmholtz solution is in error. As α increases the influence of the waves extends a smaller distance into the gas flow. Consequently the waves only see the smaller gas velocities near the surface. A consequence of this is that dependency of $\hat{P}_{SR}$ on α is less than what is indicated by (5.25).

Woodmansee and Hanratty [24] solved for $\hat{P}_{SR}$ for the range of conditions that prevailed in the experiments described in section 2. They used a quasi-laminar approximation and a Cartesian coordinate system to formulate the equations for the gas disturbances. The solution gave $\hat{P}_{SR}$ varying with $\alpha^{0.37}$ and from (5.24) a minimum critical gas velocity of 10.8 m/sec, in good agreement with experiments performed in a 2.54 cm enclosed channel. The calculations indicated that this critical condition would increase with an increase in the size of the channel.

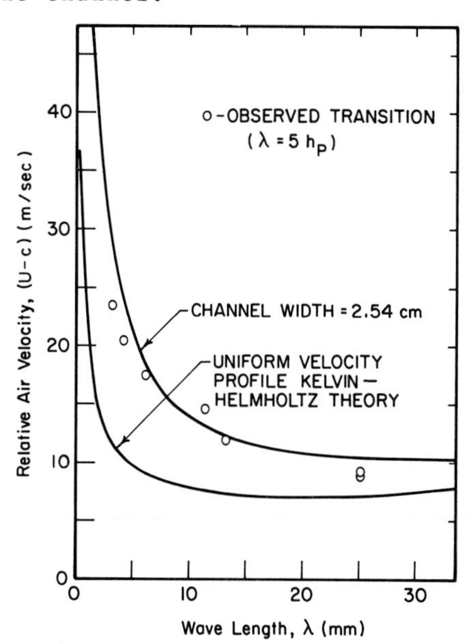

Fig. 7. Comparison of the observed initiation of atomization with Kelvin-Helmholtz theory.

A comparison of the solution of Woodmansee and Hanratty with the Kelvin-Helmholtz solution is shown in Fig. 7. The data points, representing measured critical gas velocities, are plotted against $5h_p$, where h_p the peak height of the roll waves. The use of $\lambda = 5h_p$ is in reasonable agreement with the characteristic lengths of unstable wavelets that have been observed in high speed motion pictures.

6. NOTATION.

a - wave amplitude defined by equation (1.1); a real number

A - van Driest parameter defined by equation (3.13)

$A(\mathbf{R}_L)$ - a function of liquid Reynolds number used to define the wall shear stress; see equation (4.25)

B - height of the gas space above the undisturbed liquid

C - wave velocity

D - damping coefficient defined by equation (3.14)

E_{ij} - rate of strain tensor

f_i - friction factor for gas-liquid flow

f_s - friction factor for single phase flow

F - $aF(y)$ is the amplitude of the stream function defined by equation (3.2)

g - acceleration of gravity

h - instantaneous height of the liquid film

h_x, h_y - metrics defined by equation (3.4)

k_1, k_2 - parameters defining the influence of pressure gradient on the van Driest parameter; see equation (3.15)

k_L - relaxation constant indicating the rapidity with which the turbulence adjusts to a change in pressure gradient; see equation (3.16)

ℓ - Prandtl mixing-length

ℓ_o - Prandtl mixing-length for flow over a flat plate

p - pressure in the liquid

P	-	pressure in the gas
$\tilde{P}$	-	$\frac{dP}{dx} - \rho_L g \cos\beta$
q^2	-	twice the turbulent kinetic energy per unit mass
$\mathcal{R}$	-	term representing the wave-induced Reynolds stresses, defined by equation (3.6)
R_{ij}	-	$\rho_G R_{ij}$ is a gas-phase Reynolds stress component
t	-	time
u	-	liquid velocity in the x-direction
U	-	gas velocity in the x-direction
v	-	liquid velocity in the y-direction
V	-	gas velocity in the y-direction
v_G^*	-	gas-phase friction velocity defined as $(\bar{\tau}_S/\rho_G)^{1/2}$
x	-	coordinate in the flow direction for a Cartesian coordinate system; coordinate tangent to the wave for a boundary layer coordinate system
y	-	coordinate perpendicular to x

Greek letters

α	-	wave number$=2\pi/\lambda$
β	-	inclination angle of the duct
γ_i	-	lag constant for interfacial stress, equation (5.16)
Γ	-	shape parameter for the velocity profile, equation (4.10)
λ	-	wavelength
μ	-	molecular viscosity
ν	-	molecular kinematic viscosity
ν_T	-	turbulent kinematic viscosity
σ	-	surface tension
τ	-	shear stress
τ_B	-	shear stress at the top wall
τ_C	-	characteristic shear stress, equation (4.22)
Ψ	-	stream function, equation (3.2)

Dimensionless groups

H	-	$\bar{h}/B$
P	-	$\bar{\tilde{P}}\,\bar{h}^2/\mu\bar{u}_a$
P$_1$	-	$\frac{d\bar{P}}{dx}\,\bar{h}^2/\mu\bar{u}_a$
R$_L$	-	$\bar{h}\,\bar{u}_a/\nu_L$
R$_G$	-	$B\,\bar{U}_a/\nu_G$
T	-	$\bar{\tau}_S\bar{h}/A\mu_L\bar{u}_a$
τ	-	$\bar{\tau}_S/\rho_L\bar{u}_a^{\,2}$

Superscripts

′	-	indicates a fluctuating quantity
I	-	first derivative with respect to y
II	-	second derivative with respect to y
III	-	third derivative with respect to y
IV	-	fourth derivative with respect to y
^	-	a $\hat{N}$ is the complex amplitude of a fluctuation in N
—	-	time averaged quantity

Subscripts

a	-	average $= \frac{1}{h}\int_o^h N_L dy$ or $= \frac{1}{B}\int_h^{B+h} N_G\, dy$
e	-	effective value
G	-	gas
I	-	imaginary part of a complex quantity
L	-	liquid
o	-	property evaluated at the average location of the interface
R	-	real part of a complex quantity
S	-	stress evaluated at the instantaneous location of the wave surface

7. REFERENCES.

1. Abrams, J., K. A. Frederick and T. J. Hanratty, Interaction between a turbulent flow and a wavy surface, presented at National Meeting of A.I.Ch.E., Detroit, Michigan, August, 1981; to be published in Physico Chemical Hydrodynamics.

2. Abrams, J., Turbulent flow over small amplitude solid waves, Ph.D. Dissertation, University of Illinois, Urbana, 1983.

3. Andreussi, P., J. Asali, and T. J. Hanratty, Initiation of Roll Waves in Gas-Liquid Flows, submitted for publication, 1983.

4. Asali, J. Entrainment in vertical gas-liquid annular flows, Ph.D. Dissertation, University of Illinois, Urbana, 1983.

5. Benjamin, T. B., Shearing flow over a wavy boundary, J. Fluid Mech. 6 (1959), 161.

6. Cohen, L. S. and T. J. Hanratty, Generation of waves in the concurrent flow of air and a liquid, A.I.Ch.E. Journal 11 (1965), 138.

7. Craik, A. D. D., Wind-generated waves in contaminated liquid films, J. Fluid Mech. 31 (1968), 141.

8. Craik, A. D. D., Wind-generated waves in thin liquid films, J. Fluid Mech. 26 (1966), 369.

9. Feldman, S., On the hydrodynamic stability of two viscous incompressible fluids in parallel uniform shearing motion, J. Fluid Mech. 2 (1957), 343.

10. Frederick, K. A., Ph.D. Dissertation, University of Illinois, Urbana, to be published.

11. Frederick, K. A., Wave generation at a gas-liquid interface, M.S. Dissertation, University of Illinois, Urbana, 1982.

12. Hanratty, T. J. and J. M. Engen, Interaction between a turbulent air stream and a moving water surface, AIChE J. 3 (1957), 299.

13. Hanratty, T. J. and A. Hershman, Initiation of roll waves, AIChE J. 7 (1961), 488.

14. Hanratty, T. J. and D. E. Woodmansee, Stability of the interface for horizontal air-liquid flow, Proceedings of the Conference on Two-Phase Flow, Exeter, England, 1965.

15. Henstock, W. H. and T. J. Hanratty, The interfacial drag and the height of the wall layer in annular flows, AIChE J. 22 (1976), 990.

16. Lighthill, M. J. and G. B. Whitham, On kinematic waves: I. Flood movement on long rivers, Proc. Roy. Soc. A229 (1955), 291.

17. Loyd, R. J., R. J. Moffat, and W. M. Kays, The turbulent boundary layer on a porous plate; an experimental study of the fluid dynamics with strong favorable pressure gradients and blowing Rep. No. HMT-13, Thermosciences Division, Dept. of Mechanical Engineering, Stanford University, Stanford, California, 1970.

18. Miles, J. W., On the generation of surface waves by shear flows, J. Fluid Mech. 3 (1957), 185.

19. Miya, M., D. E. Woodmansee, and T. J. Hanratty, A model for roll waves in gas-liquid flow, Chem. Eng. Science 21 (1971), 1915.

20. Phillips, O. M., On the generation of waves by turbulent wind, J. Fluid Mech. 2 (1957), 417.

21. Reynolds, W. C., Recent advances in the computation of turbulent flows, Adv. Chem. Eng. 9 (1974), 193.

22. Reynolds, W. C., Computation of turbulent flows, Annual Rev. Fluid Mech. 8 (1976), 183.

23. Thorsness, C. B., P. E. Morrisroe, and T. J. Hanratty, A comparison of linear theory with measurements of the variation of shear stress along a solid wave, Chem. Eng. Science 33 (1978), 579.

24. Woodmansee, D. E., and T. J. Hanratty, Mechanism for the removal of droplets from a liquid surface by a parallel air flow, Chem. Eng. Science 24 (1969), 299.

25. Wurz, D. E., Flussigkeits-filmstromung unter einwirkung einer uberschalluftstromung, Report Universitat Karlsruhe, Institut fur Thermische Stromungsmaschinen, February, 1977.

26. Zilker, D. P., G. W. Cook, and T. J. Hanratty, Influence of the amplitude of a solid wavy wall on a turbulent flow. Part I. Non-separated flows, J. Fluid Mech. 82 (1977), 29.

The author was partially supported in this work by the National Science Foundation Grant CPE 79-20980, by the Office of Naval Research Grant NR 657-728 and by the Shell Foundation.

Department of Chemical Engineering
University of Illinois
at Urbana-Champaign
205 Roger Adams Laboratory
Urbana, IL 61801

FILM WAVES

S. P. Lin

1. INTRODUCTION.

The wave motion in a thin film may often be observed in everyday life, as when melted ice flows down the surface of an icicle, or when a sheet of rain water runs down a sidewalk. The dynamics of film waves has received much attention from various industries because it has a dramatic effect on transport rates of mass [50], heat [16,64] and momentum [15] in devices such as distillation and adsorption columns, evaporators, condensors, nuclear reactor emergency cooling systems etc. The film waves are also of great relevance to modern precision coating of photographic emulsions, magnetic material and protective paints. A great deal has been learned about the onset of film waves and their weakly nonlinear evolution, but their often observed strong nonlinear character remains to be quantitatively understood. The problem of weakly nonlinear ripples on a viscous film has recently attracted the attention of researchers in other fields. It turns out that seemingly unrelated nonlinear phenomena are governed by the same or similar differential equations [see 25,34,54]. This article presents my personal view rather than an exhaustive review of the subject matter. Reviews of earlier works [17,29] and more recent works [56] are available.

ISBN 0-12-493220-7

2. BASIC EQUATIONS.

Electromagnetic forces, mass and heat transfer, and phase changes will be excluded from consideration. Moreover only the incompressible Newtonian fluids are considered. The governing differential equations are:

$$\rho(\partial_t \underline{V} + \underline{V} \cdot \underline{\nabla}\,\underline{V}) = -\underline{\nabla}P + \rho\underline{g} + \underline{\nabla} \cdot \underline{\underline{\tau}} - \underline{\nabla}\Pi \quad , \tag{1}$$

$$\underline{\nabla} \cdot \underline{V} = 0 \quad , \tag{2}$$

where ρ is the fluid density, t is time, $\underline{V}$ is the velocity, P is the pressure, g is the gravitational acceleration, Π a force potential which may arise, for example, from intermolecular interaction, and $\underline{\underline{\tau}}$ is the stress tensor given by

$$\underline{\underline{\tau}} = -P\underline{\underline{\delta}} + \nu(\underline{\nabla}\,\underline{V} + (\underline{\nabla}\,\underline{V})^{\dagger}) \quad ,$$

in which $\underline{\underline{\delta}}$ is the unit diadic, ν the kinematic viscosity, and $(\underline{\nabla V})^{\dagger}$ is the transpose of $(\underline{\nabla V})$.

The boundary conditions at a fluid-fluid interface $Y = H(X,Z,t)$ are

$$[\underline{\underline{\tau}}] = \underline{\nabla}_{\parallel}\sigma + \underline{n}\sigma(R_1^{-1} + R_2^{-1}) \quad , \tag{3}$$

$$[\underline{V}] = 0 \quad , \tag{4}$$

$$\dot{Y} = \partial_t H + \underline{V} \cdot \underline{\nabla}H \quad , \tag{5}$$

where (X,Y,Z) denote Cartesian coordinates, [ ] denotes the change of the quantity it brackets in the direction of the unit normal $\underline{n}$, $\underline{\nabla}_{\parallel}$ is the surface gradient, σ is the surface tension, R_1 and R_2 are the principal radii of curvature defined positive if directing from the centers of curvature in the direction of $\underline{n}$, and the upper dot denotes total time differentiation. (3) is the consequence of modeling the interface as a massless geometric surface endowed with only one mechanical property σ, and the second law of Newton. (4) and (5) result from the assumption that the interfacial velocity is the same as the fluid velocity at the interface. The boundary condition at the homogeneous solid-liquid interface is the no-slip condition

$$\underline{V} = 0 \quad . \tag{6}$$

3. SIMPLE FILM.

Consider a layer of liquid running down an inclined plane under the action of gravity as shown in Fig. 1. When the liquid layer thickness is much smaller than the linear dimensions of the incline, the film can be approximated by an infinite sheet of liquid with a smooth free surface parallel to the incline. In addition, if the viscous and inertia effects of the overlying air can be neglected, then the exact solution of (1)-(6) gives the following velocity and pressure distributions

$$U(Y) = (gd^2\sin\beta/2\nu)[1 - (Y/d)^2] \quad , \tag{7}$$

$$P(Y) = \rho gY\cos\beta - \Pi(Y) \quad , \tag{8}$$

where U is the velocity in the X-direction, Y the distance measured from the free surface into the liquid, d is the uniform film thickness, and β is the angle of inclination.

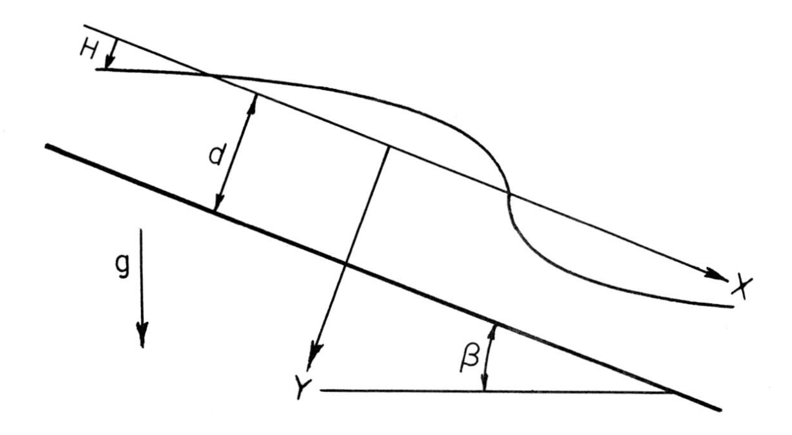

Fig. 1. Definition sketch.

The basic flow described by (7) and (8) easily becomes unstable. Two distinctively different waves, namely gravity-capillarly ripples and shear waves, have been observed as the consequence of the instability before the onset of turbulence. A spontaneous rupture has also been observed as the consequence of instability in relatively thin films. The onset and the nonlinear evolution of ripples, the onset of shear waves, and film rupture will be discussed in this section.

Before embarking on quantitative discussions, let us address the question of the origin of instability. Consider the energy contained in a given mass of fluid flowing down a film. If there is no heat transfer, the first law of thermodynamics applied to this mass is

$$\frac{d}{dt}\,(KE + PE + IE + SE) = 0 \quad ,$$

Where KE is the kinetic energy, PE the potential energy, IE the internal energy and SE is the surface energy associated with the mass, [32]. In the basic flow there are no changes in KE and SE, and the decrease in PE is balanaced exactly by the increase in IE. Thus, the above energy balance equation can be considered as applied to the disturbances. The change of IE is always positive, since the conversion of mechanical energy into internal energy through the agent of viscosity is irreversible. The change of SE is always positive, because the deformed free surface is always larger than the unperturbed flat surface. Thus if instability is defined to be a state of increasing KE of the given mass, then it follows from the above equation that both surface tension and viscous dissipation are stabilizing. Moreover, the film instability can occur only when the rate of decrease in PE exceeds the rate of increase in IE and SE. However, the instability manifests itself in so many interesting forms mentioned earlier.

3.1. Onset of Ripples.

When the ripples first appear, they have straight wave crests perpendicular to the flow direction. Thus at the onset of ripples, the basic flow is perturbed by infinitesmal disturbances which are independent of the span-wise direction.

Then (2) is the necessary and sufficient condition for the existence of the function ψ such that $u = \partial\psi/\partial y$, $v = -\partial\psi/\partial x$, where (u,v) are the components of velocity perturbation normalized by $U(0)$ in the X and Y direction respectively, and $(x,y) = (X,Y)/d$. Substituting the sum of the basic flow and the perturbation into the curl of (1) and neglecting terms which are quadratic in infinitesimal disturbances, and then taking the Fourier and Laplace transforms of the resulting equation, we have

$$\{(D^2-\alpha^2)^2-i\alpha R(\bar{u}-s/\alpha)(D^2-\alpha^2)-D^2\bar{u}\}\psi_{s\alpha} = (\alpha^2-D^2)\psi_{\alpha o} \quad , \qquad (9)$$

where $\bar{u} = U/U(0)$, $D = d/dy$, the subscripts o, s and α denote respectively initial values, the Laplace and Fourier transforms, and

$R = U(0)d/\nu \equiv$ Reynolds number,

$\alpha = 2\pi d/\lambda \equiv$ wave number of wave length λ.

Similarly the boundary conditions can be found from (3) - (5) and (1) to be

$$\bar{u}''(0)h_{s\alpha} + \psi''_{s\alpha}(0) + \alpha^2\psi_{s\alpha}(0) = 0 \quad , \qquad (10)$$

$$-(\bar{p}'(0) + \alpha^2 W)h_{s\alpha} - p_{s\alpha}(0) - i(2/R)\alpha\psi'_{s\alpha}(0) = 0 \qquad (11)$$

$$i\alpha\psi_{s\alpha} + sh_{s\alpha} + i\alpha\bar{u}(0)h_{s\alpha} = h_{\alpha o} \quad , \qquad (12)$$

$$\psi_{s\alpha}(1) = 0 \quad , \qquad (13)$$

$$\psi'_{s\alpha}(1) = 0 \quad , \qquad (14)$$

where primes denote differentiation with respect to y, $\bar{p} = P/\rho U^2(0)$, $h = H/d$ and

$W = \sigma/\rho d U^2(0) \equiv$ Weber number,

$$p_{s\alpha} = (i\alpha R)^{-1}\{\psi'''_{s\alpha}-(\alpha^2+i\alpha\bar{u}(0)R+sR)\psi'_{s\alpha}+i\alpha R\bar{u}'(0) +R\psi'_{\alpha o}-i\alpha R\pi_{s\alpha}\} \quad ,$$

$$\pi_{s\alpha} = \Pi_{s\alpha}/\rho U^2(0) \quad .$$

Note the Froude number $F^2 \equiv U^2(0)/gd = R/2$ does not appear explicitely in this system, because the viscous film flow is solely driven by gravity.

Instead of solving the initial boundary value problem (9) - (14), it is usually assumed that the stability does not depend on the initial condition and the transient effect is damped. Then one need only to consider a normal mode of the form

$$\psi = \phi(y)\exp[i\alpha(x-c\tau)] \quad ,$$

where c is the complex wave speed of the disturbances. It can be verified easily that the normal mode formulation will lead to the same system (9) - (14), except s must be replaced by $-i\alpha c$ and all quantities associated with the initial values must be put to zero. It should be pointed out that even without addressing the question of well posedness, the damping of the transient continuous spectrum of the disturbance has not been shown for this problem. For a temporally growing normal mode, α is real but c is complex i.e., $c=c_r+ic_i$. For a spatially growing normal mode, α is complex i.e. $\alpha=\alpha_r+i\alpha_i$ but the wave frequency $\omega=\alpha c$ is real. The flow is unstable with respect to temporal or spatial disturbances if $c_i>0$ or $\alpha_i<0$.

The normal mode formulation for film stability was first given by Yih [66] who neglected the surface tension and intermolecular force effects. Benjamin [6] included the surface tension and expanded the normal mode solution in power of y. He showed that the critical Reynolds number R_c below which the temporal disturbances are damped is given by

$$R_c = 5 \cot\beta/4 \quad , \tag{15}$$

which occurs at the critical wave number $\alpha_c = 0$. The same critical condition was confirmed by Yih [67] who expanded the solution in powers of small α. The solutions for finite α have been obtained by various authors [5,21,28,57,58,62] by use of different approximations with varied accuracy. No critical Reynolds number smaller than that given by (15) was found up to R=O(100). Some sample results are given in Figs. 2 and 3. As is to be expected, experimental points near $\alpha=0$ are very scarce. They require a long test section due to exceedingly

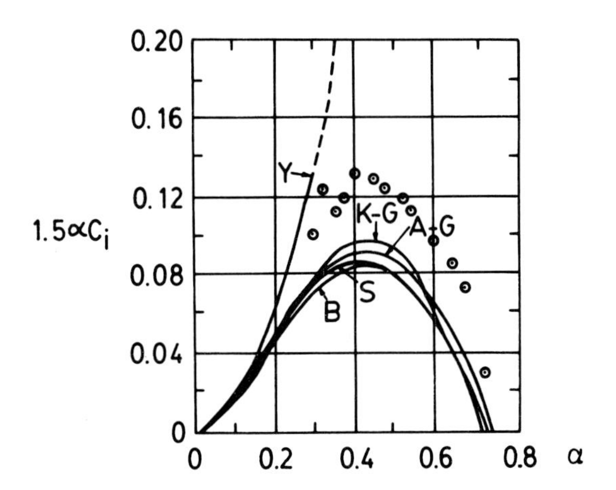

Fig. 2. Variation of the amplification factor αc_i with the wave number α; $\beta=\pi/2$, R=1.935, W=0.844. Experimental results: ⊙ Krantz and Goren [28]; ▲ Binnie [10]; ×, Kapitza and Kapitza [27]. Theoretical results: A-G, Anshus and Goren [5]; B, Benjamin [6]; K-G, Krantz and Goren [28]; S, Solesio [57]; Y, Yih [67].

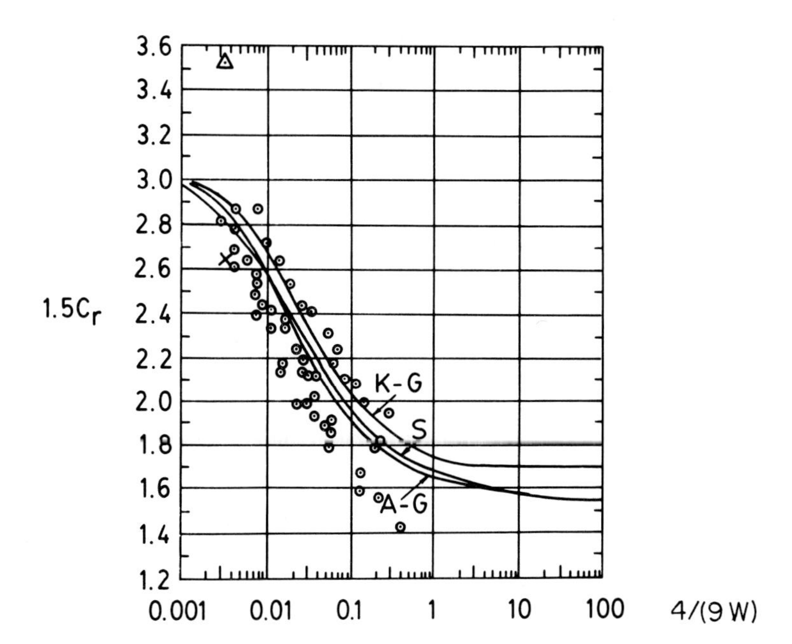

Fig. 3. Variation of the wave speed c_r corresponding to the fastest-growing wave with max. (αc_i) as a function of W; $\beta=\pi/2$, $(\sigma/\rho)(3/g\nu^4)^{1/3}=4587$. Symbols as in Fig. 2.

long wave lengths and small amplification rates. This is the inherent difficulty with the experimental verification of (15). The theoretical amplification rates in Fig. 2 are smaller than the measured ones by as much as 50% which is much larger than the claimed experimental accuracy. The difference cannot be due to the surface contamination, although a trace amount is known to have very large effects. This is because the surface contamination has been shown [7,35,63,67] to stabilize the film, and thus if the surface could be made cleaner the difference would be even greater. All wave speeds except that of Binnie [10] given in figure 3 are smaller than twice the surface velocity which is the maximum theoretical value. The scatter on wave speed data is not entirely due to experimental error. Different wave speeds for the same value of W correspond to different values of R. The larger wave speed observed by Binnie may be due to the larger R in his experiment. The larger R may correspond to larger wave amplitudes. No records of wave amplitude were given for all experimental points except that of Kapitza and Kapitza [27] entered in Figs. 2 and 3. Thus the correlations given in these figures are incomplete. This is symptomatic of existing experimental records [10,23,27,28,51,59]. Complete records must include values of α, c_r,αc_i,β,R,W, and the entire time history of wave profile evolution. Note R and W are independent parameters of the flow. Various combinations of R and W used as single parameters in some of the existing works are the consequence of particular approximation used in the theories. These combined parameters are not appropriate for use in experimental correlations.

While all of the above mentioned theories are for plane films, the measurements of Binnie and Kapitza and Kapitza were taken on annular films on the outer wall of a cylinder. The disturbances in such films indeed have higher amplification rates than those in plane films at the same flow parameters [39]. However, 50% increase in amplification rates do not occur until the film thickness and the cylinder radius become of the same order. The film thicknesses encountered in the experiments of Binnie and Kapitza and Kapitza were less than one-hundredth of the cylinder radii, and the curvature effects are not strong enough to account for the discrepancy. All evidences seem to suggest that the neglected nonlinear effects should not be overlooked.

Before looking into the nonlinear effects, we address the question of the relevance of the temporally varying disturbances to the onset of ripples [3]. Experimental observations show clearly that ripples grow or decay as they travel downstream. Thus, they are spatially varying but not temporally varying everywhere with the same rate as is assumed in the temporal formulation. However, Gaster [20] proved that the temporal and spatial eigenvalues in plane parallel flows are simply related

$$c_r = \omega/\alpha_r \tag{16}$$

$$c_i = -(\alpha_i/\alpha_r)\,\partial(\alpha_r c_r)/\partial\alpha_r = -(\alpha_i/\alpha_r)c_g \quad , \tag{17}$$

where c_g is the group velocity of the temporally varying linear waves. The Gaster theorem implies that the neutral curves obtained from two different formulations coincide, since $\alpha_i=o$ when $c_i=o$. It should be pointed out that Gaster's theorem is valid only in the region $|\alpha_i(\alpha_r c_i)\underline{max}| << c_g$. This region turns out to be quite large in the present problem. Numerical results of Solesio [57] confirm Gaster's theorem near the neutral curve. The Gaster theorem also implies that the conclusion on stability resulting from temporal formulation will not be altered by the spatial formulation results except when the group velocity of temporal disturbances changes sign at some flow parameters. The first such example was given by the stability problem of a falling liquid curtain stretched between two verticle guide wires [36] in which the group velocity changes sign at $2\sigma/\rho QU = 1$, where Q is the volumetric discharge and U is the maximum curtain velocity.

3.2 Evolution of Ripples

A brief discussion on the nonlinear film waves is in order. It was shown by Lin [31,32,33] that an infinitesimal monochromatic wave interacts nonlinearly with itself, and develops into a supercritically stable small finite amplitude periodic wave near the upper branch of the neutral stability curve (c.f. Fig. 4). He solved the Navier-Stokes equation by extending the Stuart-Watson [49] expansion formalism as refined by Reynolds and Potter [48] for parallel flows with rigid boundaries to that with a free surface. He also showed that the wave speeds decrease with amplitudes near the upper branch but increase with amplitudes near the lower branch of the neutral curve. The same conclusions were reached by Gjevik [18] and Nakaya [45,46] with different approaches. The former considered the case of large surface tension such that $\alpha^2 W = O(1)$ as Lin did and the latter studied the case of $W = O(1)$.

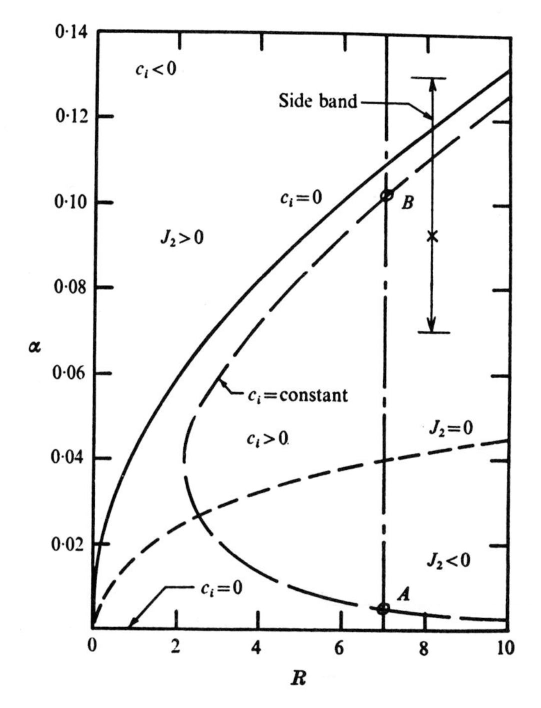

Fig. 4. Supercritically stable waves; $\beta=\pi/2$, $W=463.3$. ×, Kapitza and Kapitza's experiments. B, weak modal interaction described by Eq. (19). A, strong modal interaction described by Eq. (21).

No qualitative differences were found for the nonlinear evolution of temporally and spatially growing ripples [3,19]. Table 1 gives some typical results of supercritically stable ripples. In this table $r = (h_{max} - h_{min})/(h_{max} + h_{min})$.

TABLE 1. Supercritical Waves In Liquid Films, β=90°.

Data	Authors	c_r	r
Water, 15°C	Kapitza & Kapitza (1949)	1.76	.16
c_i = .155	Lin (1969, 1971)	1.78	.225
α = .092	Gjevik (1970)	2.21	.350
R ≈ 8.07	Gjevik (1971, Spatial)	1.71	.260
W = 463.3	Lin (1974)	1.81	.174
	Agrawal (1972, Spatial)	1.90	.218
Alcohol, 15°C	Kapitza & Kapitza (1949)	1.67	.163
c_i = .174	Lin (1969, 1971)	1.73	.221
α = .144	Gjevik (1970)	-	-
R = 5.04	Gjevik (1971, Spatial)	1.66	.270
W = 107.2	Lin (1974)	1.80	.178
	Agrawal (1972, Spatial)	1.89	.220
Water, 19°C	Binnie (1957)	2.34	Not Observed
c_i = .121	Lin (1969, 1971)	2.01	-
α = .066	Gjevik (1970)	-	-
R = 6.60	Gjevik (1971)	-	-
W = 616.7	Lin (1974)	1.72	-
	Agrawal (1972, Spatial)	2.34	-

It should be pointed out that in addition to the weakly nonlinear waves data shown in Table 1, Kapitza & Kapitza also obtained some experimental points for highly nonlinear waves the amplitudes of which are of the same order of magnitude as the film thickness. Theories of highly nonlinear waves will be mentioned at the end of this subsection.

The supercritical monochromatic waves were shown to be stable with respect to two as well as three dimensional sideband disturbances [24,34]. Sideband stability analysis was based on the nonlinear free surface displacement equation obtained from the asymptotic solution of (1) to (5) for $\alpha \to 0$, $\alpha^2 W = O(1)$, $R = O(1)$.

The obtained equation is

$$h_\tau + 2h^2h_x + \alpha[(Bh_x+Ch_{xxx}+Ch_{xxz})_x + (B_1h_z+Ch_{zzz}+Ch_{zxx})_z] + \alpha^2[\ .\ .\ .\ ] + O(\alpha^3) \quad , \tag{18}$$

where subscripts denote partial differentiations,

$$B = 8Rh^6/15 - 2\cot\beta h^3/3,\ C = 2\alpha^2Wh^3/3,\ B_1 = -(2h^3\cot\beta)/3\ ,$$

and the α^2-terms can be found in [24]. Mei [40] and Benney [9] were the first to seek the asymptotic solution of (1) to (5) for the present problem. A two dimensional version of (18) was first obtained by Benney [9] for the case of $W=O(1)$. Benney's equation was extended by Nakaya [45] to include α^3-terms. (18) was obtained earlier by Gjevik (private communication) and later by Atherton and Homsy [2] up to $O(\alpha)$ terms for the case of $\alpha^2W=O(1)$, and by Roskes [47] who neglected the surface tension. When the z dependence is neglected (18) reduces to the two dimensional results of Gjevik [19]. It should be pointed out that (18) could cover both cases of $W=O(1)$ and $\alpha^2W=O(1)$ if the $O(\alpha^3)$ terms had been included. Note that the error introduced in including higher order terms in lower order solutions is automatically corrected in the higher order regular perturbation solutions. Equations similar to (18) for cylindrical films are also available [2,39,55].

It was shown [24] that near the upper branch of the neutral curve where $c_i=O(\varepsilon^2)$, $\varepsilon\to 0$, a monochromatic wave may interact nonlinearly with itself to generate up to the third harmonics if the following solvability condition is satisfied

$$\Gamma_T - c_i'\Gamma + j_1\Gamma_{\psi\psi} + (j_2+ij_4)\Gamma^2\Gamma^* + j_5\Gamma_{\psi\phi} + j_6\Gamma_{\phi\phi} = 0\ , \tag{19}$$

where Γ is the wave amplitude of the fundamental harmonic, $T=\varepsilon^2\tau$, $c_i'=c_i/\varepsilon^2$, * denotes the complex conjugate, ψ and ϕ are the products of ε and the distances measured respectively in the flow- and spanwise- directions in a frame moving with the group velocity of the wave packet, and j's are functions of α, W, R and β. A two dimensional version of (18) was obtained earlier by Lin [34]. The two dimensional solution of the nonlinear Schrodinger equation (19) with the initial condition $\Gamma(-\infty)=0$ is

$$\Gamma = (\Gamma_1\Gamma_1^*)\exp[i\ B(T)T] \ , \ B(T) = -(j_4/T)\int_{T_o}^{T}\Gamma_1\Gamma_1^* dT \ , \quad (20)$$

$$\Gamma_1\Gamma_1^* = c_i' \exp[2c_i'(T-T_o)]/\{1+j_2\exp[2c_i'(T-T_o)]\} \ .$$

It follows that

$$\Gamma \to (c_i'/j_2)^{1/2} \text{ as } T \to \infty \quad \text{if } j_2 > 0 \ .$$

It turns out that $j_2 > 0$ is impossible if $\sigma = 0$, and possible for $\sigma \neq 0$ only in the region near the upper branch of neutral curves. This is in retrospect quite to be expected. While the higher harmonics generated by the fundamental near the upper branch fall in the stable region, those generated near the lower branch fall in the unstable region (see Fig. 4). It was also shown that the supercritical monochromatic ripple is stable with respect to side-band disturbances only if the band-width is narrower than $O(\varepsilon)$ and $j_1 < 0$. However, $j_1 < 0$ only if $\sigma \neq 0$. This may explain why the Stokes waves, contrary to the present problem, are unstable with respect to side-band disturbances [8], since there is no agent such as surface tension to store the energy generated by the side-band resonance in Stokes waves.

Far away from the upper branch, especially near the lower branch of the neutral curve, the nonlinear interaction is stronger, and (19) becomes inadequate. Krishna and Lin [24] obtained from (18) the following equation for the weakly nonlinear evolution of the long waves with strong modal interaction,

$$f_\tau = -\ 4ff_x - c_i f_{xx} + (2\alpha/3)\cot\beta f_{zz} - (2\alpha^2 W/3)(\partial_{xx}+\partial_{zz})^2 f$$
$$-\ \alpha^2\{.\ .\ .\} \ , \quad (21)$$

where x now stands for $x-2\tau$, f is the wave amplitude, and α^2-terms which contain the third and fifth derivatives dispersion terms are given in [24]. The two dimensional version of (21) was given earlier [34]. Neglecting the z-dependence of f and the α^2-terms in (21), Atherton [1] described numerically the nonlinear time evolution of ripples until they reached a stationary periodic state. The equilibrated waves were reported to be insensitive to the initial data. Similar wave forms were found by Tougou [60], Nepomnyaschchii [43,44] and Tsvelodub [61]. The same equation was recently rediscovered by Sivashinsky and Michelson [54]. They found numerically that the nonlinear evolution based on the same equation led to chaotic irregular quasi-steady wave forms. They concluded that the deterministic equation is capable of exhibiting stochastic properties without explaining the discrepancy between their finding and the findings of existing works. However, it should be pointed out that before the free surface becomes chaotic, for whatever reason, the neglected stabilizing three dimensional effects must become important. Thus, it is conjectured that if the z-dependence in (21) is retained in computation, even for the flow condition which leads to two dimensional chaos, one may still be able to find stable three dimensional ripples (roll waves) with crests wavy in the cross-stream direction [37]. It should be pointed out that all of the calculated stationary wave forms based on (21) without dispersion terms do not exhibit the characteristics of "single waves" observed by Kapitza and Kapitza. The profiles of single waves have a steep front on which small wavelets ride. This wave front is connected by a smooth and gentle slope to the rear of the long single wave ahead of it. The wavelets on the steep crest suggest that the dispersion effects represented by the third and fifth derivative terms in (21) may not be neglected in describing the "single waves". It is anticipated that (21) will be encountered in other weakly nonlinear phenomena in which nonlinear steepening, diffusion, dispersion, and restoring force (surface tension in the present problem) are of equal importance.

For large ripples whose amplitudes are of the same order as the film thickness, (18) has to be used. The possibility of associating the homoclinic trajectory of (18) with the large amplitude solitary waves exists. However (18) is valid only when the characteristic length of the wave is much larger than the amplitude. Nonlinear studies of film waves of arbitrary wave lengths have not yet appeared.

3.3 Shear Waves.

It follows from (15) that $R_c \to \infty$ as $\beta \to o$. Thus ripples will not appear on a nearly horizontal film. However, they become unstable with respect to other modes of wave motion. It turns out that relatively short shear waves (Tollmien-Schlichting waves) become the principal eigenmode, in place of ripples, when R is in the order of several thousands (c.f. Fig. 5). Lin [30] obtained the asymptotic normal mode solution of the system (9)-(14) for $\alpha R \to \infty$. He found that the shear waves travel at speeds considerably smaller than the free surface fluid velocity, and films with values of β less than 1° become unstable with respect to shear waves at finite critical wave numbers and critical Reynolds numbers of several thousands. Unfortunately, he made an error in sign at the first term of (11). This error was corrected by De Bruin [13] who ignored the surface tension. The critical Reynolds numbers predicted by De Bruin were several times smaller than those predicted by Lin, although the qualitative characteristics of shear wave in film remain the same. It should be pointed out that the shear wave results of Lin for the case of $\beta = \pi/2$ and $\sigma = o$ remain correct because the first term of (11) where a sign error was made vanishes for this case. The surface tension neglected by De Bruin was recently taken into account by Chin [12] who used liquid film to study polymer drag reduction. Chin found good agreements between the observed damping rates with the theoretical predictions based on linear theory. He also observed that turbulent spots may be made to grow at R less than 2000 when the predicted critical Reynolds number for shear waves was about 6000. This suggests that the film may become subcritically unstable with respect to nonlinear shear waves. Chin's experimental results are consistent with the earlier results of Binnie [10], in general.

Nonlinear studies of film stability with respect to shear waves have not yet appeared. Studies of nonlinear interaction between ripples and shear waves will give us entirely new experience and may lead to some happy discoveries.

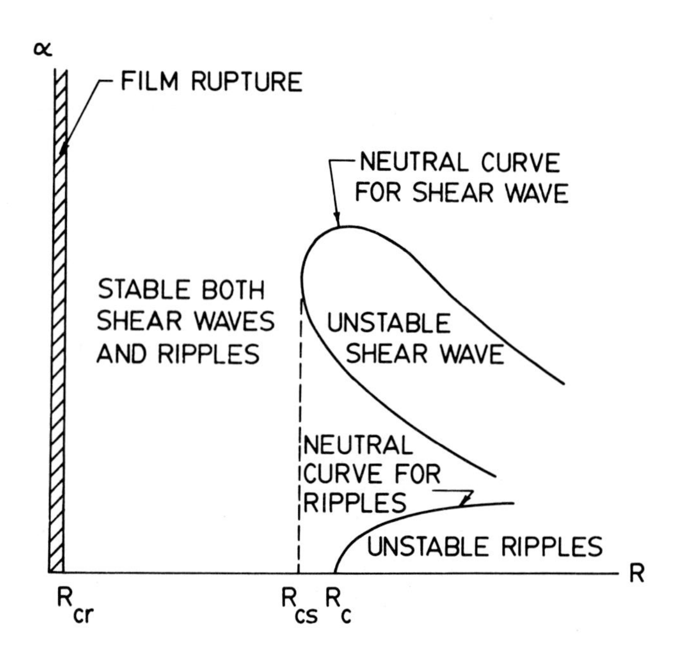

Fig. 5. Different modes of the film instability. R_c, R_{cs} and R_{cb} are respectively the critical Reynolds number of ripple, shear wave and rupture modes.

3.4 Film Rupture.

We have just seen that a thin film is stable with respect to ripple formation at small finite β and may be made stable with respect to shear waves if R is also made sufficiently small, e.g. by reducing the film thickness. However, it is our everyday experience that when a film is made too thin it will rupture. Thus, there is another mode of instability which escapes the detection of classical stability theory. (c.f. Fig. 5). In connection with the study of tear film rupture, Lin and Brenner [38] conjectured that film rupture is due to the neglected dispersion force. The dispersion force arises from dipole-dipole interaction among molecules in an inhomogeneous continuum. The interaction becomes very coherent in thin films. The formula of the dispersion force in thin planar liquid film was rigorously obtained by Dzyaloshinskii et al. [14]. A less accurate formula was obtained earlier by Hamaker [22] for a different geometry. Including Π in (1), Lin et al. obtained by use of the linear theory

$$c = ic_i = -i[(2\alpha R_1/3)(\alpha^2 W_1 + \cos\beta/F_1 + L)] \quad , \tag{22}$$

where R_1, F_1 and W_1 are the Reynolds, Froude and Weber numbers as defined in the subsection 3.1 except U(0) is now replaced by $U_s = (2\sigma/\rho d)^{1/2}$, and L is the London dispersion parameter defined by

$$L = (\partial\Pi/\partial H)_{H=d}(d/\rho U_s^2) \quad .$$

It follows from (22) that the film may become unstable if

$$\alpha^2 W_1 + \cos\beta/F_1 + L < 0 \quad ,$$

and the amplification rate is inversely proportional to the dynamic viscosity. Introducing the Hamaker approximation for "thick" films for which electromagnetic retardation is present, i.e., $\Pi = B/d^4$ where B is the Hamaker constant, the above inequality yields the critical film thickness

$$d_c = \left[\frac{4B}{(2\pi/\lambda)^2\sigma + \rho g\cos\beta}\right]^{1/5}$$

below which the film is unstable. Note that the surface tension is stabilizing just as in the cases of ripples and shear waves. Gravity is stabilizing unless the free surface faces the earth. Gravitational stabilization is due to the higher hydrostatic force acting at thicker portions of the film which tends to move the fluid toward thinner regions, where the hydrostatic pressure is smaller, to smooth out the film. On the contrary, the additional pressure created by the dispersion force is greater at thinner regions than at thicker regions. Thus if the dispersion force overpowers the stabilizing hydrostatic force, the film will continue to thin at the thinner regions until the film ruptures, to open up a bare spot on the solid surface. The maximum and minimum d_c corresponding to β=180° and 0° are given in Table 2 for various values of B.

TABLE 2. Critical Film Thickness

B(erg cm)	10^{-22}	10^{-20}	10^{-15}	10^{-14}	10^{-13}	10^{-12}
min(d_c) (μm)	.11	.28	2.8	4.4	7.0	10.9
max(d_c) (μm)	.15	.37	3.7	5.9	9.3	14.6

These predictions have not yet been borne out by experiments. In fact the predicted critical thicknesses are several orders of magnitude smaller than those observed in an isothermal film and in film flowing over a heated wall [52]. While it is certain that the dispersion force plays a dominant role during the last stage of rupture, it is not clear if it is responsible for the onset of rupture. The neglected evaporation and the associated Marangoni and viscosity stratification may play important roles in the onset of film rupture. The instability due to viscosity stratification will be discussed in the next section and more about the Marangoni effects will be mentioned in the concluding remarks. Film rupture is certainly a very nonlinear phenomenon. The linear stability theory alone may not be adequate to explain quantitatively its mechanism.

4. MULTI-LAYERED FILM.

The wave motion in a falling film of several superposed liquid layers displays some interesting new features. The wave-less basic flow of a n-layered film was obtained [4] as an exact solution of the Navier-Stokes equation. The velocity and pressure distributions are respectively parabolic and linear in each layer but their slopes are discontinuous at each interface. The stability analyses of the basic flow with respect to two dimensional normal mode ripples were carried out with the method of regular perturbation in powers of α [26 , 65]. The fourth order Orr-Sommerfeld equation must be satisfied in each of the n-layers. The 4n integration constants may be determined up to an arbitrary multiplication factor by use of the 4n boundary conditions. It follows from (3) - (6) that the boundary conditions consist of two dynamic conditions at the free surface, two kinematic conditions at the solid wall and two kinematic and two dynamic conditions at each liquid-liquid interface. These boundary conditions are expanded about the unperturbed interface by use of Taylor series. Thus there are 4n additional unknown interfacial displacements appearing explicitly in the boundary conditions. However, there are 4n additional material surface conditions relating the interfacial velocity with the unknown interfacial displacements. Thus the system is closed. It turns out that all of the non-vanishing elements in the characteristic determinant of this eigen-value problem have simple closed form expressions in terms of relevant flow parameters. This enabled us to obtain a relatively simple numerical routine for investigating the wave and stability characteristics of any n-layered films from the eigenvalue

$$c = c_{om}(\varepsilon_j, \gamma_j, \delta_j) + i\alpha c_{1m}(\alpha, \beta, \gamma_j, \delta_j, \varepsilon_j, L_j, R_j, c_{om}) \quad ,$$

where $\gamma_j = \rho_j/\rho_1$, $\delta_j = d_j/d_1$, $\varepsilon_j = \mu_j/\mu_1$, $R_j = \rho_j U_1(0) d_1/\mu_j$,

$$L_j = (\gamma_j/\varepsilon_j)(\sigma_j/\rho_1 d_1^2 g K \sin\beta), \quad K = U_1^2(0)/g d_1 R_1 \sin\beta \quad ,$$

and the subscripts denote the particular liquid layer. Since the system has n degrees of freedom, there are n independent linear modes of wave motion. All modes of long ripples are practically non-dispersive, the dispersivity being of order α^2. A wave mode which gives the j-th interface a larger wave amplitude than all other interfaces is defined as the j-th interfacial mode. The wave speeds of the free surface mode are smaller or greater than twice the free surface velocity depending on whether the composite film is bottom heavy or top heavy. Interfacial velocities of the interfacial modes differ only slightly from the interfacial fluid velocities. They may travel ahead or behind the interfacial fluid particles in the unperturbed films depending on the particular combination of the flow parameters.

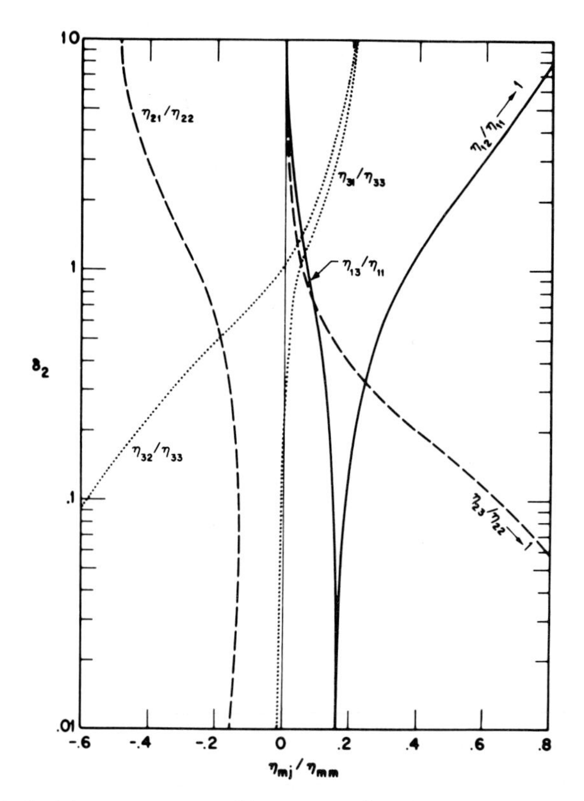

Fig. 6. Relative wave amplitudes, $\delta_1=\delta_3=1$, $\gamma_j=1,1.5,2$; $\nu_j/\nu_1=1(j=1,2,3)$; ——, free surface mode; - - -, second interfacial mode; ····, third interfacial mode.

The interfacial waves may change from varicose to sinuous mode due to change in flow parameters. An example is given in Fig. 6. Figs. 7 and 8 give two examples of stability results. Fig. 7 shows that the interfacial modes are all stable in a density stratified five-layered film at β=85° except when the stratification is top heavy. Top heavy instability is in contrast with Kao's finding [see 65]. Note, however, the small window of stability in the lower left corner of this figure indicates the difference between Lamb-Taylor instability and the present top heavy stability. Fig. 8 shows that the third mode instability associated with the step increase in viscosity at the third interface can be eliminated by an increase in thickness of the most viscous third layer. However, the film remains unstable due to the second mode. The instability due to long shear waves near the origin of the α-R plane is very typical for a viscosity stratified film. In contrast to the short shear waves in boundary layer flows, the physical characteristics of the long shear waves have not yet been experimentally determined. The instability associated with rapid viscosity variation may play more important roles than we thought in the rupture of coolant on heated walls [52]. Note that there exists a sharp viscosity increase in the liquid near the wall. This possibility is worth looking into rigorously. There is yet no satisfactory explanation for the coolant rupture which occurs at a film thickness several orders of magnitude larger than the rupture due to the dispersion force described in the last section.

The instability of multi-layered film with respect to short shear waves at small β but large R remains to be investigated. Nonlinear stability analyses of composite films are of great interest, but have not yet appeared.

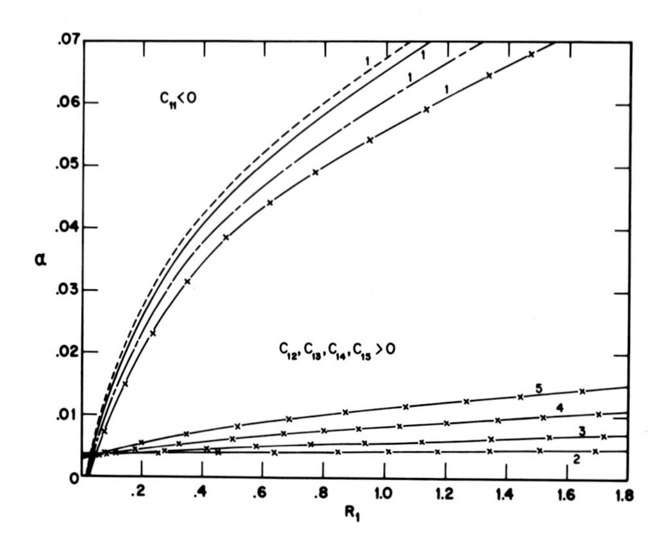

Fig. 7. Neutral stability curves, $\delta_j=1$, $\varepsilon_j=1$, $L_j=40$, $\beta=85°$. 1: $c_{11}=0$, 2: $c_{12}=0$, 3: $c_{13}=0$, 4: $c_{14}=0$, 5: $c_{15}=0$. —, $\gamma_j=1,1.1,1.3,1.6,2$; ---, $\gamma_j=1,1.2,1.4,1.6,1.8$; — — —, $\gamma_j=1,1.01,1.02,1.03,1.04$; —x—, $\gamma_j=1,0.9,0.8,0.7,0.6$. $c_{1m}<0$ (m=2 to 5) for all cases except the last one.

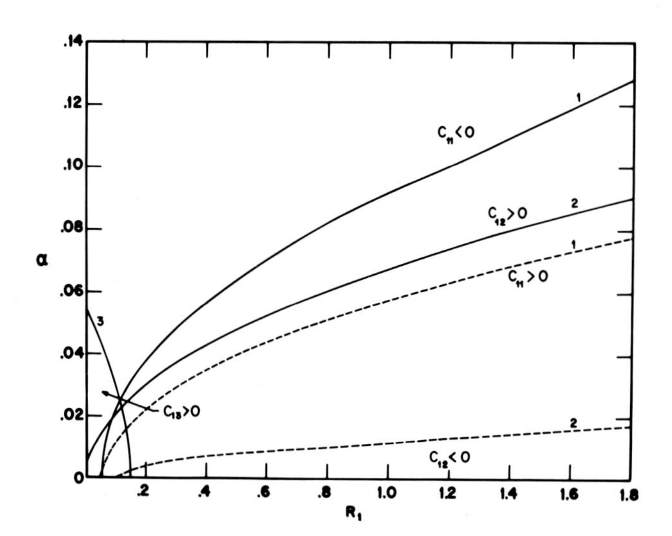

Fig. 8. Neutral stability curves, $\gamma_j=1$, $L_j=40$, $\beta=85°$, $\varepsilon_j=1,0.5,5$. 1: $c_{11}=0$, 2: $c_{12}=0$, 3: $c_{13}=0$. —, $\delta_j=1,1,2$; ---, $\delta_j=1,1,5$. $c_{13}<0$ for $\delta_j=1,1,5$.

5. CONCLUSION.

We have now a fairly good understanding of the onset of gravity-capillary waves at relatively small Reynolds numbers of the film flow. Their nonlinear evolution into supercritically stable small amplitude periodic waves is possible only when the waves introduced by the wave maker are within a narrow bandwidth and if the modal interaction is weak. The modal interaction is weak only near the upper branch of the neutral curve. The evolution of weakly nonlinear waves under strong modal interactions is currently being developed by researchers from various fields. Experiments which record the time history of nonlinear evolution for various sets of flow parameters are sorely needed at the present. Studies of highly nonlinear ripples will certainly contribute to our general understanding of nonlinear phenomena which involve competing effects of dissipation, dispersion, inertia and restoring force. The onset of short shear waves in a single-layered film has now been found both theoretically and experimentally. However, the theoretically predicted long shear waves associated with the density stratification in multi-layered films have not yet been found experimentally. The long shear waves were found to be extremely unstable at small R but stable at large R. This is very suggestive of an important role played by long shear waves at the onset of film rupture. The precise mechanism of the rupture of a thin isothermal or heated film remains very elusive. Studies on the nonlinear evolution of shear waves and on the nonlinear coupling of shear waves with ripples will hopefully, one day, help us in understanding the transition to turbulence in film flows. These studies may also shed light on other related nonlinear phenomena.

It is needless to say that I have not exhausted all of the wave phenomena in film flows. It is probably pertinent to mention that while surface tension is stabilizing in all phenomena considered in this article, it is actually the agent of instability in a liquid film with two free surfaces [36]. There is also an interesting phenomenon of the formation of Marangoni rolls parallel to the flow in a thin film which is otherwise stable to both ripples and shear waves [53].

REFERENCES

1. Atherton, R.W., Studies of the hydrodynamics of a viscous liquid film flowing down an inclined plane, Engineer's Thesis, Stanford University, 1972.
2. __________, and Homsy, G.M., On the evolution equations for interfacial waves, Chem. Eng. Comm. 2 (1976), 57-77.
3. Agrawal, S., Spatially growing disturbances in a film, Ph.D. Thesis, Part I, Clarkson College of Technology, 1972.
4. Akhtaruzzaman, A.F.M., C.K. Wang and S.P. Lin, Wave motion in multi-layered films, J. Appl. Mech. 45 (1978), 25-30.
5. Anshus, B.D. and S.L. Goren, A method of getting approximate solution to the Orr-Sommerfeld equation for flow on a vertical wall, A.I.Ch.E.J. 12 (1966), 1004.
6. Benjamin, T.B., Wave formation in laminar flow down an inclined plane, J. Fluid Mech. 2 (1957), 554.
7. __________, Effects of surface contamination on wave formation in falling liquid films, Fluid Dynamics Transaction, 2 (1965), 383.
8. __________ and J.E. Feir, The disintegration of wave trains on deep water, J. Fluid Mech. 31 (1967), 209.
9. Benney, D.J., Long waves in liquid film, J. Math. Phys. 45 (1966), 150.
10. Binnie, A.M., Experiments on the onset of wave formation on a film of water flowing down a vertical plane, J. Fluid Mech. 2 (1957), 551.
11. __________, Instability in a slightly inclined water channel, J. Fluid Mech. 5 (1959), 561.
12. Chin, R.W., Stability of flows down an inclined plane, Ph.D. Thesis, Division of Applied Sciences, Harvard University, 1981.
13. De Bruin, G.J., Stability of a layer of liquid flowing down an inclined plane, J. of Eng. Math. 8 (1974), 259.

14. Dzyaloshinskii, I.E., E.M. Lifshitz, and L.P. Pitaevskii, The general theory of Vander Waals forces, Adv. Phys. 10 (1961), 165.
15. Dukler, A.E., Characterization effects and modeling of the wavy gas-liquid interface, in Progress in Heat and Mass Transfer (G. Hetsroni, S. Sideman and J.P. Hartnet, eds.), Pergamon Press, New York, 6 (1972), 207.
16. Fedotkin, I.M. and V.R. Firisyuk, Heat transfer rate along a surface wetted by a thin liquid film, Heat Transfer Societ. Res. 1 (1969), 115.
17. Fulford, G.D., The flow of liquids in thin films, Advances in Chemical Engineering, 5 (1964), 151.
18. Gjevik, B., Occurence of finite amplitude surface waves on falling liquid films, Phys. Fluids, 13 (1970), 1918.
19. __________, Spatially varying finite-amplitude wave trains on falling liquid films, Acta Polytechnica Scandinavica, Me 61, 1971.
20. Gaster, M., The role of spatially growing waves in the theory of hydrodynamic stability, Prog. Aero. Sci., 8 (1965), 251.
21. Graef, M., Über die Eigenschaften zwei-und dreidimensionaler Störungen in Riesel Filmen an geneigten Wänden, Mitt. Max-Planck Inst.Strömungsforschung aeron. Versuchsanstalt 26, 1966.
22. Hamaker, H.C., London-Van der Waals attraction between spherical particles, Physica, 4 (1937), 1058.
23. Jones, L.O. and S. Whitaker, An experimental study Falling Liquid Films, A.I.CH.E. J. 12 (1966), 525.
24. Krishna, M.V.G. and S.P. Lin, Nonlinear stability of a viscous film with respect to three-dimensional side-band disturbances, Phys. Fluids, 20 (1977), 1039.
25. Kuramoto, Y., Instability and turbulence of wavefronts in reaction-diffusion systems, Prog. Theor. Phys. 63 (1980), 1885.
26. Kao, T.W., Role of viscosity stratification in the stability of two-layer flow down an incline, J. Fluid Mech. 33 (1968), 561.

27. Kapitza, P.L. and S.P. Kapitza, Collected works. Wave flow of thin layers of a viscous fluid, Pergamon Press, New York, 1965, 690.
28. Krantz, W.B. and S.L. Goren, Stability of thin liquid films flowing down a plane, Ind. Eng. Chem. Found. 10 (1971), 91.
29. Levitch, V.G. and V.S. Krylov, Surface-tension-driven phenomena, Annual Review of Fluid Mech. 1 (1969), 293.
30. Lin, S.P., Instability of liquid film flowing down an inclined plane, Phys. Fluids, 10 (1967), 308.
31. __________, Finite amplitude stability of a parallel flow with a free surface, J. Fluid Mech. 36 (1969), 113.
32. __________, Roles of surface tension and Reynolds stress on the finite amplitude stability of a parallel flow with a free surface, J. Fluid Mech. 40 (1970), 307.
33. __________, Profiles and speed of finite amplitude waves in a falling liquid layer, Phys. Fluids, 14 (1971), 263.
34. __________, Finite amplitude side-band stability of a viscous film, J. Fluid Mech. 63 (1974), 417.
35. __________, Finite amplitude stability of a contaminated liquid film, in Progress in Heat and Mass Transfer (G. Hetsroni, S. Sideman and J.P. Hartnet, eds.), Pergamon Press, New York 6 (1972), 263.
36. __________, Waves in viscous liquid curtain, J. Fluid Mech. 112 (1981), 443.
37. __________, and M.V.G. Krishna, Stability of a liquid film with respect to initially finite three dimensional disturbances, Phys. Fluids, 20 (1977), 2005.
38. __________, and H. Brenner, Tear film rupture, J. Colloid and Interface Sci. 89 (1982), 226.
39. __________, and W.C. Liu, Instability of film coating of wires and tubes, A.I.Ch.E. J. 21 (1975), 775.
40. Mei, C.C., Nonlinear gravity waves in a thin sheet of viscous fluid, J. Math. Phys. 45 (1966), 266.
41. Manneville, P., Statistical properties of chaotic solutions of a one-dimensional model for phase turbulence, Phys. Lett. 84 A (1981) N3.

42. Marschall, E. and R. Salazar, Linear stability analysis of falling liquid films, Ind. Eng. Chem. Fundam. 13 (1974), 289.
43. Nepomnyashchii, A.A., Stability of wave regimes in a film flowing down an inclined plane, Izv. Akad. Nauk SSSR, Mekh. Zhidk. Gaza, No. 3 (1974).
44. __________, Stability of wave motions in a layer of viscous liquid on an inclined plane, in Nonlinear Wave Process in Two-Phase Media [in Russian], Novosibirsk (1977).
45. Nakaya, C., Long waves on a thin fluid layer flowing down an inclined plane, Phys. Fluids, 18 (1975), 1407.
46. __________, Waves of large amplitude on a fluid film down a vertical wall, J. Phys. Sci. Jpn. 43 (1977), 1821.
47. Roskes, C.J., Three dimensional long waves on a liquid film, Phys. Fluids, 13 (1970), 1440.
48. Reynold, W.C. and M.C. Potter, Finite-amplitude instability of parallel shear flows, J. Fluid Mech. 27 (1967), 465.
49. Stuart, J.T., On the nonlinear mechanics of wave disturbances in stable and unstable parallel flows. Part I: The basic behavior in plane Poiseuille flow, J. Fluid Mech. 9 (1960), 353.
50. Stainthorp, F.P. and G.J. Wild, Film flow - the simultaneous measurement of wave amplitude and the local concentration of a transferable component, Chem. Engng. Sci. 22 (1967), 701.
51. __________, and J.M. Allen, The development of ripples on the surface of a liquid film flowing inside a vertical tube, Trans. Inst. Chem. Eng. (London) 43 (1965), 185.
52. Simon, F.F. and Y.Y. Hsu, Thermocapillary induced breakdown of a falling liquid film, NASA TN D-5642, Jan. 1970.
53. Sreenivasan, S. and S.P. Lin, Surface tension driven instability of a liquid film flow down a heated incline, 21 (1978), 1517.
54. Sivashinsky, G.I. and D.M. Michelson, On irregular wavy flow of a liquid film down a vertical plane, Prog. Theor. Phys. 63 (1980), 2112.

55. Shlang T. and G.I. Sivashinsky, Irregular flow of a liquid film down a vertical column, J. Physique 43 (1982) 459.
56. Solesio, J.N., Instabilites des films liquides isothermes, Rapport CEA-R-4835, Centere d'Etudes Nucleaires de Grenoble, 1977.
57. __________, La methode de quadrature par differentiation appliquee a l'hydrodynamique des films liquides, Rapport CEA-R-4888, ibid, 1977.
58. __________, Stabilite des films liquides en ecoulement cylindrique vertical, Rapport CEA-R-4887, ibid, 1977.
59. Tailby, S.R. and S. Portalski, Determination of the wave length on a vertical film of liquid flowing down a hydrodynamically smooth plate, Trans. Inst. Chem. Eng. (London) 40 (1962), 114.
60. Tougou, H., Deformation of supercritically stable waves on a viscous liquid film down an inclined plane wall with the decrease of wave number, J. Physical Soci. Japan, 50 (1981), 1017.
61. Tsvelodub, O.Yu., Stationary traveling waves in a film flowing down an inclined plane, Fluid Dynamics (1981) 591-594. Translated from Izvestiya Akademii Nauk SSSR, Mekhanika Zhidkosti i Gaza, No. 4 (1980), 142.
62. Whitaker S., Effects of surface-active agents on the stability of falling liquid films, Ind. Eng. Chem. Fundam. 3 (1964), 132.
63. __________, and L.O. Jones, Stability of falling liquid films, Effect of interface and interfacial mass transport, A.I.CH.E. J. 12 (1966), 421.
64. Williams, A.G., S.S. Nandapurkar and F.A. Holland, A review of methods for enhancing heat transfer rates in surface condensors, Chem. Engng. 223, CE367-CE373, 1968.
65. Wang, C.K., J.J. Seaborg and S.P. Lin, Instability of multi-layered liquid films, Phys. Fluids 21 (1978), 1669.
66. Yih, C.S., Stability of parallel laminar flow with a free surface, Proc. 2nd U.S. Congr. Appl. Mech., Amer. Soc. Mech. Engrs, (1954), 623.

67. __________, Stability of liquid flow down an inclined plane, Phys. Fluids, 6 (1963), 321.

The author's works on various aspects of film waves were in part supported by National Science Foundation Grant ENG 74-12442, Kodak Company Grants and a Baush and Lomb Grant.

Mechanical and Industrial Engineering Department
Clarkson College
Potsdam, New York 13676

RUPTURE OF THIN LIQUID FILMS

S. Davis

1. INTRODUCTION

The rupture of thin liquid films occurs in many colloid systems. The coalescence of droplets and bubbles, the creation of dry patches, and the breakdown of films into rivulets all involve such rupture.

The spontaneous rupture of a film often occurs at a critical thickness h_c. Scheludko [7] proposes that negative disjoining pressures cause thin regions to thin further. Surface tension resists these tendencies but loses this competition for thin enough films. Scheludko [7] uses a static stability analysis to obtain for a free film an expression for h_c. This expression involves the surface tension, the wavelength of the surface corrugations, and the Hamaker constant involved in a simple model for the long-range molecular forces.

Ruckenstein and Jain [6] study spontaneous rupture of a liquid film on a planar solid. They model the liquid as a Navier-Stokes continuum having an extra body force. This body force has a potential, due to van der Waals interactions, inversely proportional to the third power of the film thickness, Ruckenstein and Jain [6] use a lubrication approximation (though not in a formal manner) to obtain dynamic linear instability results. These include the maximum disturbance growth rate and its corresponding wavelength. From this theory one can also obtain a rough estimate for the

ISBN 0-12-493220-7

rupture time, the time needed for the film to attain zero thickness at some point. This estimate, however, requires that the linearized theory be taken seriously up to rupture when disturbances have amplitudes comparable to the undisturbed layer depth. Linear theories can follow unstable disturbances for only a short time and so cannot indicate whether small unstable disturbances result in rupture.

The present work aims at examining nonlinear effects in film rupture. We follow previous work by applying to the liquid the Navier-Stokes equations modified by an extra body force as described above. London/van der Waals forces are included but double-layer forces are neglected. We formalize a long-wave theory for the nonlinear dynamic instabilities of a static film on a planar solid and derive a <u>nonlinear evolution equation</u> for h(x,t), the film thickness as a function of the lateral space dimension x and time t. The method we use is an adaptation of those used by Benney [2] and Atherton and Homsy [1] for falling films.

Since the evolution equation is a strongly nonlinear partial differential equation, we use numerical methods to solve it as part of an initial-value problem for periodic boundary conditions in x. We obtain true rupture (given our model) in the sense that the film thickness becomes zero in a finite time. We obtain rupture characteristics and find how nonlinearities affect the rupture properties. This work follows that of Williams and Davis [12].

Finally, we discuss limitations of the modeling and extensions to the theory to non-isothermal situations in which thermocapillary (Marangoni) effects are present.

2. FORMULATION

Consider a thin liquid layer bounded below by a rigid plane and above by an interface separating the liquid from a passive gas. The liquid layer is assumed thin enough (say, a few hundred angstroms) that van der Waals forces are effective and thick enough that a continuum theory of the liquid is applicable.

We assume that the liquid is a Newtonian viscous fluid having kinematic viscosity ν and constant density ρ. We nondimensionalize the governing equations and boundary conditions using the following scales: length $\sim h_0$, time $\sim h_0^2/\nu$, velocity $\sim \nu/h_0$, and pressure $\sim \rho\nu^2/h_0^2$, where h_0 is the mean thickness of the layer. For two-dimensional motions of the liquid, we have the (modified) Navier-Stokes equations,

$$u_t + uu_x + wu_z = \nabla^2 u - p_x - \phi_x \quad , \tag{1}$$

$$w_t + uw_x + ww_z = \nabla^2 w - p_z - \phi_z \quad , \tag{2}$$

and the continuity equation

$$u_x + w_z = 0 \quad . \tag{3}$$

Here we have used a Cartesian coordinate system (x,z) with corresponding velocity components (u,w). The origin lies on the solid plane, z increases upward into the liquid while the x-axis lies along the solid plane. The pressure is p and the van der Waals forces are represented through the potential function ϕ which depends on the layer thickness. We follow Ruckenstein and Jain [6] and write

$$\phi = Ah^{-3} \quad , \tag{4a}$$

where we omit the usual additive constant and the (dimensional) Hamaker constant A' is related to A by

$$A = A'/6\pi h_0 \rho \nu^2 \quad . \tag{4b}$$

Thus, we have a continuum description of the liquid dynamics modified only by the presence of an extra distributed body force due to long-range molecular forces.

On the rigid bottom plate there is no slip:

$$u = w = 0 \quad \text{at} \quad z = 0 \quad . \tag{5}$$

Let $z = h(x,t)$ denote the (nondimensional) position of the interface whose mean position is $z = 1$. The kinematic boundary condition has the form

$$w = h_t + uh_x \quad \text{at} \quad z = h \quad . \tag{6}$$

The surface tension σ is assumed to be constant; therefore, the shear stress on the interface vanishes

$$(u_z + w_x)(1 - h_x^2) + 2h_x(w_z - u_x) = 0 \quad \text{at} \quad z = h \quad . \tag{7}$$

The normal stress boundary condition balances the normal stress with the product of surface tension times curvature of the interface; in nondimensional form this takes the form

$$-p + 2\left[(1 - h_x^2)w_z - h_x(u_z + w_x)\right](1 + h_x^2)^{-1}$$

$$= 3Sh_{xx}(1 + h_x^2)^{-3/2} \quad \text{at} \quad z = h \quad . \tag{8}$$

The parameter S,

$$S = h_0\sigma/3\rho\nu^2 \quad , \tag{9}$$

is a nondimensional measure of surface tension. In Eq. (8) we have taken the pressure in the gas to be zero.

3. LONG-WAVE THEORY

In the linear theory analysis of Ruckenstein and Jain [6] when the layer is thinner than a critical value, infinitesimal disturbances begin to grow. These waves have wavelengths much larger than the mean thickness of the layer. We can formalize this notion in the present nonlinear analysis by defining a small parameter k that is directly related to the (small) wavenumber of such disturbances. We then rescale the governing system consistent with linear theory by writing

$$\xi = kx \quad , \quad \zeta = z \quad , \quad \tau = kt \tag{10}$$

and assuming that

$$\frac{\partial}{\partial\xi}, \frac{\partial}{\partial\zeta}, \frac{\partial}{\partial\tau} = O(1) \quad \text{as} \quad k \to 0 \quad . \tag{11}$$

Given that $u = O(1)$, the preservation of the continuity equation (3) requires that

$$w = O(k) \quad \text{as} \quad k \to 0 \quad . \tag{12}$$

We can then examine the x component (1) of the equation of motion. For small k, the inertia is order k, the viscous term is $u_{\zeta\zeta} + k^2u_{\xi\xi}$, and the modified pressure gradient is $k(\partial/\partial\xi)(p + \phi)$. If this pressure gradient were negligible for $k \to 0$ compared to $u_{\zeta\zeta}$, then our approximate system would govern the instability of a static layer subject to no extra body force due to van der Waals attraction. Clearly, such a layer is absolutely stable. Thus, in order to examine spontaneous rupture, we must take this term comparable to

$u_{\zeta\zeta}$. Therefore, we let both

$$p, \phi = O(1/k) \quad \text{as} \quad k \to 0 \quad . \tag{13a,b}$$

We can now obtain approximate solutions to system (1)-(8) by substituting forms (10) and writing

$$u = u_0 + ku_1 + k^2 u_2 + O(k^3) \quad , \tag{14a}$$

$$w = k\{w_0 + kw_1 + k^2 w_2 + O(k^3)\} \quad , \tag{14b}$$

$$p = (1/k)\{p_0 + kp_1 + k^2 p_2 + O(k^3)\} \quad ,$$

with

$$\phi_0 \equiv k\phi = O(1) \quad \text{as} \quad k \to 0 \quad . \tag{15}$$

We equate to zero like powers of k in each equation and boundary condition and obtain a sequence of problems to solve. In this process h is considered an arbitrary O(1) function and the u_j, w_j, and p_j obtained as functionals of h. Given these, we have u, w, and p that are substituted into the kinematic condition (6). This yields the sought after evolution equation that determines h.

At leading order in k, one obtains the system

$$0 = u_{0_{\zeta\zeta}} - p_{0_\xi} - \phi_{0_\xi} \quad , \tag{15a}$$

$$0 = - p_{0_\zeta} - \phi_{0_\zeta} \quad , \tag{15b}$$

$$u_{0_\xi} + w_{0_\zeta} = 0 \quad , \tag{15c}$$

$$u_0, w_0 = 0 \quad , \quad \zeta = 0 \quad , \tag{15d}$$

$$u_{0_\zeta} = 0 \quad , \quad \zeta = h \quad , \tag{15e}$$

$$- p_0 = 3\bar{S}h_{\xi\xi} \quad , \quad \zeta = h \quad . \tag{15f}$$

In Eq. (15f),

$$\bar{S} = k^3 S \tag{16}$$

and we assume that $\bar{S} = O(1)$ as $k \to 0$ in order to retain the effects of surface tension in the problem. This is the typical assumption used to retain surface tension effects in gravity-driven film flows [1].

The solutions of this system are

$$p_0 + \phi_0 = p_0(\xi,\tau) + \phi_0(\xi,\tau) \tag{17}$$

and

$$u_0 = \left(\frac{1}{2}\zeta^2 - h\zeta\right)\Phi \quad , \tag{18}$$

where

$$\Phi \equiv \phi_{0_\xi} - 3\bar{S}h_{\xi\xi\xi} \quad . \tag{19}$$

The solution for w_0 from Eqs. (15c) and (15d) is

$$w_0 = -\frac{1}{2}\left\{\left(\frac{1}{3}\zeta^3 - h\zeta^2\right)\Phi_\xi - \Phi\zeta^2 h_\xi\right\} \quad . \tag{20}$$

We can go on and obtain u_1, w_1 plus further corrections. However, we obtain the first nontrivial nonlinear information on the evolution of nonlinear waves from forms (18) and (20). To see this we take the kinematic condition (6), rescale according to (10) and (11), and use expansions (14a,b) to obtain

$$h_\tau + h_\xi u_0 - w_0 = O(k) \quad \text{at} \quad \zeta = h \quad . \tag{21}$$

If we now substitute forms (18) and (20), we obtain from form (21)

$$h_\tau - \frac{1}{3}(\Phi h^3)_\xi = O(k) \quad . \tag{22}$$

If we now replace Φ in Eq. (22) by its definitions (19) and (4a), we obtain at leading order in k

$$h_\tau + \bar{A}(h^{-1}h_\xi)_\xi + \bar{S}(h^3 h_{\xi\xi\xi})_\xi = 0 \quad . \tag{23}$$

Note that

$$\bar{A} = kA = O(1) \quad \text{as} \quad k \to 0 \tag{24}$$

due to assumption (13b) and recall that the undisturbed thin film has $h \equiv 1$. Eq. (23) is the leading order evolution equation for thin film rupture.

Now that we have derived this equation, it is possible to replace the physical variables x, t, and h so that

Eq. (23) has the form

$$h_t + A(h^{-1}h_x)_x + S(h^3h_{xxx})_x = 0 \quad .$$

The parameters S and A of this system can now be removed by rescaling. We write

$$X = (A/S)^{1/2}x \quad , \quad T = (A^2/S)t \tag{25a}$$

and obtain the final form of the system:

$$h_T + (h^{-1}h_X)_X + (h^3h_{XXX})_X = 0 \tag{25b}$$

with initial condition

$$h(X,0) = g(X) \quad . \tag{25c}$$

Note that by combining the new scales in (25a) with those used previously that the variables X and T are nondimensional lengths and times related to their dimensional counterparts by scales $(A/S)^{1/2}h_0$ and $(A^2/S)(\nu/h_0^2)$, respectively.

4. LINEAR STABILITY THEORY

The evolution equation (25) governs long-wave interfacial disturbances to the static film (having h = 1) subject to van der Waals attractions. The disturbance amplitude is, however, unrestricted so strongly nonlinear interactions are possible. System (25) should also govern infinitesimal disturbances and we examine linear stability theory as a check on the theory.

We formally obtain the linearized problem from system (28) by writing $h = 1 + H(X,T)$ and linearizing in H. We obtain

$$H_T + H_{XX} + H_{XXXX} = 0 \quad . \tag{26}$$

If we abandon the initial condition and instead use normal modes in X and T, we write

$$H(X,T) = H_0\exp(\omega T + iqX) \quad . \tag{27}$$

Here ω is the disturbance growth rate and q is the wavenumber. If we substitute form (27) into Eq. (26), we find the characteristic equation

$$\omega = q^2(1 - q^2) \quad . \tag{28}$$

Small disturbances grow if $\omega > 0$ and decay if $\omega < 0$. The cutoff wavenumber $q = q_c$ is given by $\omega = 0$,

$$q_c = 1 \quad , \tag{29}$$

and there is growth for all $0 < q < 1$. In addition, the small disturbance having maximum growth rate ω_M occurs at $q = q_M$, where $(d\omega/dq)(q_M) = 0$. This computation yields

$$q_M = 2^{-1/2} \tag{30a}$$

and

$$\omega_M = 1/4 \quad . \tag{30b}$$

Results (29) and (30) are equivalent to those of Ruckenstein and Jain [6] taking into account their different notation. Hence, our evolution system (28) reduces to the proper linear theory.

Before we go on, let us define a rupture time T_L based on the linearized result. Even though linear theory is valid only for infinitesimal departures $|h - 1|$ from the planar interface, we presume that we take seriously the linear theory prediction up to rupture. We define T_L so that $1 + H(X, T_L) = 0$ for at least one value of X. If we use form (27), we have for some X, $1 + H_0 \exp(\omega T_L + iqX) = 0$. Now the most dangerous linearized disturbance has $q = q_M$ and $\omega = \omega_M$ given by Eqs. (30). We define T_L as that time when $1 - H_0 \exp(\omega_M T_L) = 0$, so that

$$T_L = -\frac{1}{\omega_M} \ln (H_0) \tag{31}$$

Clearly, the estimate T_L of Eq. (31) pushes the linear theory far beyond its expected range of validity. It is the object of the analysis that follows to obtain the predicted rupture time T_N of the nonlinear system (25) and compare it to the rough value given in Eq. (31).

5. RESULTS

We use finite difference methods to solve system (25) with initial condition

$$q(X) = 1 + B \sin q_M X \tag{32}$$

Centered differences in space are employed from outside in while the midpoint rule is used in time. The difference equations are then solved iteratively. Details are given in Williams [11].

The rupture time T_N is the first time for which h(X,T) vanishes. T_N depends only upon initial amplitude B. When the computed value of T_N is compared with T_L, we see how the actual rupture time is affected by nonlinearity. Williams and Davis [12] find that T_N/T_L decreases monotonically with B from $T_N/T_L \approx 0.45$ at B = 0.2 to $T_N/T_L \approx 0.2$ at B = 0.8. Hence, the nonlinearities of the system accelerate the rupture process (compared to the linearized theory estimate) by up to a factor of 10.

Initial condition (32) presumes a periodicity in X given by the maximizing wavenumber of linearized theory. We can seek that initial wavenumber q_* for which the nonlinear system displays the mininal rupture time; q_* depends on B. Williams and Davis [12] find for B = 0.1 that $q_*/q_M \approx 1.2$ so that shorter waves are preferred than in linearized theory.

6. DISCUSSION

The above work, based on Williams and Davis [12], concentrates on the effect of nonlinearity on the mechanics of film rupture. A strongly nonlinear evolution equation is derived for a highly idealized model system in which a static liquid film of infinite horizontal extent spontaneously ruptures due to long range forces modeled by pair-wise interactions of molecules. The formalism presented, however, is applicable to more complex systems using more realistic models.

The model film is assumed charge-neutralized so that electrical double-layers are absent. The inclusion of double layers [3] could lead to the film seeking a non-ruptured finite equilibrium balancing electrical and surface tension forces against van der Waals attractions. Thus, so called "black films" could be studied.

The model film is assumed non-draining so that the basic state is static. Gumerman and Homsy [4] examine effects of draining in the linearized theory. Although the draining has an effect, its magnitude is not substantial.

This is, perhaps, due to the fact that long-wave instabilities perform an effective vertical averaging on the layer; long waves do not depend strongly on the vertical dynamical structure.

The model film is assumed to be of infinite horizontal extent. Gumerman and Homsy [4] considered the effects of boundary sidewalls in the linearized theory. Since the present instabilities have long waves, the effects are substantial. The present nonlinear theory for the unbounded layer serves as an "outer" solution to that of a film in a finite container. The disturbance flows must "turn around" in end regions near the sides; these flows should asymptotically match the outer flow, the composite fields giving rise to the sidewall-modified evolution equation. In this case, $h(X,T)$ would have boundary conditions at $X = \pm L$ rather than periodicity conditions in X.

The model film is assumed to be uncontaminated. Insoluble surfactants on the interface [6] can alter the rupture process by modifying the surface tension and giving rise to solutocapillary effects.

The model film is assumed to be isothermal. Thermocapillary effects can also lead to alteration of the rupture process. If we imagine the model layer to be placed on a "hot plate" while the surrounding gas is cool, then an instantaneous corrugation of the layer causes the thin region to heat up as compared to the thick region. Since surface tension σ decreases with temperature T for most common liquids, thermocapillarity causes surface and hence bulk flow away from the thin region enhancing the corrugation. Thus, thermal variations should promote rupture. An analysis of such a generalized model leads to the following evaluation equation:

$$h_t + \{[MP^{-1}h^2 - Gh^3 + Ah^{-1}]h_x\}_x + S\{h^3 h_{xxx}\}_x = 0 \quad . \quad (33)$$

Here the Marangoni number M is a non-dimensional measure of $\left|\frac{d\sigma}{dT}\right|$ and G is a non-dimensional measure of gravity; buoyancy is considered negligible. P is the Prandtl number.

If we examine the linearized instability of the flat interface at $h = 1$, we obtain growth for wavenumber q in the range $(0,q_c)$ where

$$q_c^2 = \left[MP^{-1} - G + A\right]/S \quad . \tag{34}$$

It is of interest to specialize this case to pure Marangoni instability of static layers i.e. $A = 0$. Such a mechanism was explained by Pearson [5] in a layer where the interface is non-deformable. From Eq. (34), the limit $S \to \infty$ gives $q_c \to 0$ and recovers Pearson's result that no long-wave convection is possible. If the interface is allowed to deform, Scriven and Sternling [8] conclude that (for $G = 0$) $q_c > 0$ and hence that the layer is always unstable to long waves. Smith [9] clarifies the situation by including gravity, $G \neq 0$, and thus allowing gravity waves on the interface. Since for most common situations $G > MP^{-1}$, long wave instabilities are again absent and Pearson's result is essentially regained.

Finally, we turn to the most drastic assumption in the model film, the form of the van der Waals effect. We followed previous authors and postulated a potential function ϕ for a body force that represents the effect of the long range molecular action through a single Hamaker constant and a single exponent. This is clearly an oversimplified model containing no variations across the layer and presuming a scalar effect of the interactions. In reality the thinness of the layer should give rise to non-isotropies in the mechanical behavior of the material which in a continuum representation might be represented in terms of a non-isotropic stress tensor. If one could formulate such a model, one would then again have a static-state solution subjected to long-wave instabilities. One could again adapt the present formalism to this system and obtain an evolution equation. However, if as discussed above, the long-wave theory effectively averages over the depth, then the more elaborate evolution equation obtained may, indeed, give results similar to those of the model system already addressed. This remains, however, only a conjecture.

We have set up a formalism for the study of film rupture which in principle includes interface departures comparable to the mean film thickness. Hence, it is possible to predict rupture in a self-consistent manner and the way is open for the inclusion of those physical effects essential to many situations.

REFERENCES

1. Atherton, R. W. and Homsy, G. M., Chem. Eng. Commun. 2 (1975).
2. Benney, D. J., J. Math. Phys. 45 (1966).
3. Clunie, J. S., Goodman, J. F. and Ingram, B. T., Surface Colloid Sci. 3 (1971).
4. Gumerman, R. J. and Homsy, G. M., Chem. Eng. Commun. 2 (1975).
5. Pearson, J. R. A., J. Fluid Mech. 4 (1958).
6. Ruckenstein, E. and Jain, R. K., Chem. Soc. Faraday Trans. 70 (1974).
7. Scheludko, A., Advan. Colloid Interf. Sci. 1 (1967).
8. Scriven, L. E. and Sternling, C. V., J. Fluid Mech. 19 (1964).
9. Smith, K. A., J. Fluid Mech. 24 (1966).
10. Vrij, A., Discuss. Faraday Soc. 42 (1966).
11. Williams, M. B., Ph.D. dissertation, Johns Hopkins University, Baltimore, MD (1981).
12. Williams, M. B. and Davis, S. H., J. Colloid Interf. Sci. 90 (1982).

The author gratefully acknowledges the support of the National Science Foundation, Fluid Mechanics Program.

Department of Engineering Sciences and Applied Mathematics
Northwestern University
Evanston, IL 60201

THE MOVING CONTACT LINE

E. B. Dussan V.

1. INTRODUCTION

Many hydrodynamical analyses primarily concerned with the behavior of fluid interfaces formed by two immiscible fluids are restricted to systems infinite in extent in the horizontal direction. Such approaches cannot be used to assess the influence of side bounding walls. On the other hand, including the entire solid boundary requires taking into account the presence of contact lines, refer to Figure 1. Unfortunately, the appropriate equations and boundary

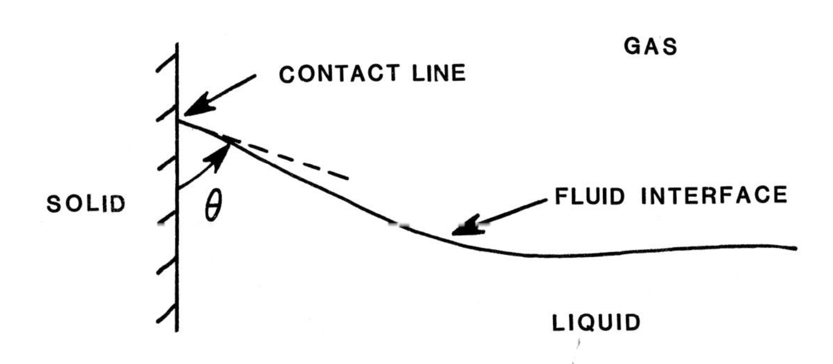

Figure 1: The contact line is formed at the intersection of the fluid interface and the surface of the solid. The immiscible fluids are referred to as a liquid and a gas for convenience. The contact angle, θ, is the angle formed between the plane tangent to the fluid interface at the contact line and the surface of the solid.

ISBN 0-12-493220-7

conditions which describe the dynamics of the fluids in their immediate neighborhood are as yet unknown. The aim of this presentation is threefold. It will begin with an identification of the mechanism by which the dynamics of the fluids in the immediate vicinity of the moving contact line influences the dynamics of the entire fluid body. Awareness of this mechanism would enable one to assess the necessity for including these regions when analyzing the influence of side bounding walls. Next, the nature of the failure of the Navier-Stokes equation and the usual boundary conditions to describe the dynamics of the fluids in this region will be reviewed from both a theoretical and experimental point of view. Finally, a well-posed boundary-value problem will be presented that bypasses this difficulty, along with an illustration of a procedure by which its validity can be tested experimentally.

2. MECHANISM

There are many problems of practical concern in which moving contact lines abound. Most processes whose objective is to apply a protective coating to the surfaces of solids begin by laying down a thin uniform layer of liquid. Contact lines are present during the initial phase when the dry solid is wetted by the liquid. The formation and growth of bubbles of vapor on hot surfaces, or the condensation of vapor in the form of droplets on cold surfaces, contain moving contact lines. Contact lines are also usually present during immiscible fluid displacement through solid porous media. This could represent water displacing oil through a porous rock, or a liquid such as water displacing air through cloth or paper.

It is evident from these examples that contact lines can be present under a variety of conditions; heat and mass transfer may be taking place; the temperature field within the solid may not be constant, its value depending explicitly on the location of the contact line; and surface-active-agents, chemicals that adsorb preferentially onto fluid interfaces, may be present. However, most efforts focused on understanding the dynamics of the fluids within the immediate region surrounding the moving contact line have been restricted to chemically pure systems in which the temperature is held constant with no mass transfer. We will restrict our attention to such systems.

Under these conditions, the mechanism by which the dynamics of the fluids in the immediate vicinity of the moving contact line influences the entire body of fluid can easily be identified. It is through the shape of the fluid interface by way of the contact angle that this occurs†, refer to Figure 1. The contact angle in some sense serves as a bridge across which phenomena occurring on a very small scale can travel and influence the rest of the fluid. On the one hand, there is little argument that the contact angle is controlled by the behavior of the system in the region surrounding the contact line. Its value is determined by both the nature of the materials, that includes the identity and finish of the solid as well as the identity of the two fluids, and the dynamics of the fluids. On the other hand, the contact angle is the natural boundary condition for the shape of the fluid interface in hydrodynamical analyses. Hence, only for those fluid mechanical problems for which the contact angle plays an important role must one be concerned with the dynamics of the fluids immediately surrounding the moving contact line. A general rule of thumb for identifying such problems is to examine the relevant length scale; the smaller its value the greater the likelihood that the contact angle cannot be ignored.

Unfortunately, very little is known about the contact angle under either static or dynamic conditions [1]. Young's equation is often referred to when describing the static contact angle, θ_s,

$$\sigma \cos \theta_s = \sigma_{gs} - \sigma_{\ell s}$$

Here, each phase boundary has been modeled as having a constant energy per unit area; the quantities σ, σ_{gs}, and $\sigma_{\ell s}$ denoting its value on the gas-liquid, gas-solid, and liquid-solid interface, respectively. However, there are two fundamental weaknesses with this equation. To begin with, one would like

†This conclusion is based upon both the theoretical and experimental results presented in section 3.

to use it as a means for predicting the value of the static contact angle. Unfortunately, this cannot be done since measurements of σ_{gs} and $\sigma_{\ell s}$ are impossible to make. Consequently, one often finds in the literature Young's equation used for determining the value of $\sigma_{gs} - \sigma_{\ell s}$ from experimental measurements of θ_s and σ. The second and more telling weakness comes from the fact that Young's equation implies that the value of the static contact angle is unique. This is infrequently observed experimentally. Instead, it is found that two angles, θ_A and θ_R, often referred to as the advancing and receding contact angles, are used for characterizing a static system. As long as $\theta \varepsilon [\theta_R, \theta_A]$ the contact line will not move [2, 3, 4]. One of the remaining problems in the theory of capillarity is to identify the model which gives rise to this characteristic, frequently referred to as contact angle hysteresis.

Very little effort has been directed towards predicting the value of the contact angle under dynamic conditions. A few theories exist [5], however they all suffer from the inability to measure experimentally all the parameters upon which they depend. The only firm information about the dynamic contact angle comes from experimental measurements [6, 7, 8].

3. INADEQUACY OF THE USUAL HYDRODYNAMIC MODEL

a. Theory

Analyzing the motion of the fluids in the immediate vicinity of the moving contact line is not a straightforward procedure. This is due to the fact that the validity of several assumptions frequently used in the field of fluid mechanics has been brought into question. The most controversial is the no-slip boundary condition, a condition always assumed at surfaces formed between real (viscous) fluids and solids.

There are various levels at which the compatibility of the no-slip boundary condition and the existence of a moving contact line have been addressed. The most fundamental concerns its kinematics. It has long been thought that a physically acceptable velocity field could not exist which simultaneously describes one immiscible fluid displacing another from the surface of a solid and fluids adhering to the solid. (The term "adherence" has often been used in the literature interchangeably with "no-slip" [8, 9].) However, this is not the case. A generalized rolling type of motion has been identified which establishes their compatibility [11]. In addition, such a motion serves as an illustration of the lack of equivalence between the terms adherence and no-slip. Unfortunately, velocity fields which give rise to such motions must be multivalued at the moving contact line†. It is this characteristic that is responsible for many of the difficulties which arise when attempting to determine explicit forms of the velocity field. We will return to this point below.

An interesting offshoot of the study of kinematically permissible velocity fields associated with fluids containing moving contact lines has been an investigation of the <u>kinematic boundary condition</u>††. This is a necessary restriction imposed on the velocity field at the boundary of all material bodies. It has frequently been assumed that surfaces bounding material bodies must always consist of the same material points. Either it has been assumed in the derivation of the kinematic boundary condition, or it has been said to be implied by the kinematic boundary condition. In general, neither statement is correct. It has been demonstrated that a velocity field need not obey the kinematic boundary condition at every point on the surface of a material body, and even if it does, the bounding surface need not be a material surface. The key characteristic is the differentability of the velocity field at the bounding surface. It has been shown

---

†The velocity vector, $\underset{\sim}{u}(\underset{\sim}{x},t)$, may be continuous everywhere except at the contact line, $\underset{\sim}{x}_{c\ell}$, in which case the value of $\lim_{\underset{\sim}{x}\to\underset{\sim}{x}_{c\ell}} u(\underset{\sim}{x},t)$ depends upon the particular path used to evaluate the limit as $\underset{\sim}{x}$ approaches $\underset{\sim}{x}_{c\ell}$.

††Let $F(x,t)=0$ describe the location of the surface bounding a material body. The mathematical form of the kinematic boundary condition is:

$$\frac{\partial F}{\partial t} + \underset{\sim}{u}\cdot\nabla F = 0 \text{ at } F(\underset{\sim}{x},t) = 0$$

that in order for surfaces bounding material bodies not to consist of the same material, the velocity field must not be Lipschitz continuous[†] at some point on the boundary [12]. Whether or not such velocity fields are physically realizable depends on the constitutive nature of the fluids involved.

The usual approach taken over the years for obtaining the velocity field of the fluids near the moving contact line has been to assume a model in which the fluids are incompressible, Newtonian, and obey the no-slip boundary condition at a rigid, smooth solid surface. Such approaches have always lead to velocity fields which are singular at the moving contact line. This singularity is responsible for the fluids exerting an infinite drag on the solid, dissipating mechanic energy at an infinite rate, and for an inability to specify the contact angle as a boundary condition for the shape of the fluid interface [1]. One should not lose sight of the fact that the contact angle represents an important boundary condition, reflecting the physio-chemical nature of the two fluids and solid surface. By not being able to specify its value, the shape of the entire fluid interface cannot be determined.

For a long time the existence of this singularity was attributed to the seemingly inherent kinematic incompatibility created by simultaneously demanding the velocity field to obey the no-slip boundary condition and to contain a moving contact line; however, it has already been pointed out that this is not the case. Rather, the singularity arises from the lack of compatibility between the multivalued velocity field demanded by the no-slip boundary condition, and the other elements of the model specified above [11]. Hence, the velocity field in the immediate vicinity of the moving contact line cannot be obtained by solving an already well-established boundary-value problem. The correct model,

---

†The velocity field, $\underset{\sim}{u}(\underset{\sim}{x},t)$, is Lipschitz continuous if there exists a constant, A, independent of $\underset{\sim}{x}$ and t such that

$$|\underset{\sim}{u}(\underset{\sim}{x}_1,t) - \underset{\sim}{u}(\underset{\sim}{x}_2,t)| < A|\underset{\sim}{x}_1 - \underset{\sim}{x}_2|$$

for any choice of $\underset{\sim}{x}_1$ and $\underset{\sim}{x}_2$ as long as they lie within either fluid.

i.e. constitutive relation and boundary conditions, is as yet unknown.†

b. Experiment

It is also of interest to note the manner in which the lack of knowledge of the appropriate model of the fluids in the immediate vicinity of the moving contact line manifests itself experimentally. By far, the most popular reported measurable quantity has been the contact angle. Various configurations have been used to measure its value, some of which are wires or tapes entering or leaving large baths of liquids, drops spreading on horizontal solid surfaces, and liquids flowing through capillaries or slots [1]. For the remainder of this section, attention will be focused on one such study [15] whose techniques and accuracy typify others. In section 4 it will be shown how these results can be used to test the validity of a method which has been suggested for eliminating the singularity at the moving contact line.

The contact angles were measured in a small chamber formed by two parallel glass microscope slides separated from each other, a distance 2a, in a sandwich-like configuration by narrow pieces of metal placed along their two longer edges. The slides were held together with Teflon tape applied to their two longer and one of the two shorter edges, thus creating a narrow slot through which silicone oil was forced to flow. The oil entered the chamber through a small hole in the piece of tape running along the shorter edge. The chamber was maintained in a vertical position causing the air-oil interface to have a horizontal orientation which traveled in an upward direction as the oil entered at various controlled rates.

---

†For the special case in which the contact angle for the liquid is zero and the gas is replaced by a void, it is possible for a velocity field to exist which is not multivalued. The dynamics of the liquid can then be studied within the context of the usual hydrodynamic model [13, 14]. However, this represents a singular limit for the case when $\mu_g/\mu_\ell \to 0$. Regardless of the size of ρ_g and μ_g, as long as they are nonzero the velocity at the contact line must be multivalued.

The key measurement in such a configuration is the apex height, h, refer to Figure 2. It was impossible to measure the value of the contact angle directly due to the presence

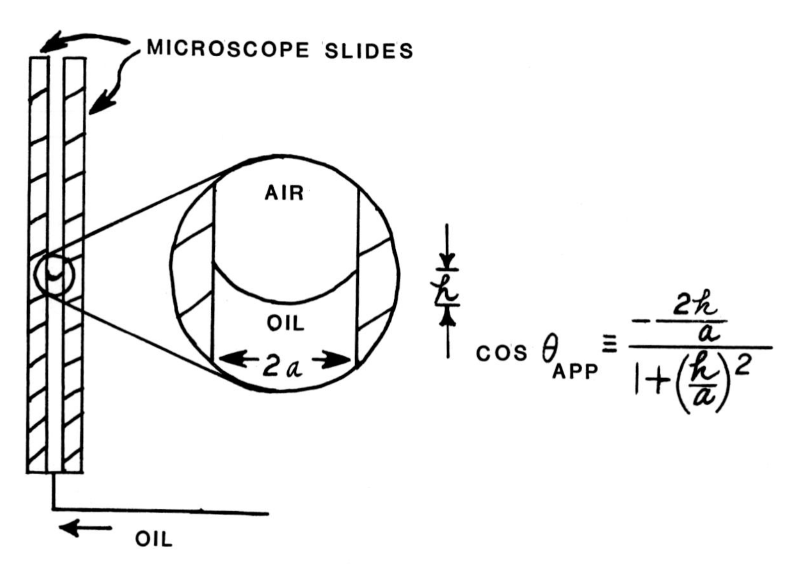

Figure 2: Side view of the chamber. This view is impossible to achieve experimentally due to the presence of the two strips of metal along the vertical edges needed to maintain a distance of 2a between the microscope slides. The quantity which is measured is the apex height, h. Values of the contact angle are obtained by the above formula which assumes the shape of the air-oil interface is that of an arc of a circle.

of the metal strips along the vertical edges of the chamber, so the contact angles were calculated on the assumption that the fluid interface had the same shape as a segment of a horizontal circular cylinder. This was justified by the fact that in a set of separate experiments of oil displacing air through a glass capillary of diameter approximately equal to 2a at the same speeds, it was impossible to distinguish the shape of the fluid interface from that of a segment of a sphere. It should be noted that values reported by others of contact angles measured in capillaries are frequently calculated by the same formula indicated in Figure 2 (in this case 2a represents the inner diameter of the capillary).

The subscript APP, short for apparent, refers to the fact that the value of the contact angle is calculated from the formula in Figure 2.

Three sets of experiments were performed, the results of which appear in Figure 3, for values of 2a corresponding to 0.01, 0.07 and 0.12 cm. This differs from studies by

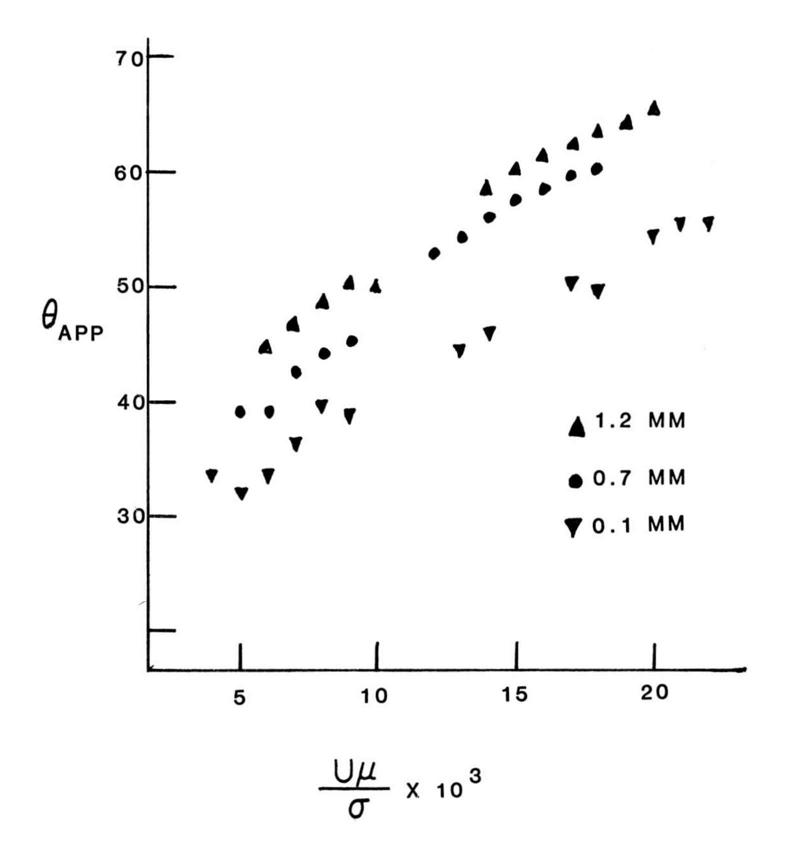

Figure 3: The displacement of air by 1000 cstk Union Carbide L-45 silicone oil through slots formed by microscope slides, at 25$\pm$ 0.2°C. (This is Figure 4 in [15].)

others in that it is usual for the identity of the materials to vary keeping the geometry of the apparatus the same. Two conclusions were drawn from the results. The first simply acknowledged the fact that the measured contact angle cannot be a property solely of the materials of the system since its value depends on 2a; hence, it cannot represent the actual contact angle. It is the actual contact angle which serves as the boundary condition for the shape of the fluid interface. The second conclusion was that the usual hydrodynamic model identified earlier in this section cannot be correct for systems containing moving contact lines. This conclusion was arrived at independent of any interpretation of θ_{APP}, i.e., it relied solely upon the fact that the results in Figure 3 indicated h/a depends on 2a as well as $U\mu/\sigma$.

The details behind the second conclusion were straightforward making use of dimensional analysis. Assuming the usual model was valid, an analysis of the motion of the oil and air would yield:

$$\frac{h}{a} = f\left(\frac{\Delta\rho g a^2}{\sigma}, \frac{\mu U}{\sigma}, \frac{Ua\rho}{\mu}, \frac{\mu_{air}}{\mu}, \frac{\rho_{air}}{\rho}, \theta\right)$$

where U denotes the speed of the contact line; σ denotes the surface tension; μ and μ_{air} denote the viscosity of the oil and air, respectively; ρ and ρ_{air} denote the density of the oil and air, respectively; g denotes the gravitational constant, θ denotes the value of the actual contact angle which itself can depend on the dynamics of the system such as U, and $\Delta\rho \equiv \rho - \rho_{air}$. Since the experiments were performed at small values of $\Delta\rho g a^2/\sigma$ and $Ua\rho/\mu$, both can be ignored in the above relationship. The values of μ_{air}/μ and ρ_{air}/ρ were the same in all the experiments. The value of the actual contact angle, $\theta(U)$, though unknown, must be independent of 2a because the model assumes it is a material property. Hence, if the usual hydrodynamic model were adequate, then h/a should only depend on the value of $U\mu/\sigma$, i.e., all of the results in Figure 3 should lie on the same curve! Thus, it appears that something is missing from the model. Perhaps another length scale should emerge from the region surrounding the moving contact line which could combine with 2a to create a dimensionless parameter upon which h/a can depend.

4. WELL-POSED BOUNDARY-VALUE PROBLEM

It is clear that in order for a physically acceptable analysis to be performed which includes the region surrounding a moving contact line, one must be willing to use a model for the fluids other than that typically employed in the field of fluid mechanics. By far the most popular approach has been to disregard the no-slip boundary condition.

It seems at first glance somewhat surprising to suggest a model which would allow a viscous fluid to slip at the surface of a solid. However, when one stops to think about it, maybe we should be astonished by the success the no-slip boundary condition has enjoyed over the years. Goldstein [16] points out in his historical review of the subject that other boundary conditions such as the one introduced by Navier have been used. Navier assumed that

$$\underset{\sim}{\tau} \cdot [\underset{\sim}{T}(\underset{\sim}{n})] = \beta\, \underset{\sim}{\tau} \cdot \underset{\sim}{u}$$

where $\underset{\sim}{n}$ and $\underset{\sim}{\tau}$ denote the unit normal and tangent vectors to the fluid-solid interface, respectively; $\underset{\sim}{T}$ denotes the stress tensor; and β is a constant, often referred to as the slip coefficient. The no-slip boundary condition was gradually adopted during the mid-eighteen hundreds only after much debate. Its justification seems to rest principally upon the fact that it accurately predicts experimentally observed laminar motion of fluids regardless of the finish of the fluid-solid interface. The most familiar exception, as demonstrated by Maxwell, occurs when the fluid is a rarefied gas, in which case the condition suggested by Navier is in agreement with experiments. Here, $1/\beta$ approaches zero, i.e., the no-slip boundary condition is retrieved, in the limit as the ratio of the mean free path of the gas to the characteristic length scale of the geometry of the system approaches zero.

It is of interest to note that Goldstein's review addresses systems consisting of only one component of fluid. Apparently, the appropriate boundary condition at the fluid-solid boundary of a multicomponent system is as yet unknown. Jackson [17] points out in a review restricted to viscous gases that using the no-slip boundary condition on the mass-averaged velocity, the velocity appearing in the Navier-Stokes equation, can lead to substantial error especially for the case of isobaric diffusion. Theoretical predictions of the motion of the fluids disagree with experimental observations. The inappropriateness of the no-slip boundary condition on the mass-averaged velocity is further strengthened by the presentation of a generalization of Maxwell's derivation to a multicomponent system of gases. Whereas, Maxwell retrieves the no-slip boundary condition in the limit as the ratio of the mean free path of the gas to the characteristic length scale of the geometry of the system approaches zero, a slip boundary condition results in the multicomponent case. However, it does reduce to the no-slip boundary condition when applied to a system consisting of only one component. Jackson presents experimental evidence which generally supports predictions based upon this derived slip boundary condition.

The results of a recent study by Richardson [18] gives some insight into the success enjoyed by the no-slip boundary condition for single component systems. The motion of fluid flowing over a solid surface with periodic irregularities was analyzed. Far above the surface of the solid the velocity field approached a specified gradient. At the surface of the irregularities the velocity obeyed either the no-slip boundary condition or the sheer-free condition. That is to say, the slip boundary condition given by Navier was used in which the value of β was assumed to be either infinite or zero. Two length scales characterize such flows. The smaller one is associated with the size of the irregularities while the larger one characterizes the overall geometry responsible for creating the specified gradient in the velocity field far above the surface. Richardson found that the velocity field was insensitive to the value of β when viewed from the larger length scale in the limit as the size of the irregularities approached zero while keeping the ratio of their amplitude to wavelength fixed[†]. Hence, fluids can be slipping at a solid boundary; however, the presence of roughness microscopic in size can make it appear on the larger length scale, that from which almost all hydrodynamic observations are made, as if the no-slip boundary condition is obeyed.

It is conceivable that the geometric configuration and the requirements placed on fluids in the immediate vicinity of the moving contact line, responsible for creating a very large gradient in the velocity field, can cause the dynamics of the fluids to be highly sensitive to the value of slip coefficient. Analyses have been performed using the boundary condition introduced by Navier assuming an absolutely flat solid surface[††]. [19, 20, 21] Indeed, it has been found that such a boundary condition eliminates the singularity at the moving contact line. An attempt has been made to justify its use based upon an analysis similar to Richardson's [22].

Other boundary conditions which allow the fluids to slip at the surface of the solid have been investigated. In an analysis of a thin drop of liquid spreading over a horizontal surface, Greenspan [23] found that the singularity could conveniently be removed by assuming a slip coefficient which is directly proportional to the local thickness of the layer of liquid. Huh & Mason [20] have pursued a model in which they assumed $\underset{\sim}{t} \cdot [\underset{\sim}{T}(\underset{\sim}{n})] = 0$ within a small specified distance of the moving contact line, and the no-slip boundary condition applies elsewhere. Realizing

---

†This represents an important restriction. If instead only the wavelength is held fixed, then the results are entirely different.

††One should keep in mind that the value of the slip coefficient used when modeling the surface as flat probably differs considerably from that which might be appropriate when the roughness of the surface is not ignored, such as in Richardson's analysis. It is not even obvious if Naviers slip boundary condition evaluated on the actual roughened surface would give rise to a similar boundary condition with different slip coefficient for a model in which the roughened surface is assumed to be flat.

that the singularity would be removed by any slip boundary condition which eliminated the multivalued velocity field at the moving contact line, Dussan V. [24] analyzed the motion of a fluid resulting from three diverse slip boundary conditions. The purpose of the study was to assess the sensitivity of the velocity field away from the contact line to various aspects of the models of the slip boundary condition. She found that the only property affecting the entire body of fluid was their effective slip length, i.e., the length scale associated with the size of the region surrounding the moving contact line within which the velocity of the fluid in contact with the solid differed substantially from that of the solid. Recasting the three boundary conditions into the form $\underset{\sim}{\tau}\cdot[\underset{\sim}{T}(\underset{\sim}{n})] = f(\underset{\sim}{\tau}\cdot\underset{\sim}{u})$, where f denotes a nonlinear function of the velocity of the fluid at the surface of the solid, representing in some sense a generalization of Navier's model, $(f(\underset{\sim}{\tau}\cdot\underset{\sim}{u})=\beta\underset{\sim}{\tau}\cdot\underset{\sim}{u})$, revealed that quite a diverse family of functions can give rise to the same observable motion. Thus, her results suggest that different slip models could not be distinguished through observations based solely upon the dynamics of the fluids away from the contact line.

All of the analyses mentioned above involving slip boundary conditions give rise to fluid motions having the following common structure. Two length scales emerge. The smaller length scale characterizes the size of the region neighboring the contact line within which the velocity of the fluid at the solid surface differs substantially from that of the solid. Within this region, often referred to as the "inner region", the dynamics of the fluids is very sensitive to the form of the slip boundary condition, however, analyses are simplified by the fact that the motion is locally two-dimensional. The larger length scale characterizes the size of the entire fluid body. For example, in the experiments described in the previous section it is given by the distance between the two parallel microscope slides, or the diameter of the capillary. It also represents the characteristic length scale of the "outer region". This region includes all the fluid except that in the immediate vicinity of the moving contact line. Hence, the fluid within the outer region obeys the no-slip boundary condition, however, its geometrical configuration may cause the analysis of its velocity field to be quite complex. The location of these regions is illustrated in Figure 4.

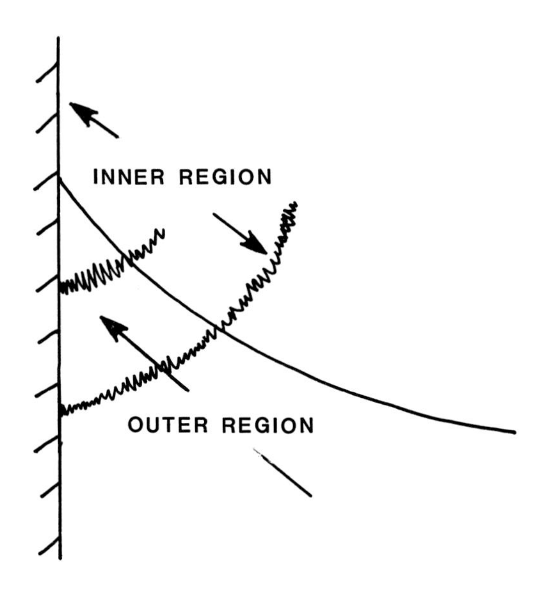

Figure 4: The entire fluid body consists of the inner and outer regions. In order for the solutions of the dynamics of the fluids in these regions to completely describe the entire fluid body, the regions must overlap. Within the overlap region the dynamics of the fluids has characteristics consistent with that found in each region separately. Hence, the dynamics must be locally two-dimensional since it is part of the solution of the inner region; in addition, it must satisfy the no-slip boundary condition, since it is a property of the solution of the outer region.

Although the behavior of the fluids in the inner region is of great scientific interest, it is not necessarily of direct major concern to most fluid mechanicians. Rather, their prime objective is to determine the dynamics of the fluids in the outer region. As a consequence, it is important to identify the manner in which the dynamics of the fluids in the outer region is affected by that which occurs in the inner region. All of the above cited analyses which use a slip boundary condition have the following two common characteristics. (i) The velocity field in the outer region does not depend explicitly on the form of the slip boundary condition. (ii) The slope of the fluid interface can be completely determined except for a constant of integration. This constant of integration represents the extent to which the solution in the outer region depends upon the behavior of the fluids in the inner region. It should be emphasized that all of the analyses upon which these statements are based assume a vanishingly small value of the Reynolds number, and a small but non-zero value of $U\mu/\sigma$.

While no experiments presently exist which can confirm the fact that fluids actually slip at the surface of a solid, there does exist evidence illustrating its consistency with experimental observation. Expressions for θ_{APP} derived from analyses which assume a slip boundary conditions depend upon the value of length scale associated with the outer region, in addition to the values of $\mu U/\sigma$ and θ [20, 25]. This is in agreement with the data reported in Figure 3 of Ngan & Dussan V. Admittedly, the former refers to analyses of flow in a capillary, while the latter were obtained from experiments performed with a slot.

Based upon the above-cited characteristics common to analyses using the slip boundary condition, formulating a well-posed boundary-value problem for the dynamics of the fluid in the outer region is a relatively straightforward procedure (a detailed description appears in [25]). The entire solution in the outer region is completely determined upon knowledge of the constant of integration appearing in the solution for the shape of the fluid interface. The constant would be determined by the value of the actual contact angle if the solution could be evaluated at the contact line. However, this cannot be done. Instead, the constant can be evaluated using the angle θ_I, refer to Figure 5. Since this angle is assumed to lie close enough to the contact line, in the region where the dynamics is locally two-dimensional, it can be considered to be a

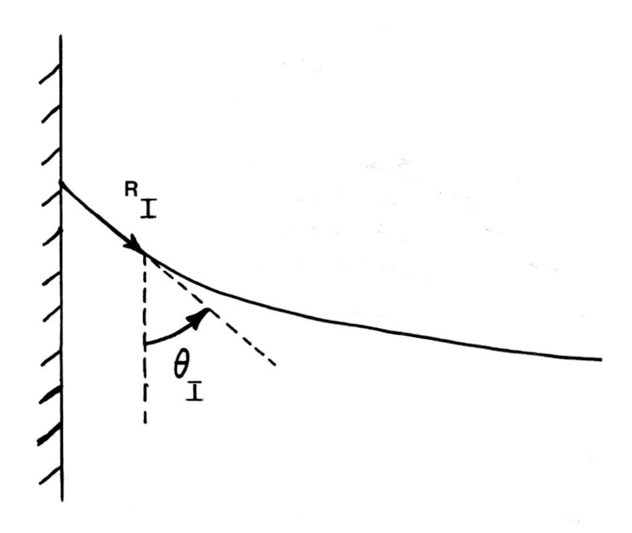

Figure 5: The angle formed between the local tangent plane to the fluid interface a distance R_I from the contact line and the plane tangent to the surface of the solid is given by θ_I.

material property of the system. On the other hand, since it is assumed to be located at a sufficient distance away from the contact line, it can be used as a boundary condition on the dynamics of the fluids in the outer region. From a cause-and-effect point of view, the value of θ_I is controlled by both the dynamical behavior of the actual contact angle and the physical model of the fluids in the immediate vicinity of the contact line through the deformation of the fluid interface. If the correct model of the fluids within the inner region were known, then θ_I could be calculated. However, this seems unlikely in the near future. Fortunately, it can be measured using existing experimental techniques.

There is no question that the above-specified boundary-value problem is self-consistent with all existing analyses which assume a slip boundary condition. However, it is not at all obvious that fluids actually slip. By specifying the dynamics of the fluids in the outer region in this way, not only are we freed from having to use a specific slip boundary condition whose form cannot be verified, fortuitously, we might be using the proper outer solution corresponding to an inner solution whose model involves other assumptions besides slip.

On the other hand, if it can be shown that solutions to the above boundary-value problem are <u>inconsistent</u> with experimental observations then it can be definitely concluded that allowing fluids to slip on the surface of the solid is <u>not</u> the correct modeling assumption.

In order to test the validity of this approach, we† are currently solving the boundary-value problem appropriate for the flow field investigated experimentally by Ngan & Dussan V. Correct to $0(U\mu/\sigma)$ as $U\mu/\sigma$ approaches zero it has been found that

$$\cos \theta_{APP} \sim \cos \theta_I + \frac{U\mu}{\sigma} [F_1(\theta_I) \ln \frac{R_I}{a} + F_2 \ (\theta_I)]$$

where

$$F_1(\theta_I) \equiv \frac{4 \sin^2 \theta_I}{2 \ \theta_I - \sin^2 \theta_I}$$

and the value of $F_2(\theta_I)$ is presently being calculated numerically. The experimental results appearing in Figure 3 can be used to calculate the function F_1 without knowledge of θ_I through

$$F_1 = \frac{\cos\theta_{APP}(a_1) - \cos\theta_{APP}(a_2)}{\frac{\mu U}{\sigma} \ln \frac{a_2}{a_1}}$$

Since data is presented in Figure 3 for the dependence of θ_{APP} on $\frac{U\mu}{\sigma}$ for three values of a, three curves can be calculated for the dependence of F_1 on $\frac{\mu U}{\sigma}$, refer to Figure 6 [26]. If the boundary-value problem is appropriate, then all three curves for F_1 should coincide. If any two of the curves differ beyond that which can be anticipated from

---

†Joint work with C. G. Ngan

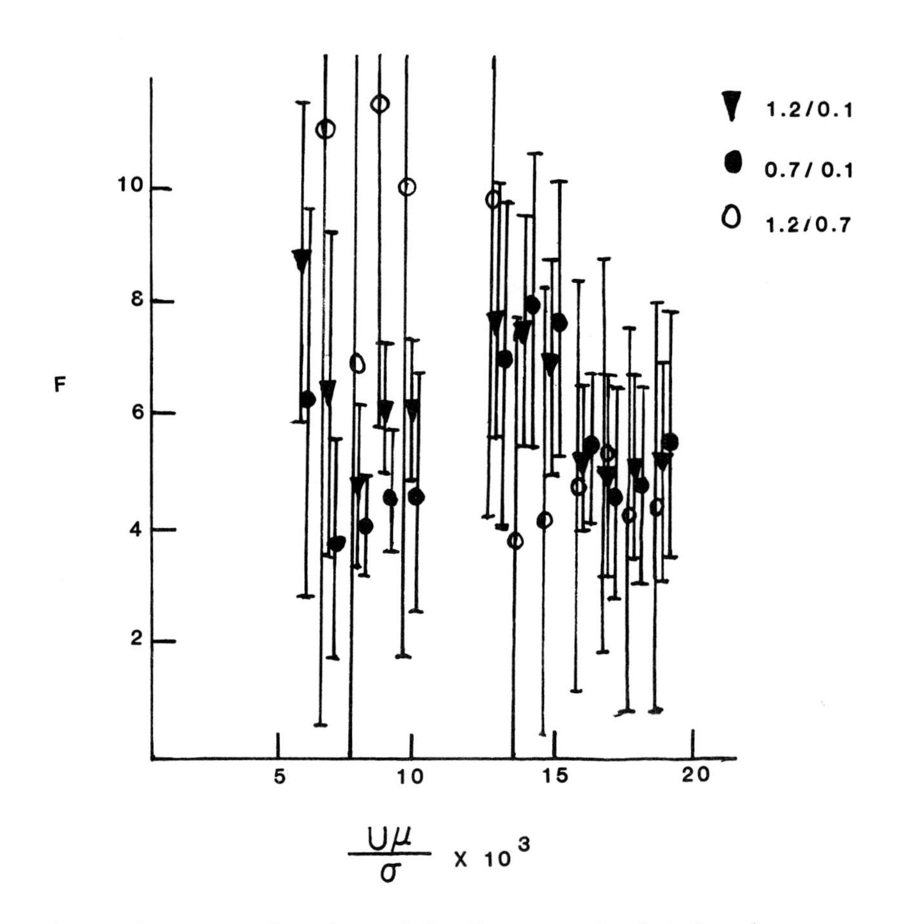

Figure 6: At each value of U, F_1 was calculated using every combination of the data whose average is presented in Figure 3. The average and standard deviation of the calculated values of F_1 for the three possible combinations of gap widths appears above. The large standard deviation associated with 0 can be anticipated due to the close proximity of the data corresponding to gap widths 1.2 and 0.7 mm. In addition, large standard deviations would always occur for data taken at small speeds because all three curves in Figure 3 converge.

experimental error, then assuming the fluids slip at the solid is not appropriate. It is evident that the results reported in Figure 6 are inconclusive.

Currently, experiments are being performed in which the contact line is permitted to move at one of two precise speeds in order to generate significantly more data in an attempt to improve the statistical accuracy. In addition, the analysis is being extended to include $0([\mu U/\sigma]^2)$ effects.

REFERENCES

1. Dussan V., E. B., On the spreading of liquids on solid surfaces: static and dynamic contact lines, Ann. Rev. Fluid Mech. 11 (1979), 371-400.
2. Penn, L. S., and B. Miller, A Study of the Primary Cause of Contact Angle Hysteresis on Some Polymeric Solids, J. Colloid Interface Sci. 78 (1980), 238-241.
3. Johnson, R. E., R. H. Dettre and D. A. Brondreth, Dynamic Contact Angles and Contact Angle Hysteresis, J. Colloid Interface Sci. 62 (1977), 205-12.
4. Morrow, N. R. and M. D. Nguyen, Effect of Interface Velocity on Dynamic Contact Angles at Rough Surfaces, J. Colloid Interface Sci. 89 (1982), 523-31.
5. Blake, T. D. and J. M. Haynes, Kinetics of Liquid/Liquid Displacement, J. Colloid Interface Sci. 30 (1969), 421-23.
6. Hoffman, R., A study of the Advancing Interface. I. Interface Shape in Liquid-Gas Systems, J. Colloid Interface Sci. 50 (1975), 228-41.
7. Inverarity, G., Dynamic Wetting of Glass Fibre and Polymer Fibre, Br. Polym. J. 1 (1969), 245-51.
8. Schwartz, A. M. and S. B. Tejada, Studies of Dynamic Contact Angles on Solids, J. Colloid Interface Sci. 38 (1972), 359-75.
9. Truesdell, C. A. and W. Noll, Classical Field Theories, Handbuck der Physik vol. 3/1. Springer (1960).
10. Coleman, B. D., H. Markovitz and W. Noll, Viscometric Flows of Non-Newtonian Fluids, Springer, 1966.
11. Dussan V., E. B. and S. H. Davis, On the motion of a fluid-fluid interface along a solid surface, J. Fluid Mech. 65 (1974), 71-95.

12. Dussan V., E. B., On the difference between a bounding surface and a material surface, J. Fluid Mech. 75 (1976), 609-23.
13. Pismen, L. M. and A. Nir, Motion of a contact line, Phys. Fluids 25 (1982), 3-7.
14. Benney, D. J. and W. J. Timson, The Rolling Motion of a Viscous Fluid on and off a Rigid Surface, Stud. Appl. Mech. 63 (1980), 93-98.
15. Ngan, C. G. and E. B. Dussan V., On the nature of the dynamic contact angle: an experimental study, J. Fluid Mech. 118 (1982), 27-40.
16. Goldstein, S., Modern Developments in Fluid Mechanics, Dover, 1965 (see pages 676-80).
17. Jackson, R., Transport in porous catalysts, Elsevier, 1977.
18. Richardson, S., On the no-slip boundary condition, J. Fluid Mech. 59 (1973), 707-19.
19. Hocking, L. M., A moving fluid interface. Part 2. The removal of the force singularity by a slip flow, J. Fluid Mech. 79 (1977), 209-29.
20. Huh, C. and S. G. Mason, The steady movement of a liquid meniscus in a capillary tube, J. Fluid Mech. 81 (1977), 401.
21. Lowndes, J., The numerical simulation of the steady movement of a fluid meniscus in a capillary tube, J. Fluid Mech. 101 (1980), 631-46.
22. Hocking, L. M., A moving fluid interface on a rough surface, J. Fluid Mech. 76 (1976), 801-17.
23. Greenspan, H. P., On the motion of a small viscous droplet that wets a surface, J. Fluid Mech. 84 (1978), 125.
24. Dussan V., E. B., The moving contact line: the slip boundary condition, J. Fluid Mech. 77 (1976), 665-84.
25. Kafka, F. Y. and E. B. Dussan V., On the interpretation of dynamic contact angles in capillaries, J. Fluid Mech. 95 (1979), 539-65.
26. Ngan, C. G., private communication, 1982.

The author was partially supported by the National Science Foundaton Fluid Mechanics Program.

Department of Chemical Engineering
Towne Building/D3
University of Pennsylvania
Philadelphia, PA 19104

THE ENDOTHELIAL INTERFACE BETWEEN TISSUE AND BLOOD

S. Weinbaum

1. INTRODUCTION

The movement of water, solutes, proteins and other macromolecules across the vascular interface are vital physiological processes governing transcapillary fluid exchange in the microcirculation and the homeostasis of the larger vessels. The transport of macromolecules is thought to play an important role in such disease states as edema and atherosclerosis. The vascular interface of both the larger vessels and the microcirculation is composed of a single layer of cells called the endothelial lining. Before electronmicroscopic studies could identify the ultrastructural pathways for the transendothelial passage of large and small molecules, microcirculatory physiologists [1] had discovered that there was at least a small pore system which provided separate pathways for the passage of water and small ions and the large molecules. Electronmicroscopic studies starting in the early 1960's have since shown that there are four, if not more, passive pathways by which molecules can cross the endothelial lining. The smallest pathway is the 3-4 Å hydrophilic pore which is an integral part of the phospholipid bilayer membrane of the cell. This pathway is crucial in the movement of water and small ions into and between (communicating junctions) the endothelial cells themselves. This small hydrophilic pore system is probably of lesser importance in transendothelial water transport than the larger pathways described next because of its larger hydraulic

ISBN 0-12-493220-7

resistance. The next larger pathway is the intercellular cleft between adjacent endothelial cells. Electronmicroscopic studies with labeled tracer molecules have shown that the largest molecules which can effectively pass through the narrowest constrictions of the intercellular channels (frequently called tight junctions) under normal conditions are approximately 20-40 Å. Larger molecules, such as albumin (42 Å) and low density lipoproteins carrying cholesterol (150-200 Å), are transported either through plasmalemma vesicles, as first proposed by Palade [2], or through widened cell junctions and fenestra. It is now thought likely that a widened intercellular space might exist during cell turnover (death and replacement of endothelial cells) when the integrity of the tight junctions is possibly disrupted. Fenestra are larger circular pores of approximately 1000 Å diameter which have thus far been observed only in post capillary venules where the endothelial cells have become greatly attenuated (transendothelial cell dimensions are 1000 Å or less).

Figure 1 is an electronmicrograph of a capillary showing both a thin section (1a) of an endothelial cell in transmission and a freeze replica (1b) of an endothelial cell in transverse fracture. The important ultrastructural features for transport are the numerous attached and free plasmalemma vesicles v and the intercellular cleft (normal spacing approximately 200 Å) which is partially occluded on the tissue side by the tight junction j. Plasmalemma vesicles are spherically shaped membrane bound bodies about 700 Å in diameter which are founded in large concentrations especially in the peripheral zone pz of the cell. For the relatively thin capillary endothelium shown roughly 50 percent of the vesicles are free in the cells cytoplasm, whereas our own studies with arterial endothelium have shown that for the larger vessels only about 10 percent of the vesicle population is free. The attached vesicles are equally distributed on the lumen (pl) and tissue sides (c) of the cell. It was first hypothesized by Palade [2] that the vesicles in their attached state fill through diffusion, then break off from their attaching stalks, migrate across the endothelial cell by Brownian diffusion and reattach at the opposite plasmalemma where they unload their contents.

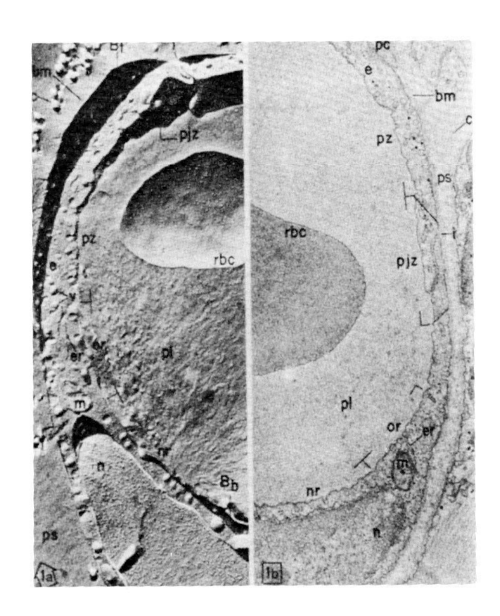

Fig. 1. Comparison between a replica of a nearly transverse fracture through a blood capillary of rat diaphragm (1a, x 22,600) and a transmission section (1b, x 20,600). The fracture exposes the B face of the plasmalemma on the tissue front at B_t and the A and B faces of the plasmalemma on the blood front at A_b and B_b, respectively. Note: red blood cell rbc in lumen, vesicle v and intercellular cleft with junction j. (Courtesy of Simionescu, N. Simionescu, M. and Palade, G. E.)

The process proceeds in both directions, but, if the concentration of proteins in the luminal fluid is greater than that in the tissue, there will be a net transport from blood to underlying tissue. The dimension of the transported molecule is limited in this process by the internal diameter of the vesicle attachment stalk, which is typically 200 Å. The process just described has been experimentally documented in mammalian endothelium using time dependent labeled tracer techniques [3,4]. An alternate hypothesis, which has recently been advanced by Michel and coworkers [5,6] based on their experiments with frog capillary endothelium, is that the vesicles do not undergo a transendothelial migration but fuse with one another forming transient patent channels across the cell. These investigators also observe that the attached vesicles do not fill immediately suggesting that there is a diffusion barrier at the mouth of the vesicle. It is not known at present which hypothesis is correct or if both occur and the difference is due to species differentiation. The vesicular transport process just described should not be confused with endocytosis or exocytosis. The latter are active transport processes involving either the encapsulation or expulsion of material required for the internal metabolism of the cell rather than transport across the cell.

A problem of great current interest because of its implications for arterial disease is why in certain regions of the larger arterial vessels the uptake of lipoproteins is 50 to 100 percent larger than neighboring areas. Since fenestra are not observed in arterial endothelium, the two pathways that might be involved in such regional variations are the intercellular cleft and the vesicular transport pathways. The theory and experiment described in Chien and Weinbaum [7] suggest that such large regional differences would be hard to explain on the basis of vesicular transport, since vesicle attachment rather than diffusion is the rate limiting process and the latter does not appear to be significantly affected by hemodynamic factors. The alternate explanation is that the variations in transport are mediated by subtle changes in the permeability of the endothelial cell layer as might occur during cell turnover. Two possibilities have been examined, either the dying cell becomes leaky over its entire surface or only its junctional complexes become leaky. In either case one has to explain quantitatively how a twofold increase in transport can occur when less than one percent of the endothelial surface is involved in cell turnover at any given instant. Although the endothelial cell layer is recognized as the principal barrier to passage of proteins and lipoproteins into the arterial media (the underlying tissue of the artery wall composed of elastic connective tissue and smooth muscle cells) it can not be treated as an isolated diffusion barrier. The three-dimensional interaction between the cell layer and the underlying tissue and the filtration flow through the intercellular cleft can both have a very substantial effect on the transport and uptake by the arterial wall. Small damages in the integrity of the endothelial lining involving a minute fraction of its surface can produce dramatic changes in the permeability of the wall to macromolecules.

In this short survey paper we shall briefly summarize recent theoretical models of (a) the vesicular transport process, (b) the three-dimensional diffusional interaction between the endothelial cell layer and the underlying tissue and the effect of minute endothelial injuries, (c) the filtration flow across the arterial wall and (d) a hydrodynamic interaction theory for determining the osmotic reflection and diffusion coefficients for spherical molecules passing through the intercellular cleft.

2. VESICULAR TRANSPORT

(a) Background. In the past decade a variety of theoretical models have been introduced to analyze quantitatively the steady state vesicle transport process [8-12]. A more detailed summary of these basic models and their extensions appears in Weinbaum and Chien [13]. The earliest steady state models were based on one-dimensional continuum diffusion theory or random walk computer simulation experiments [9] in which the interaction of the vesicles with the boundary plasmalemma membranes of the cell was neglected entirely. These models were later modified to treat the plasmalemmas as imperfectly absorbing barriers with an empirically chosen elastic reflection coefficient [8,10]. The first dynamic model of the steady state transport process which attempts to describe in an approximate manner the hydrodynamic and electrodynamic force interactions between the vesicle and the boundary plasmalemma is presented in Weinbaum and Caro [11]. This basic model and its extension to time dependent vesicle transport is outlined below and described in greater detail in Arminski et al [14].

(b) Mathematical Model. Figure 2 is a schematic illustration showing the simplified geometry and typical dimensions for the diffusion of a vesicle across an endothelial cell. The vesicles are assumed to be released from the lumen at a finite distance y = 200 Å equal to the length of the vesicle

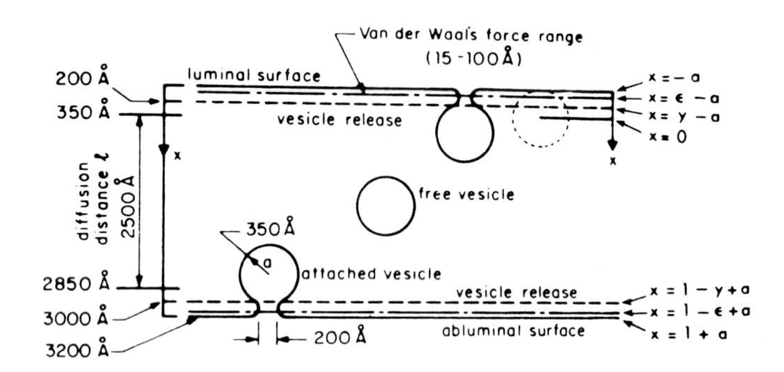

Fig. 2. Sketch of mathematical model showing typical dimensions for the transendothelial motion of a vesicle between parallel plasmalemmas. From Weinbaum and Chien [13].

attachment stalk. The distance between the plasmalemmas is equal to the diffusion distance of the center of the vesicle L plus the vesicle diameter 2a representing the excluded distance of the vesicle center from the two boundary membranes. In the vicinity of each plasmalemma narrow regions are hypothesized to exist where electrodynamic forces are important. The effective center of interaction, ε, of these layers lies from 15 to 100 Å from the plasmalemma membrane although the long-range effect of the van der Waals forces can be felt as far as 400 - 500 Å from this surface. The interior surface of the plasmalemma and the exterior surface of the vesicle are believed to contain little negative charge. For this reason electric double layer forces have been neglected in the model.

The conservation equation for the vesicle number density c on each side of the release plane x = y can be described by a continuum equation (1) of the form (all quantities in dimensional form)

$$\frac{\partial c}{\partial t} + \frac{\partial (c u_{vw})}{\partial x} = \frac{\partial}{\partial x}\left[D(x) \frac{\partial c}{\partial x} \right] \qquad \begin{vmatrix} 0 \le x < y \\ y < x \le L \end{vmatrix} \tag{1}$$

where u_{vw} can be thought of as a convective velocity for the vesicles created by the macroscopic London-van der Waals force and D(x) is a spatially dependent diffusion coefficient that describes the fluid resistance a spherical vesicle experiences in undergoing a one-dimensional Brownian motion between parallel plasmalemmas. All distances in equation (1) will henceforth be scaled relative to the diffusion distance L.

The expression for D(x) is obtained by introducing an effective viscosity $\mu(x) = \lambda(x)\mu_o$ into the Stokes-Einstein relation for the diffusion coefficient where μ_o is the unbounded fluid viscosity and λ is a hydrodynamic interaction parameter describing the variation in fluid resistance for perpendicular motion of a spherical particle between the plasmalemmas. Thus,

$$D(x) = D_o/\lambda(x) \ , \tag{2}$$

where D_o is the vesicle diffusion coefficient in an infinite medium.

The expression for u_{vw} is obtained by taking a force balance on the vesicle in which the sum of van der Waals, concentration gradient and hydrodynamic resistance forces is set equal to zero.

$$0 = F_{vw} + 6\pi a\mu\lambda u_D - 6\pi a\mu\lambda u \ . \tag{3}$$

The total velocity u is the sum of the Brownian diffusion velocity u_D and u_{vw}. An approximate expression for the macroscopic force of attraction F_{vw} between the vesicle and the adjacent plasmalemma can be obtained by integrating the non-retarded binary interaction potential between all surface elements, neglecting all interference effects. This resultant force is

$$F_{vw} = \frac{3}{2}\pi^2 k \left[\frac{1}{x^4} - \frac{1}{(x+a)^4}\right] , \tag{4}$$

where k is proportional to the surface density of molecules and the difference in the polarization properties of the bilayer membrane and the intervening fluid. In Weinbaum and Caro [11] it is shown that the effective center of the van der Waals layer is defined by the condition $u_{vw} = u_D$. This is the apparent location $x = \varepsilon$ at which the free vesicle concentration in the cell interior would vanish if electrodynamic forces were not present. Since $\varepsilon << 1$ it is convenient to develop a separate analysis for the thin van der Waals force layers adjacent to each plasmalemma. In these regions it is shown in

[14] that one can solve a quasisteady approximation to equation (1) where an approximate expression for u_{vw} is derived from equations (3) and (4) in which the effect of the more distant plasmalemma is neglected.

In the interior of the cell, outside the van der Waals layers, equation (1) is given by another approximation in which the unsteady term must be retained but $u_{vw} = 0$ since the van der Waals forces vanish. In this region the simplified dimensionless form of equation (1) is

$$\frac{\partial c}{\partial t} = \frac{\partial}{\partial x}\left[\frac{1}{\lambda(x)}\frac{\partial c}{\partial x}\right] \qquad \begin{array}{l} \varepsilon \leq x < y \\ y < x \leq 1 - \varepsilon \end{array} \tag{5}$$

where the characteristic reference time is L^2/D_o and the reference vesicle concentration is $\phi L/D_o$. ϕ is the number of vesicles released per unit time, per unit area at the luminal surface.

As discussed in the foregoing, the effect of the van der Waals forces on the solution in the cell interior can be approximated by simply relocating the position at which the free vesicle concentration vanishes. Thus

$$c(\varepsilon,t) = 0, \tag{6a}$$

$$c(1 - \varepsilon,t) = 0 \tag{6b}$$

The dimensionless matching conditions at the release plane $x = y$ require that the concentration be continuous and that the difference in diffusive flux on each side of the plane of release be equal to the vesicle release rate:

$$c(y^-,t) = c(y^+,t) \tag{7a}$$

$$1 = \frac{1}{\lambda(y)}\left[\frac{\partial c(y-,t)}{\partial x} - \frac{\partial c(y^+,t)}{\partial x}\right]. \tag{7b}$$

Laboratory experiments to study vesicle transport are conducted by introducing at some time $t = 0$ a labeled marker molecule and then studying the subsequent advance of the labeled vesicles at different time intervals across the cell. The initial condition for these experiments is

$$c(x,0) = 0 . \tag{8}$$

An important feature of the boundary and initial value problem defined by equations (5) through (8) is the specification of the diffusional resistance function $\lambda(x)$. Freeze fracture electron micrographs indicate that there is sufficient crowding of attached vesicles to affect substantially the hydrodynamic interaction of a free vesicle moving toward either plasmalemma. Since this interaction is too difficult to treat by analytical methods, a laboratory model experiment was constructed to measure empirically the additional resistance due to this steric hindrance effect [15]. A convenient mathematical form was then introduced for the function $\lambda(x)$.

$$\lambda(x) = \beta + \frac{\theta a}{x} + \frac{\theta a}{1 - x} \quad (9)$$

Equation (9) provides for a steric hindrance layer near each plasmalemma described by the second and third terms with an empirically determined constant θ and a second constant β to account for the additional resistance in the interior of the cell due to the presence of both boundaries. The latter constant was determined from the recent exact solutions in [16] for the perpendicular motion of a sphere between two parallel boundaries.

(c) Results and Discussion. The solution of equation (5), with λ given by equation (9), which satisfies conditions (6) through (8) is not straightforward because both the concentration and its gradient at $x = y$ are unknown functions of time. A novel approximate integral solution technique was developed and is presented in detail in [14]. This solution describes the spreading of two coupled concentration boundary layers from the vesicle plane of release toward the plasma membranes on each side of the cell.

In figure 3 we have compared the predictions of the theoretical model for the time dependent labeled vesicle concentration profiles (C/C_T is the instantaneous fraction of labeled vesicles) with experimental data in [4] for mouse heart endothelium. Although there is a large standard error deviation because of the difficulty of the experiment and reliably identifying free and attached vesicles, reasonable agreement is obtained for both short and long times.

The vesicle diffusion model just described can also be used to predict the probability that a vesicle released at the luminal front will eventually migrate to the opposite plasma-

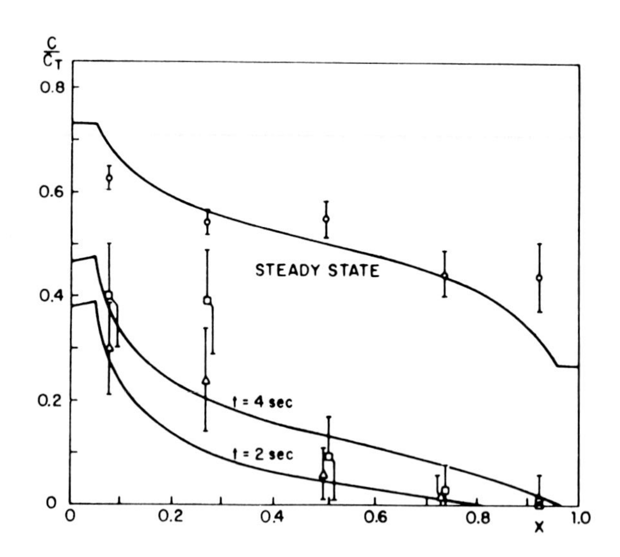

Fig. 3. Theoretical and experimental time dependent proportion of labeled vesicles c/c_T as a function of distance x for an effective van der Waals cut-off distance ε of 15 Å, a transendothelial diffusion distance L of 4300 Å and a vesicle center-to-center distance of 1100 Å. Theoretical curves were based on the expression for λ(x) including the steric hindrance correction for attached vesicles. Experimental data: Δ, 2 s; □, 4 s; o, $\geq$ 16 s (Casley-Smith and Chin [4]: vesicles labeled with horseradish peroxidase). Vertical bars represent ± 1 standard error. From Arminski, et al [15].

lemma. Thus if the vesicle attachment time and surface number density of attached vesicles is known one can predict the transendothelial vesicle transport rate. The model can also be used to predict the changes in vesicular transport that might be expected due to mechanical disturbances such as increased luminal pressure (hypertension) and blood pulse. These extensions of the theory and corroborating experiments are described in [7].

3. MARCOMOLECULE TRANSPORT ACROSS THE ARTERY WALL

(a) Background. In vivo experiments with pig, canine and rabbit aorta have shown that there are localized regions of enhanced macromolecule permeability. These regions which stain more readily with protein binding Evans blue dye than unstained white areas have also been shown to exhibit a significant increase in cell turnover as indicated by silver staining or thymidine labelling techniques. Figure 4 taken from Gerrity et al [17] is a Häutchen preparation of endothelial cells from pig aorta showing two such damaged cells. Although macromolecules are too large to pass through the intercellular clefts of adjacent undamaged cells and transport across these cells takes place by the vesicular process described in the previous section, it is hypothesized that during the process of cell death and endothelial healing either the dying cell itself becomes permeable across its entire surface or the integrity of the junctional complexes in the intercellular cleft between the dying cell and its neighbors is disrupted. Mathematical models have recently been developed for quantitatively examining the effect on transport and uptake of each of these two types of endothelial injury. Before presenting these models we shall briefly describe the theoretical studies that provide the basis for the multiple damage site models described in part (b) of this section.

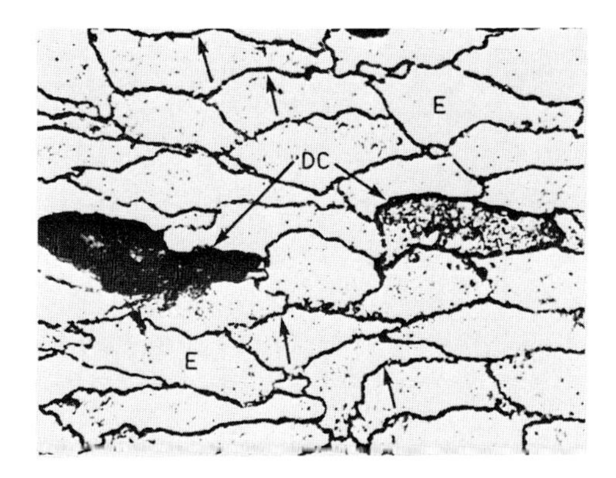

Fig. 4. Light micrograph of silver-stained endothelial cell; injured and dead cells show intense uptake of silver stain. (Courtesy R. G. Gerrity and C. J. Schwartz.)

The basic model for macromolecule transport in arterial vessels is presented in Weinbaum and Caro [11]. In this model the arterial wall is considered to be a medium consisting of two phases, an interstitial fluid phase and a uniformly dispersed cellular phase representing the large number of smooth muscle cells which are present in the media. The entire interstitial space is described by an effective diffusion coefficient which takes into account the presence of collagen fibers, elastin and other intercellular structure which is present in the interstitial space. The smooth muscle cells are assumed to be uniformly distributed across the wall and act as a continuous distribution of sinks since the dimension of the cells is small compared to the thickness of the arterial media. In the original model [11] the endothelial cell layer is treated as an intact endothelium without damage sites in which macromolecule transport occurs only through vesicle diffusion. The endothelial cell layer is typically less than one percent of the arterial wall thickness for most major vessels yet represents more than 80 percent of the total resistance to macromolecule transport if the endothelium is intact.

Nir and Pfeffer [18] have extended the basic two phase arterial wall model just described to take into account the presence of a single local endothelial injury in an otherwise undamaged endothelial cell layer. By neglecting the interaction between multiple dispersed dying cells these authors obtained analytic solutions for the time-dependent and steady-state concentration distribution, flux and uptake in the arterial wall as a function of damage size, fraction of damaged surface and thickness of the artery wall. The results of this model suggested that only a minute fraction of the endothelial cell layer needs to be damaged, if the thickness of the wall is large compared to the distance between the damaged endothelial cells, for the resistance of the endothelium to be significantly impaired. The theory, however, could not be directly applied to the physiological problem since cell turnover studies have shown that the spacing between dying cells is too close for the interaction between damage sites to be neglected.

(b) Mathematical Model for Interacting Endothelial Injury. The simplified models for multiple interacting injuries assume that the dying cells are arranged in some periodic two-dimensional array as shown in figure 5, where ξ is the effective radius of a unit cell. The dying cell in figure 5 is modelled as a circular hole of radius ε and thus depicts the case where the damaged cell is permeable over its entire surface. The second case, where the damaged cell is leaky only through its junctional complex, is schematically shown in figure 6. In the latter case, the vesicular transport boundary condition is assumed to apply everywhere except at the junction $R_1 < r < R_2$.

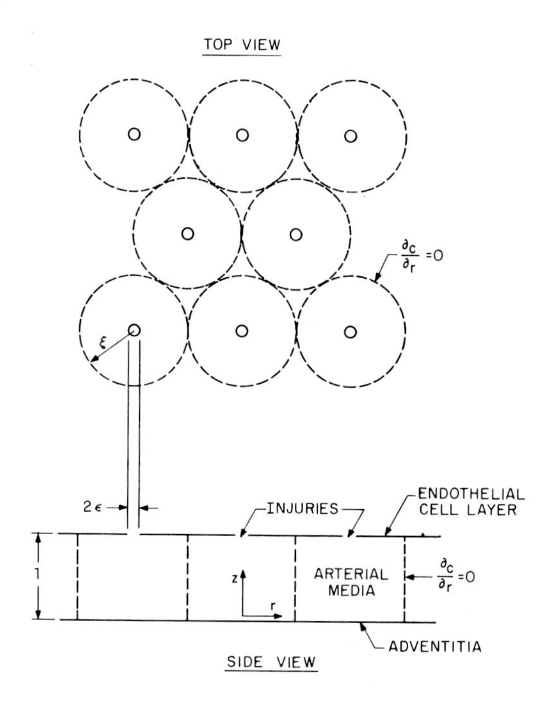

Fig. 5. Schematic diagram of mathematically idealized model of artery wall with periodically distributed dying endothelial cells of diameter 2ε.

The simplified models depicted in figures 5 and 6 treat the artery wall as a slab of uniform thickness with a vanishingly thin endothelial layer at $z = 1$ (all lengths have been scaled by the arterial wall thickness L). The adventitial surface is $z = 0$. The arterial media consists of a dispersed cellular array of smooth muscle cells in a continuous interstitial fluid phase. The equations describing the change in concentration of macromolecules in the interstitial fluid

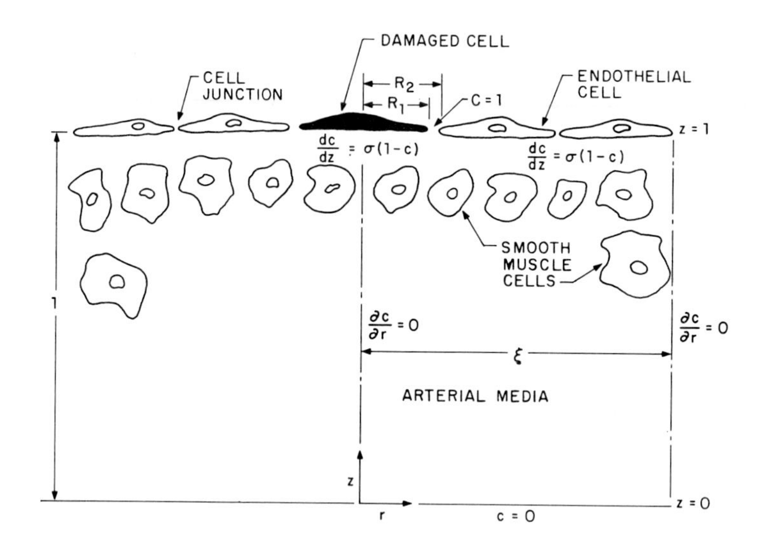

Fig. 6. Schematic diagram of unit cell in periodic model artery for case where dying cell (shown in black) becomes leaky to macromolecules only through its junctional complex.

phase and the cellular phase of the arterial are given by

$$\nabla^2 c_1 + \beta(c_2 - c_1) = \alpha_1 \frac{\partial c_1}{\partial t} \tag{10}$$

$$\beta(c_1 - c_2) = \alpha_2 \frac{\partial c_2}{\partial t} \tag{11}$$

Here c_1 and c_2 are the dimensionless concentrations of macromolecules in the interstitial fluid phase and the dispersed cellular phase, respectively, defined by $c_1 = (C_i - C_A)/(C_L - C_A)$, $i = 1, 2$, where C_A is the concentration at the adventitial surface and C_L is the concentration in the lumen. The term $\beta(c_2 - c_1)$ accounts for the transport of macromolecules from the interstitial phase into the dispersed cellular phase by endocytosis or other mechanisms. The coefficients α_1 and α_2 are the volume fractions of the interstitial fluid phase and the dispersed cellular phase, respectively. The mathematical model assumes a uniform effective diffusion coefficient throughout the artery wall structure.

At steady-state conditions (long times) $c_1 = c_2 = c$, i.e., the concentration distribution in the two phases is identical.

Thus (10) reduces to

$$\nabla^2 c = 0 \tag{12}$$

For both the leaky junction and the permeable cell model we assume that the macromolecules are effectively removed at the adventitial surface by the lymphatic system

$$c(r, 0) = 0, \qquad 0 < r < \xi \tag{13}$$

and that the concentration is bounded at r = 0. At the edge of the unit cell we require that there be no mass transfer

$$\frac{\partial c(\xi,z)}{\partial r} = 0. \qquad 0 < z < 1 \tag{14}$$

At the luminal surface for the permeable cell model we assume unobstructed diffusion of the macromolecules to the underside of a dying cell

$$c(r,1) = 1. \qquad 0 < r < \varepsilon \tag{15}$$

In the area where the endothelium is intact, the net flux of macromolecules across the endothelial cell layer by vesicular transport must equal the net flux by molecular diffusion in the intima directly beneath the endothelium. Thus, the appropriate boundary condition for this area is

$$\frac{\partial c}{\partial z}(r,1) = \sigma[1 - c(r,1)]. \quad \varepsilon < r < \xi \tag{16}$$

Here σ is a dimensionless Biot number which is shown in Weinbaum and Caro [11] to represent physically the ratio of the resistance to macromolecule transport of the arterial media to that of the intact endothelium. For the leaky junction model the boundary conditions at the luminal surface are given by

$$c(r,1) = 1, \qquad R_1 < r < R_2 \tag{17}$$

$$\frac{\partial c(r,1)}{\partial z} = \sigma[1 - c(r,1)]. \qquad 0 < r < R_1,\ R_2 < r < \xi \tag{18}$$

Equation (17) neglects the transport resistance of the intercellular cleft around the perimeter of the dying cell. In the more detailed model of the intercellular cleft presented in section 4 we shall examine how this resistance changes as a function of the relative dimensions of the diameter of the diffusing molecule to the width of the cleft. The mathematical solution of the boundary valve problems just outlined is not straightforward because of the complexity of the mixed boundary condition at the endothelial surface. Special analytic solution procedures had to be developed and are described in Ganatos et al [19] and Tzeghai et al [20].

(c) Results and Discussion. We shall first examine the results for the flux and uptake for the permeable cell model as a function of ϕ, the fraction of dying cells. Figure 7 shows the average axial flux normalized by the flux for an intact endothelial cell layer as a function of damage size and fraction of damaged area for a Biot number of 0.2. Since $\psi/\psi_i = u/u_i$ these curves also represent the uptake per unit volume ratio. Superimposed on this figure are the results of Nir and Pfeffer [18] for noninteracting damages. For large values of ξ (damages spaced far apart) the results of the present study approach those of Nir and Pfeffer. As $\phi \to 0$ the flux reduces to that for an intact endothelium. As $\phi \to 1$ the solution approaches the one in which the endothelium is totally removed, i.e., $\psi/\psi_i = 6.0$. It is interesting to note that if the fraction of damaged surface ϕ is held constant and the damage size is increased the flux approaches that for an intact endothelial layer. As the damage size is decreased (but the number of damages are increased) the flux approaches that for a denuded endothelium even though only a small fraction of the total surface area may be damaged.

A deeper insight into the mechanism by which a small fraction of damaged surface can effectively destroy the diffusional resistance of the endothelium can be obtained by examining the axial and radial concentration profiles in figures 8a,b. In each figure the dashed lines are the limiting curves for the axial concentration for an intact and completely denuded endothelium. Figures 8(a) and 8(b) are for a small damage $\varepsilon = 0.01$ and fraction of damaged surface of 0.1 percent and 3 percent, respectively. Due to the small damage size, large concentration gradients are found only in the vicinity of the damage site. The axial concentration profiles are nearly linear over most of the range of z. The concentration is nearly independent of r away from the damage site. In figure 8(a) the damages are spaced relatively far apart and the axial concentration profile approaches that for an undamaged endothelium while at closer damage spacing (figure 8(b)) the concentration

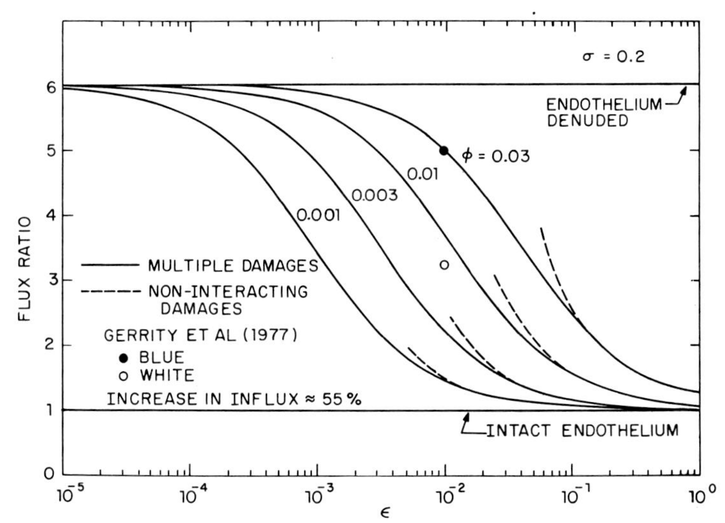

Fig. 7. Relative enhancement of flux or uptake as a function of damage size and fraction of damaged surface for case where dying cell is permeable across its entire surface. Dashed curves theory of Nir and Pfeffer [18] for isolated damages. Also shown experimental results Gerrity et al [17].

approaches that for a completely denuded endothelium even though only 3 percent of the endothelial surface is damaged.

The equivalent results to figure 7 for the leaky junction model are shown in figure 9 for a 200 Å cleft and a wall thickness of 1 mm. The results for this mode of endothelial injury are even more dramatic since the exposed area of the intercellular cleft is less than .001 the surface area of a single dying cell when the width of the junctional complex is 200 Å. Thus, the leaky junction region is less than 10^{-5} of the endothelial surface area when $\phi < .01$. One notes that as the number of damaged cells or turnover rate increases from 1/1000 to 5/1000, the flux or uptake ratio increases by 67 percent clearly indicating the rate controlling effect of the endothelial cell monolayer. Thus large changes in the equilibrium balance of cholesterol carrying LDL in the arterial wall can occur at very small cell turnover rates. The theory also shows that these results are relatively insensitive to the width of the junctional complex provided its dimensions are sufficiently large to permit the rapid passage of the diffusing molecule.

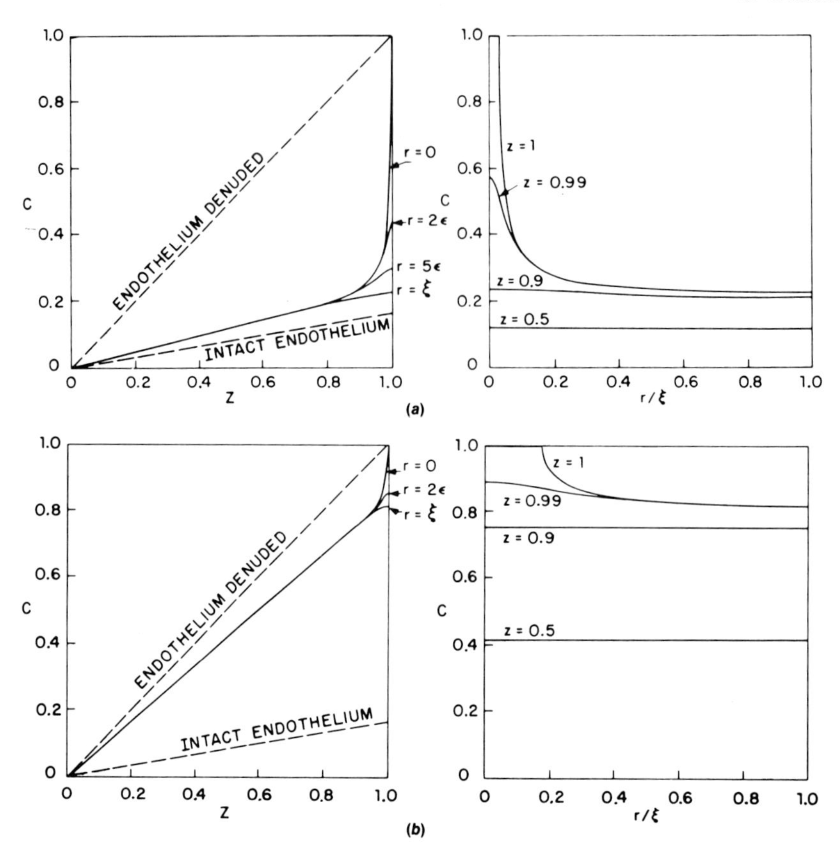

Fig. 8a,b. Axial and radial concentration profiles for case where dying cell is permeable to macromolecules across its entire surface ($\sigma = 0.2$). (a) $\varepsilon = 0.01$, $\phi = 0.001$; (b) $\varepsilon = 0.01$, $\phi = 0.03$.

This occurs because the transport rate is controlled by the perimeter of the dying cell rather than its area. This last conclusion is evident when one compares the results of the leaky junction and permeable cell models for the same value of ϕ.

4. FILTRATION ACROSS THE ARTERY WALL

(a) Background. Experimental studies on the movement of water across the arterial wall due to a pressure driving force and the hydraulic conductivity of the various layers of the wall suggest that as much as half of the total hydraulic resistance occurs across the endothelial cell layer [21]. This finding is rather surprising considering that the thickness of

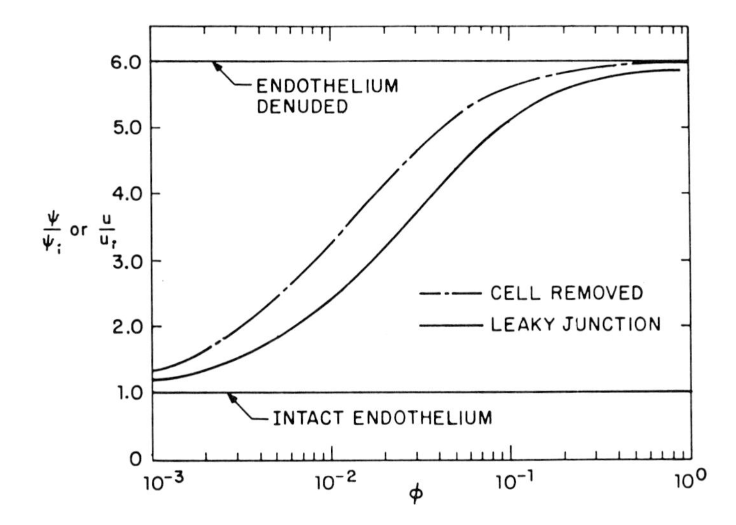

Fig. 9. Relative enhancement of flux or uptake as a function of cell turnover for the leaky junction model; width of intercellular junction 200 Å, wall thickness 1 mm. Dashed curve equivalent result if entire endothelial cell removed.

the endothelial cell layer is less than 1/1000 the thickness of the arterial media for the larger arteries of most mammals and humans. A mathematical model has been proposed to provide fundamental insight into the fine structure of the water movement and pressure distribution in the arterial wall and, in particular, to show quantitatively how the dimensions of the intercellular cleft between adjacent endothelial cells are related to both the water movement and pressure distribution in the subendothelial space. The movement of water across the artery wall is also important in studying the effect of convection on macromolecule transport in the arterial media. The extension of the models in section 2 to include convective transport is currently in progress.

(b) Mathematical Model. Figure 10 is a schematic diagram showing our mathematical model (not to scale) for pressure filtration across the artery wall. The important unknown about which there has been much discussion is the steady state pressure distribution in the tissue and in particular in the subendothelial space P_2. In modeling the problem, we first solve a porous media diffusion equation for the pressure field, and then use Darcy's Law to obtain an expression for the volumetric flow across the arterial media in terms of the unknown pressure P_2 just beneath the endothelium.

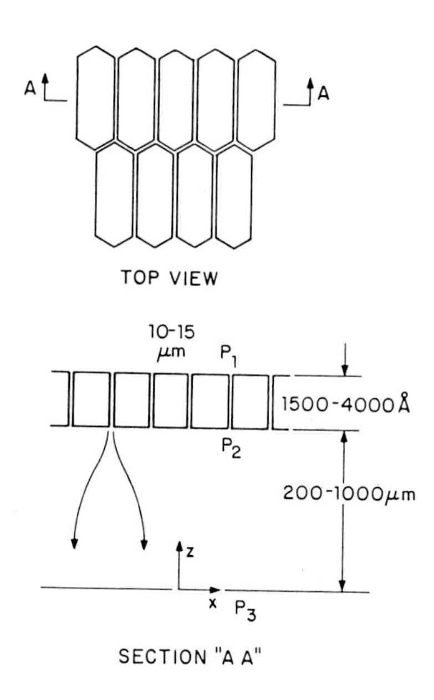

Fig. 10. Schematic diagram showing typical dimensions and simplified geometry of mathematical model for water filtration across arterial endothelium and media.

The boundary value problem for determining the steady state pressure distribution in the media involves a diffusion equation with a novel boundary condition for the endothelial cell layer. The model assumes that there is negligible water movement across the surface of the cell and that the filtration is due the transendothelial pressure difference $P_1 - P_2$ across the intercellular cleft. A second equation is obtained by applying lubrication theory analysis to the fluid motion within the intercellular channels. At the channel exit we require that the pressure be continuous and that the volumetric flow entering the media be compatible with the total pressure drop across the intercellular channels. These equations are solved simultaneously to yield the flow across the arterial wall and the pressure distribution beneath the endothelium and in the arterial media. The intercellular space is treated as a parallel channel with a localized constriction of length w and sinusoidal shape which narrows from 200 Å to $2\varepsilon_{min}$ in the region of the junctional complex. As shown by the right hand ordinate in figure 11, the hydraulic resistance of the intercellular cleft can be expressed in terms of an equivalent parallel wall channel of the same length but width $2\varepsilon_0$.

(c) Results and Discussion. Figure 11 shows the solutions for the pressure P_2 at the entrance to the subendothelial space when the luminal pressure is 100 cm. of water. One observes that when the effective intercellular cleft width $2\varepsilon_0$ is roughly 35 Å, more than one half of the total transmural pressure drop can be sustained by an endothelial cell layer for a vessel whose wall thickness is less than 1 mm. In general, the total flow into the arterial media will be the sum of the filtration flow just described and the flow due to the transendothelial colloidal osmotic pressure difference.

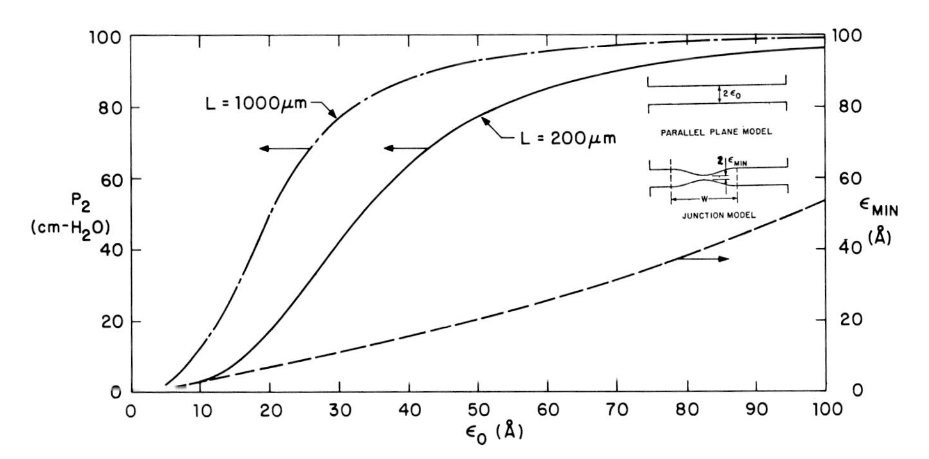

Fig. 11. Subendothelial pressure as a function of effective intercellular cleft width for 0.2 and 1.0 mm. thick arterial walls. $2\varepsilon_{min}$ is the minimum width of junctional complex. Insert shows model of intercellular junction and right hand coordinate, the lubrication theory result for determining the equivalent width of a parallel walled channel of the same length and same hydraulic resistance.

In figure 12 the axial velocity profiles are plotted as a function of depth in the arterial media. The striking feature of these curves is that the fluid jets entering the arterial media decay very rapidly and asymptotically approach a uniform flow after penetrating a few percent of the thickness of the arterial media. This behavior will permit a significant simplification of the convective -diffusion model for arterial wall mass transport.

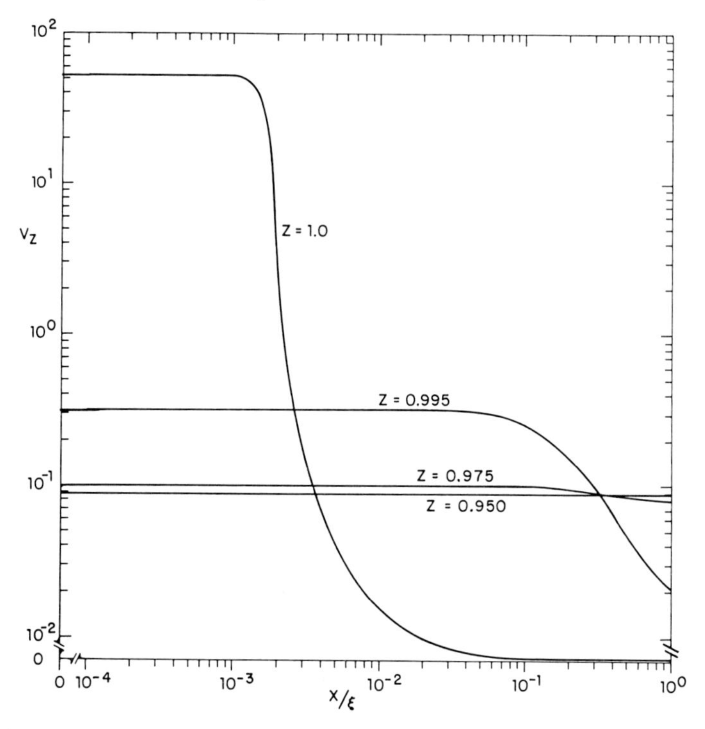

Fig. 12. Velocity profiles in z direction at increasing depths below endothelial surface. Logarimthmic scale z direction. Width intercellular cleft 200 Å, arterial wall thickness 0.2 mm.

5. TRANSPORT THROUGH THE INTERCELLULAR SPACE (OSMOSIS)

(a) Background. The movement of macromolecules and solutes through the intercellular cleft is crucial both in understanding the osmotic effects produced by the endothelial cell layer and the mechanism by which macromolecules will pass through the cleft if the junctions become leaky as hypothesized in section 2. Most existing analyses of the reflection of nonelectrolytes by biological membranes are based on the Kedem-Katchalsky equations derived from the theory of irreversible thermodynamics. The two important phenomenological coefficients that appear in the equation for the solute flux are the osmotic reflection coefficient σ and the diffusive permeability ω. Bean [22] and Levitt [23] have shown how both these coefficients can be related using continuum hydrodynamic theory to low Reynolds number hydrodynamic parameters that describe the arbitrary motion of a particle through a biological pore or channel. Electron microscopic studies strongly suggest that the biological pores are actually two-dimensional slits representing the intercellular space between adjacent cells.

(b) Mathematical Model. Figure 13 is a schematic showing the idealized geometry for the arbitrary motion of a spherical particle in a parallel walled channel representing the intercellular cleft. The hydrodynamic force and torque

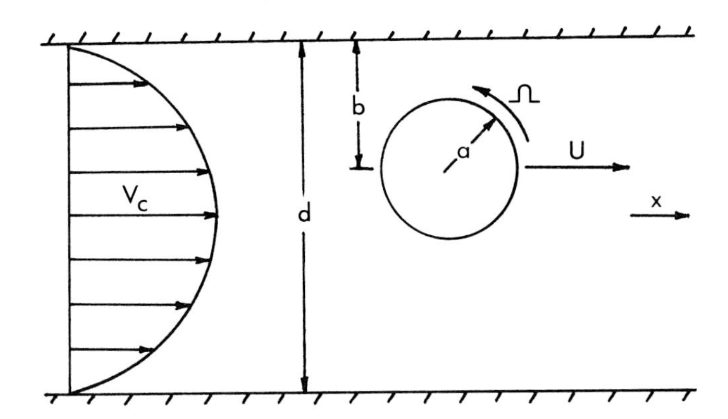

Fig. 13. Schematic diagram showing flow geometry for arbitrary motion of a sphere in a parallel walled channel flow with a Poiseuille profile at infinity.

on a neutrally buoyant spherical particle translating parallel to two plane walls with velocity U at low Reynolds number and/or rotating with angular velocity Ω about an axis which is parallel to the walls and perpendicular to the direction of motion can be represented by the superposition of three motions; (i) a pure translation, (ii) a pure rotation in a stationary fluid and (iii) a Poiseuille flow past a stationary sphere:

$$D = 6\pi\eta a[UF_x^t + a\Omega F_x^r + V_c F_x^p] \qquad (19)$$

$$T = 8\pi\eta a^2[UT_y^t + a\Omega T_y^r + V_c T_y^p] \qquad (20)$$

Here η is the fluid viscosity, a is the particle radius and V_c is the centerline velocity of the undisturbed Poiseuille profile. The force and torque coefficients F_x^t, F_x^r, T_y^t, T_y^r and T_y^p account for wall effects on the particle for each of the three motions. These functions, which depend on relative particle size to channel width ($\alpha = 2a/d$) and particle position ($s = b/d$) (see figure 1), have been evaluated to a high degree of accuracy in [24] using a strong interaction theory in which the no-slip boundary conditions are first satisfied exactly along the channel walls and a boundary collocation procedure then used to satisfy the no-slip boundary conditions on the surface of the sphere.

Equating the hydrodynamic force to the gradient of the chemical potential and the net torque to zero yields

$$6\pi\eta a[UF_x^t + a\Omega F_x^r + V_c F_x^p] = -\frac{1}{N_A}\left[\frac{RT}{c}\frac{dc}{dx} + \bar{V}_s\frac{dP}{dx}\right] \qquad (21)$$

$$8\pi\eta a^2[UT_y^t + a\Omega T_y^r + V_c T_y^p] = 0, \qquad (22)$$

where N_A is Avogadro's number, R is the universal gas constant, T is absolute temperature, c(x) is the concentration of solute, $\bar{V}_s$ is the molar volume of solute and dc/dx and dP/dx are concentration and pressure gradients parallel to the slit walls. Solving (22) for the angular velocity Ω and substituting into (21) allows (21) to be recast in the form

$$\frac{6\pi\eta a}{F(\alpha,s)}[U - G(\alpha,s)V_c] = \frac{1}{N_A}\left[RT\frac{dc}{dx} + \bar{V}_s\frac{dP}{dx}\right] \qquad (23)$$

where

$$F(\alpha,s) = \frac{-T_y^r}{F_x^t T_y^r - F_x^r T_y^t} \qquad (24)$$

$$G(\alpha,s) = \frac{F_x^r T_y^p - F_x^p T_y^r}{F_x^t T_y^r - F_x^r T_y^t} \qquad (25)$$

The F and G functions defined by equations (24) and (25) have a simple physical meaning. If there was no filtration ($V_c = 0$) the function 1/F would describe the hydrodynamic correction to the Stokes drag due to wall effects for a particle that is free to translate and rotate under the action of an external force in a stationary fluid. If no external forces are present one obtains the zero drag motion of a neutrally buoyant particle. In this case the translation velocity of the particle is given by $U = GV_c$ where the function G describes the slip velocity of the sphere relative to the Poiseuille profile.

For $b/a \geq 1.1$ the force and torque coefficients in the expressions for F and G can be obtained from the highly accurate solutions presented in [24]. At closer particle-to-wall spacing, the numerical solution procedure used in [24] becomes too time consuming and an approximate lubrication theory analysis described in [25] is more expedient. The variation of F and G with particle position s and relative particle size α is shown in figure 14. The solid lines are the collocation theory results from [24] and the dashed curves are the approximate lubrication solutions from [25].

To determine the solute permeability and reflection coefficents in terms of the F and G functions we follow the procedure used by Levitt [23]. If molecular level interactions between the diffusing molecule and the membrane boundaries of the intercellular cleft are neglected, Brownian motion forces will produce a uniform distribution of particles across the width of the channel except for a particle exclusion layer. This theory can be modified as shown in Anderson and Quinn [26] to include molecular level wall interactions by introducing a probability density distribution function for predicting the position s of the particle in the channel.

For the simple case where the concentration is uniform across the channel except the exclusion layer, the solute flux is given by

$$J_s = 2 \int_{a/d}^{1/2} c(x)\ U(x,s)\ ds = \text{const.} \tag{26}$$

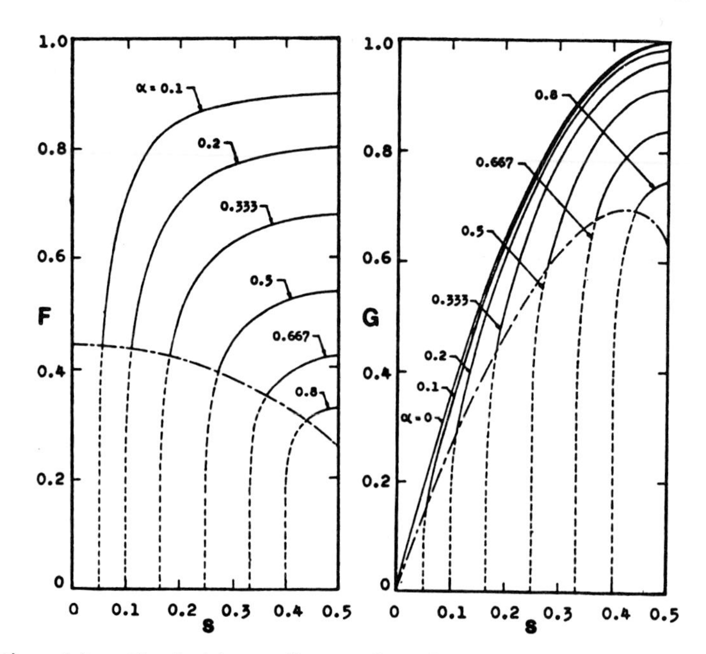

Fig. 14. Variation of F and G drag function with particle position s. ———— collocation theory, Ganatos, Pfeffer & Weinbaum [24]; ----- lubrication theory,; —— · —— b/a = 1.1.

(26) is integrated along the length of the channel L,

$$J_s = \frac{2}{L} \int_0^L \int_{a/d}^{\frac{1}{2}} c(x)\ U(x,s) ds\ dx. \tag{27}$$

Solving for U(x,s) and substituting into (27) gives

$$J_s = \omega[RT\Delta c + \overline{V}_s\overline{c\Delta P}] + (1 - \sigma)\overline{V}_w J_w \overline{c}\ , \tag{28}$$

where

$$\omega = \frac{L_p}{\overline{V}_s} \cdot \frac{2}{3}\alpha^2(1 - \alpha)\overline{F} \tag{29}$$

and

$$\sigma = 1 - \frac{3}{2}(1 - \alpha)\overline{G} \tag{30}$$

are the permeability coefficient and reflection coefficient, respectively, and

$$L_p = \frac{d^2}{12\eta L}\ ,\ \overline{V}_s = N_A \cdot \frac{4}{3}\pi a^3 \tag{31}$$

$$\Delta c = \int_0^L \frac{dc}{dx} dx,\ \overline{c\Delta P} = \int_0^L \frac{dP}{dx} dx,\ \overline{c} = \frac{1}{L}\int_0^L c dx \tag{32}$$

and

$$\overline{F} = \frac{2}{1 - \alpha} \int_{a/d}^{\frac{1}{2}} F ds, \qquad \overline{G} = \frac{2}{1 - \alpha} \int_{a/d}^{\frac{1}{2}} G ds \tag{33}$$

(c) Results and Discussion. It is evident from equations (29) and (30) that ω is related to the average value of the F function whereas σ is related to the average value of the G function integrated across the width of the channel. Table 1 shows the numerical results for the diffusive permeability $\omega\overline{V}_s/L_p$ and the reflection coefficient σ as a function of the particle to channel diameter ratio α. Prior to the theory presented in [24,25] there were no accurate solutions for the hydrodynamic interaction of a spherical particle in a parallel walled channel for $\alpha > 0.2$. Thus, while the expressions for ω and σ given by equations (29) and (30) have been available since 1975, ad hoc methods had been previously used for estimating $\overline{F}$ and $\overline{G}$. The results of one widely used approximation Curry [27] are also shown in Table 1.

Table 1...Comparison of "exact" theory of the present study with the approximate results of Curry [27] for the reflection σ and diffusive permeability ω.

	σ		$\omega\overline{V}_s/L_p$	
α	Present Study	Curry (1974)	Present Study	Curry (1974)
0	1	0	0	0
0.1	0.0253	0.015	0.00501	0.0054
0.2	0.0898	0.048	0.0155	0.0170
0.333	0.215	0.142	0.0302	0.0336
0.5	0.413	0.313	0.0407	0.0454
0.667	0.628	0.531	0.0377	0.0428
0.8	0.797	0.719	0.0259	0.0328
1	1	1	0	0

One result of particular importance in the interpretation of the leaky junction model presented in section 3 is the resistance that a molecule of given size experiences as the width of the junctional complex increases. Examination of the values for σ in Table 1 provides an approximate measure of this change in resistance. Thus, when $\alpha = 0.8$ the results of the model indicate that the average velocity of a molecule transported through the cleft is roughly 20 percent of the average water velocity and that this increases to nearly 60 percent when $\alpha = 0.5$. This indicates that the hydrodynamic resistance of the intercellular cleft rapidly diminishes once the width of the junction becomes larger than the diameter of the transported molecule. In the leaky junction model the assumption used in equation (18) is that the concentration in the lumen is the same as in the subendothelial space at the exit of the intercellular channel. The intercellular cleft, even when leaky, can offer a significant fraction of the total wall resistance if the thickness of the media is small. This refinement will need to be included in future models of macromolecule transport across thin walled arteries.

REFERENCES

1. Pappenheimer, J. R., E. M. Renkin and L. M. Borrero, Am. J. Physiol., 167:13, 1951.
2. Palade, G. E., J. Appl. Phys. 24:1424, 1953.
3. Simionescu, N., M. Simionescu and G. E. Palade, J. Cell Biol. 57:434-452, 1973.
4. Casley-Smith, J. R. and J. C. Chin, J. Microscopy, 93:167-189, 1971.
5. Loudon, M. F., C. C. Michel and I. F. White, J. Physiol., 296:97-112, 1979.
6. Clough, G. and C. C. Michel, J. Physiol.. 315:127-142, 1981.
7. Chien, S. and S. Weinbaum, ASME J. Biomech. Eng. 103:176-186, 1981.
8. Green, H. S. and J. R. Casley-Smith, J. Theor. Biol., 35:103-111, 1972.
9. Shea, S. M., M. J. Karnovsky and W. H. Bossert, J. Theor. Biol. 24:30-42, 1969.
10. Shea, S. M. and W. H. Bossert, Microvasc. Res. 6:305-315, 1973.
11. Weinbaum, S. and C. G. Caro, J. Fluid Mech. 74:611-640, 1976.
12. Rubin, B. T., J. Theor. Biol., 64:619-647, 1977.

13. Weinbaum, S. and S. Chien, "Vesicular Transport of Macromolecules Across Arterial Endothelium," Mathematics of Microcirculation Phenomena, eds., J. F. Gross and A. Popel, Raven Press, New York, 109-131, 1980.
14. Arminski, L., S. Weinbaum and R. Pfeffer, J. Theor. Biol., 85:13-43, 1980.
15. Arminski, L., S. Chien, R. Pfeffer and S. Weinbaum, Biorheol., 17:431-444, 1980.
16. Ganatos, P., S. Weinbaum and R. Pfeffer, J. Fluid Mech., 99:739-753, 1980.
17. Gerrity, R. G., M. Richardson, J. B. Somer, F. P. Bell and C. J. Schwartz, J. Amer. Path., 99:131-134, 1977.
18. Nir, A. and R. Pfeffer, J. Theor. Biol., 81:685-701, 1979.
19. Pfeffer, R., P. Ganatos, A. Nir and S. Weinbaum, ASME J. Biomech. Eng., 103:197-203, 1981.
20. Tzeghai, G., P. Ganatos, R. Pfeffer and S. Weinbaum, "Effect of Cell Turnover and Leaky Junctions on the Transport of Macromolecules Across the Arterial Wall," to be presented in the Symposia "Fundamentals in the Life Sciences" A.I.Ch.E. Annual Meeting Nov. 14-18, Los Angeles, 1982.
21. Vargas, C. B., F. F. Vargas, J. G. Pribyl and P. L. Blackshear, Amer. J. Physiol., 236:H53-H60, 1978.
22. Bean, C. P., "The Physics of Porous Membranes-Neutral Pores," In Membranes - A Series of Advances, G. Eisenman, ed. Marcel-Dekker, Inc., New York, 1:1, 1972.
23. Levitt, D. G., Biophys. J. 15:533-551, 1975.
24. Ganatos, P., R. Pfeffer and S. Weinbaum, J. Fluid Mech., 99:755-783, 1980.
25. Ganatos, P., S. Weinbaum and R. Pfeffer, "Strong Interaction Solutions for the Gravitational and Zero Drag Motion of a Sphere in an Inclined Channel," J. Fluid Mech., in press, 1982.
26. Anderson, J. L. and J. A. Quinn, Biophys. J., 14:130, 1974.
27. Curry, F. E., Microvasc. Res., 8:236-253, 1974.

This research was supported by National Science Foundation Grant ENG 78-22101 and Subcontract Award 5R 01 HL 19454 from the National Institute of Health.

Department of Mechanical
Engineering
The City College of New York
New York, New York 10031

INDEX

A

Accelerating fluid layers, 71
Air flow over waves
  linear theory for, 226–232
  shallow gas analysis of, 231–233
  turbulence models for, 229–231
  wave-induced pressure variation in, 223, 226–228
  wave-induced shear variation in, 223, 226–228
Alternating direction implicit algorithm, 103
Apex height of meniscus, 310, 312
Atomization
  by air flow over waves, 226, 253–255
  critical gas velocity for, 223–224

B

Bernoulli constant, 3
Bifurcation, 23
Boundary conditions
  electrohydrodynamic, 180
  fluid–solid
    adherence, 307
    generalization of Navier, 316
    kinematic, 307
    multicomponent system, 313
    Navier, 313
    no-slip, 306, 314, 317
    slip, 313–316, 318–320
  gas–liquid, 223–224
Boundary-fitted coordinates, 98
Boundary-integral techniques, 55, 89
Boussinesq waves, 3
Breaking water waves, 54
Bubble deformation, 106

C

Capillary, 310
  constant, 114
  length, 123
  number, 311, 312, 318, 320, 321
  surfaces, disappearance of, 121
  waves, 123, 134, 153, 158
    at high gas velocities, 226, 247–248
    caused by air flow, 233–235
Cleft, intercellular, 326–329, 335, 337–339, 341, 343–347, 349, 352
Cell turnover, 328, 335, 341
Characteristic numbers, table of electrohydrodynamic, 186
Characteristic time
  charge relaxation, 173, 186
  electrohydrodynamic, 186
  electroviscous, 186
  gravity-viscous, 186
  surface tension-viscous, 186
  viscous diffusion, 186
Charge conservation, interfacial, 180
Charge density
  surface free, 173
  surface polarization, 174
Charge relaxation time, 173
Classification of waves on thin liquid layers, 224–226
Conductivity, electrical, 170
Constitutive law
  electrical conduction, 170
  polarization, 170
Contact angle, 114, 116, 117, 119, 121, 303, 305, 309
  actual, 312, 318, 319
  advancing, 306
  apparent, 310–312, 318, 320
  dynamic, 306
  hysteresis, 306
  intermediate θ_I, 318–320
  measured, 311
  receding, 306
  static, 305
Contact line, 303, 304, 307
Contact model
  Dussan V., 316
  Greenspan, 315
  Huh and Mason, 315
  Navier, 313, 316
  usual hydrodynamic, 308, 312
Contamination, 300
Continuous spectrum, 266
Continuum electromechanics, 198
Coriolis, 210

Corner phenomenon, 116, 117
Correlation length, 152, 155, 162, 163
Creeping motion, 88
Current jump, 17
Cylindrical film, 268

D

Damped transients, 266
Density
 correlations, 131
 functional theory of capillary waves, 142
 stratification, 281
Dilatations, surface charge, 175
Dipole
 distributions, 58
 force on, 169
Disjoining pressure, 291
Dispersion, 274
 equation, electrohydrodynamic, 184
 force, 277
Dissipation, 234, 241, 242, 274
Disturbances
 spatially growing, 266–269
 temporally growing, 266–269
Domain perturbations, 94
Double layer, 168, 299
 nonequilibrium dynamics of, 194
 standing wave sensor probe of, 195
 traveling-wave pumping of, 196
 potential, 96
Drag
 effect of waves, 232, 252, 253
 by gas at an interface, 252
Draining, 299
Dynamical limit, 4, 9

E

Eigenfunction expansions, 88
Eigenvalue problem, 20
Electrohydrodynamics, 167
Endothelial injury, 329, 335–338, 340, 341
Energy density, 5, 6
Equilibrium free surface, 113
Evolution equation, 292, 296, 297, 298

F

Film
 rupture, 277
 waves, 261
Filtration, 329, 342–345, 349
Finite-amplitude waves, 17
Finite-difference approximations, 84
Finite-element
 approximations, 83
 method, 119
Flat interface, 88
Flux-potential relations, electric, 170
Force density
 free charge, 168
 Korteweg–Helmholtz, 169
 polarization, 169
 surface and volume, 168
 tenuous dipole, 169
Free
 boundary, 83
 surface mode, 280
Fundamental solutions, 89

G

Generalised vortex method, 61
Generation of wave
 by air flow, 224–226, 241–255
 patterns on thin layers, 224–226
Geometrical limit, 4, 9, 14
Gradient theory, 124
Gravity, 300
 current, internal, 211
 stabilization of liquid layers, 242, 246, 247, 249
 waves, 1, 17
Gravity-capillary waves, 43, 49, 264
 field coupled, 181
Green's formula, 57
Growth
 of unstable waves in gas-liquid flows, 240, 245, 249
 rates, electrohydrodynamic instability, 187

H

Hamaker constant, 291, 293, 301
Height of flowing liquid layers, 238, 239

I

Imposed ''ω–k'' technique, 167
Inertia destabilization of a liquid layer, 234, 237, 238, 249
Influence parameter, 124
Ink, 168
Inner region, 316
Instability
 electrohydrodynamic, 185
 used to print, 184

Integral equations, stability of liquid layers, 235–241
Integro-differential equation, 44
Intensity correlation, 154, 158–160
Interdigital electrodes, 176
Interface
  charge at solid–liquid, 195
  internal electrohydrodynamics of, 194
Interfacial
  conditions, electrohydrodynamic, 180
  density profile, 125
  fluctuations, 134
  modes, 280
  pumping, electrohydrodynamic, 188
  tension, 126, 152, 160–162, 164
  waves, 1, 10, 17, 22, 30, 39
Interstitial fluid, 335, 337, 338
Isobaric diffusion, 314

K

Kelvin waves, 202, 217
  solitary, 210
Kelvin–Helmholtz instability, 17, 21
  for thin liquid layers, 226, 253–255
Kinematic waves in gas–liquid flows, 240

L

Ladyzhenskaya, 95
Lagrangian, 4
Laplace–Young equation, 113
Leaky junction, 328, 335, 337–339, 341–343, 346, 351, 352
Lee wave, 203
Light scattering, 153, 154, 157–159, 162
Linear momentum equations for gas–liquid flows, 221–224, 226–228
London/van der Waals interactions, 291–293, 297, 301
Long-range molecular forces, 291–293
Long-wave theory, 292, 294, 300, 301

M

Marangoni
  effect, 292, 300
  instability, 301
  rolls, 283
Maxwell, 313
Mean curvature, 114, 115
Mixing lengths for flow over waves, 229
Modes, surface-dilatation, 167
Monolayers, charge, 175
Multilayered film, 279

N

Navier, 313
Negative energy, 6, 9
Neutral curves, 269
Newton's method, 7
Nonlinear
  mode jumping, 30, 39
  Schroedinger equation, 273
  steepening, 274
Normal mode, 266
Numerical methods, 6, 47

O

Origin of instability, 264
Orr–Sommerfeld equation for flow over small amplitude waves, 227
Orthogonal mapping, 99
Osmosis, 329, 346–352
Outer region, 316, 318, 319
Overhanging waves, 4, 12

P

Pade approximants, 3
Permittivity, electric, 169
Phase lock amplifier, 176
Phase speed effect of air flow, 235, 240, 242, 245–249
Point force, 89
Polarization density, 169
Potential
  dipole, 90
  electric, 170
Printing, 167
Pumping
  traveling-wave, 196
  traveling-wave induced, 188

R

Rayleigh–Taylor instability, 70
Resistance at a wall, 231, 232, 236, 238, 240
Resonance, 22
  frequency, charge monolayer, 183
  frequency, gravity-capillary, with electric field, 181
Reynolds stresses
  mixing length theory, 229
  wave-induced variation of, 227–231
Ripples, 264

Roll waves, 274
generation by air flow, 226
at high gas velocities, 251–253
on thick films, 248–251
Rossby
number, 210
radius, 210, 214–216
Rupture
of films, 291, 292, 294, 296, 299, 300
time, 292, 298, 299

S

Salinity, 152, 160–163
Second Landau coefficient, 273
Shallow water theory, waves on thin layers, 236
Shear
flow instability, 203
stress, electrical, 167
waves, 275
Side-band stability, 273
Sidewall effects, 300
Silicone oil, 311
Single-layer potential, 96
Single waves, 274
Singularity methods, 84
Sinuous wave, 281
Slender bodies, 89
Slip
coefficient, 313, 315
length, 316
Slow waves, 224–225, 245–247
Solitary wave, 43, 48, 49
internal, 202, 203, 209
Kelvin, 210
Solution methods, 83
Spacelab, 121
Special solutions, 8
Stability, 17, 291, 292, 297, 301
analysis, large Reynolds numbers, 233–235
analysis, shallow liquids, 235–241
effect of gas flow, 221–224
on inclined planes, 246
surface stress, influence of, 241–255
Standing-wave
response, electrohydrodynamic, 177
sensor, electrohydrodynamic, 195
Stokes expansions, 3
Stokeslet, 90
Stratified layered flow, 54
Stream function, 171
Streaming, traveling-wave
charge-relaxation induced, 191
field-coupled gravity-capillary induced, 192
induced, 188
trapped-charge induced, 192
Stress
at wavy interface
pressure and shear stress, 223, 224, 226–233
relaxation of, 251, 252
balance, interfacial electromechanical, 180
tensor
electric, 169
Newtonian fluid, 171
Subcritically unstable, 275
Supercritically stable, 270
Superposition of singularities, 88
Surface
contamination, 268
dilatation, 167
energy, 113, 114, 121, 305
force, average, 190
force density, 173
tension, 41, 114
stabilization of liquid layers, 247–254
waves, 2, 167
electric potential conserving, 173
electrohydrodynamic, 167
field coupled, 173
polarization, 173
Surfactant, 152, 156

T

Thermal fluctuations, 130
Thermocapillarity, 292, 300
Thermodynamic stability, 129
Three-dimensional waves, 1
Tide, barotropic, 202
Transverse correlations at an interface, 136
Traveling-wave, electric, 168
Turbulence
wave-induced variation of, 227–231
spots, 275

V

van der Waals forces, 330, 331
Varicose waves, 281
Velocity
Lipschitz continuous, 308
multivalued, 307, 308, 316

profiles in liquid layers, 238, 239
singularity, 308, 315
Vesicle
diffusion, 330–333
plasmalemma, 326–334
Viscosity stratification, 281
Viscous skin depth, 173
Vortex
point method, 206
sheet, 7, 9
sheet motion, 54

W

Water waves, 22, 41
Wave
evolution equation, 273
speed, 6, 9
trapped-charge electrohydrodynamic, 183
weakly nonlinear, 4

Y

Young's equation, 305